CLONING: CHRONOLOGY, ABSTRACTS AND GUIDE TO BOOKS

CLONING: CHRONOLOGY, ABSTRACTS AND GUIDE TO BOOKS

STEPHEN D. FAIRBANKS
EDITOR

Nova Science Publishers, Inc.
New York

Senior Editors: Susan Boriotti and Donna Dennis
Coordinating Editor: Tatiana Shohov
Office Manager: Annette Hellinger
Graphics: Wanda Serrano
Editorial Production: Marius Andronie, Alexandra Columbus, Maya Columbus,
Alexis Klestov, Vladimir Klestov,
Matthew Kozlowski and Lorna Loperfido
Circulation: Ave Maria Gonzalez, Vera Popovic, Luis Aviles, Sean Corkery,
Raymond Davis, Melissa Diaz, Meagan Flaherty, Magdalena Nuñez,
Marlene Nuñez, Jeannie Pappas and Frankie Punger
Communications and Acquisitions: Serge P. Shohov
Marketing: Cathy DeGregory

Library of Congress Cataloging-in-Publication Data
Available Upon Request

ISBN: 1-59033-879-0.

400 Oser Ave, Suite 1600
Hauppauge, New York 11788-3619
Tele. 631-231-7269 Fax 631-231-8175
e-mail: Novascience@earthlink.net
Web Site: http://www.novapublishers.com

Printed in the United States of America

CONTENTS

FOREWORD

Chapter one is a select chronology of the events surrounding the cloning of a sheep from a single adult sheep cell by Scottish scientists, announced in February 1997. It also addresses subsequent reports of other cloning experiments, including the first one using human cells. Information on presidential actions and legislative activities related to the ethical and moral issues surrounding cloning is provided, as well as relevant web sites.

Chapter two provides a summary and comparison of existing state laws on human cloning, as well as a discussion of similar legislation currently under consideration in Congress.

In the 107th Congress, the House passed H.R. 2505 (Weldon) which would have banned the process of human cloning as well as the importation of any product derived from an embryo created via cloning. In the 108th Congress, H.R. 534 (Weldon) has been introduced and is similar to H.R. 2505 from the 107th Congress, but lacks the importation ban on products derived from cloned embryos. S. 303 (Hatch), the Human Cloning Ban and Stem Cell Research Protection Act of 2003, was introduced on February 5, 2003. It would prohibit human reproductive cloning while allowing cloning for medical research purposes, including stem cell research. Chapter three discusses these issues.

The terms "recombinant DNA technology," "DNA cloning," "molecular cloning" or "gene cloning" all refer to the same process: the transfer of a DNA fragment of interest from one organism to a self-replicating genetic element such as a bacterial plasmid. The DNA of interest can then be propagated in a foreign host cell. This technology has been around since the 1970s, and it has become a common practice in molecular biology labs today. Reproductive cloning is a technology used to generate an animal that has the same nuclear DNA as another currently or previously existing animal. Dolly was created by reproductive cloning technology. In a process called "somatic cell nuclear transfer" (SCNT), scientists transfer genetic material from the nucleus of a donor adult cell to an egg whose nucleus, and thus its genetic material, has been removed. The reconstructed egg containing the DNA from a donor cell must be treated with chemicals or electric current in order to stimulate cell division. Once the cloned embryo reaches a suitable stage, it is transferred to the uterus of a female host where it continues to develop until birth. Therapeutic cloning, also called "embryo cloning," is the production of human embryos for use in research. The goal of this process is not to create cloned human beings, but rather to harvest stem cells that can be used to study human development and to treat disease. Stem cells are important to biomedical researchers because they can be used to generate virtually any type of specialized cell in the human body.

This new book presents an up-to-date Chronology of Cloning along with current and selected abstracts dealing with cloning as well as a guide to books on the topic. Access to the abstract and books sections is provided by title, subject and author indexes.

Chapter 1

CHRONOLOGY OF CLONING*

Mary V. Wright

INTRODUCTION

This is a selected chronology of the events surrounding and following the cloning of a sheep from a single adult sheep cell by Scottish scientists, which was announced in February 1997. The project was cosponsored by PPL Therapeutics, Edinburgh, Scotland, which has applied for patents for the techniques used. This chronology also addresses subsequent reports of other cloning experiments, including the first one using human cells. Information on presidential actions and legislative activities related to the ethical and moral issues surrounding cloning is provided, as well as relevant Web sites.

CHRONOLOGY

1997

February 23, 1997

Dr. Ian Wilmut, a Scottish embryologist and his colleagues at the Roslin Institute, Edinburgh, Scotland, succeeded in cloning a sheep from a single adult sheep cell. *Dolly,* the sheep that was created in this manner, is genetically identical to the adult sheep from which she was cloned.

February 24, 1997

President Clinton asked the 18-member National Bioethics Advisory Commission to study the ethical and legal implications of cloning. Human cloning is not currently regulated by law in the United States.

* Excerpted from CRS Report RL31211 Updated July 9, 2003.

March 2, 1997

Scientists at the Oregon Regional Primate Research Center in Beaverton, OR, reported cloning two monkeys. The monkeys, born in August 1996, were cloned from monkey embryo cells, not cells from an adult monkey. The cloned primates were not genetically identical to any adult monkey.

March 4, 1997

President Clinton issued a memorandum for the heads of executive departments and agencies entitled *Prohibition on Federal Funding for Cloning of Human Beings*. This memorandum may be accessed on the Internet at [http://frwebgate.access.gpo.gov/cgi-bin/getdoc.cgi?dbname=1997_public_papers&docid= pap_text-144]. Also, the President issued remarks entitled *Remarks by the President on Cloning*, which may be accessed at [http://frwebgate.access.gpo.gov/cgi-bin/getdoc.cgi?dbname=1997_public_papers&docid=pap_text-143].

The House Committee on Science's Subcommittee on Technology held a hearing entitled *Bio technology and the Ethics of Cloning: How Far Should We Go?*

March 12, 1997

A hearing entitled *Scientific Discoveries in Cloning: Challenges for Public Policy* was held by the Senate Labor and Human Resources Committee's Subcommittee on Public Health and Safety.

June 9, 1997

The National Bioethics Advisory Commission presented its report, *Cloning Human Beings*, to President Clinton. The report may be accessed on the Web at [http://www.georgetown.edu/research/nrcbl/nbac/pubs/cloning1/cloning.pdf]. President Clinton's remarks, entitled *Remarks by the President at Announcement of Cloning Legislation*, may be accessed at [http://frwebgate.access.gpo.gov/cgibin/ getdoc.cgi?dbname=1997_public_papers&docid=pap_text-418].

June 17, 1997

The Senate Labor and Human Resources Committee's Subcommittee on Public Health and Safety held a hearing entitled *Ethics and Theology: A Continuation of the National Discussion on Human Cloning.*

July 22, 1997

The House Committee on Science's Subcommittee on Technology held a hearing entitled *Legislative Hearing on the Prohibition of Federal Funding for Human Cloning Research.*

August 7, 1997

Researchers at ABS Global, Inc., DeForest, WI, announced that they had succeeded in cloning a Holstein bull from fetal stem cells. Stem cells are "blank slate" cells that have not yet specialized their function, such as a liver or muscle cell. The stem cell has the potential to be any part of the mature animal. Gene, the bull, was created in this manner. Dolly, the sheep, was cloned from the genetic material of an adult cell.

December 18, 1997

Roslin Institute scientists reported data showing the production of the world's first lambs that carry a human gene (transgenic lambs) created by nuclear transfer. To produce the lambs, they first exposed skin cells (fibroblast) to DNA that included a human gene and a marker gene. Then they took the cells that contained both the marker gene and the human gene and followed the same cloning technique they used to make Dolly. Both lambs contained the transgenic gene in their cells.

1998

January 7, 1998

Dr. Richard Seed, a physicist from Riverside, IL, on National Public Radio discussed his plans to open a clinic to clone humans before Congress outlawed the procedure. He stated he had already assembled a team of doctors, and four couples volunteered to participate. His objective was to provide childless couples with children. He first announced his intention on December 5, 1997, at a symposium on reproductive technologies in Chicago.

January 10, 1998

President Clinton, speaking in his weekly radio address, urged swift action by Congress to ban the cloning of humans. He reiterated his support for related areas of cloning research, which might lead to medical breakthroughs.

January 20, 1998

Michael A. Friedman, Acting Commissioner of the Food and Drug Administration (FDA), stated that the FDA has the authority under the Food, Drug, and Cosmetic Act to regulate human cloning, since it is a form of cellular or genetic therapy. Since such therapies require prior approval by FDA reviewers, anyone planning to legally attempt human cloning would have to file a formal application with the agency, which would then undertake a lengthy review of the proposal. Friedman also said the FDA would initiate legal action against anyone who failed to file such an application.

July 22, 1998

Researchers at the University of Hawaii announced that they had created dozens of mice by cloning, using a new technique in the most commonly-used laboratory animal. A paper detailing their method was published in the July 23 issue of the science journal *Nature*. The researchers were able to reprogram nuclei from cells taken from ovaries of adult mice. Known as cumulus cells, these differentiated cells surround the eggs of mice, as well as humans, and are shed with eggs during ovulation. For the cloning experiment, nuclei from cumulus cells were inserted directly into egg cells whose nuclei had been removed. The combination was then activated with chemicals prompting the eggs to start dividing and form embryos. The embryos were transferred to the wombs of surrogate mice, and some resulted in the birth of mice clones (identical to mice from which the cumulus cells were taken). This was the first published documentation that adult animals could be cloned since researchers in Scotland announced the birth of a cloned sheep named Dolly in February 1997. Some scientists questioned the validity of that study. In the same issue of *Nature*, two other papers were published where researchers in England documented that Dolly is indeed a clone.

August 19, 1998

Scientists at the Ruakura Research Center in Hamilton, New Zealand, announced the cloning of the lone survivor of a rare breed of cow. The cloned cow is the last member of a herd which had lived in isolation on Enderby Island, a barren, subantarctic section of the Auckland Islands. From the cow's ovaries, the scientists took several granulosa cells, which nourish egg cells in the ovaries. The granulosa cells are similar to the cumulus cells that Hawaiian researchers used to clone mice. The nuclei from the granulosa cells were inserted directly into egg cells from which the nuclei had been removed. This combination was then activated with electrical shocks. The resulting fused cells divided and developed into embryos, which were transferred into the wombs of surrogate Angus cows. On July 31, 1998, the first clone was born, with slightly different markings than the adult cow. However, the results of a DNA fingerprinting test proved that the clone and the cloned cow were genetically identical.

December 11, 1998

Japanese scientists published data in the journal *Science* reporting the cloning of eight calves from an adult cow's oviduct and cumulus cells. The success rate was higher than that of any other group that cloned large mammals. Ten embryos, derived from differentiated (oviduct and cumulus) cells of one adult cow, resulted in the birth of eight calves, but four calves died at or soon after birth.

1999

April 26, 1999

An announcement was made that a collaboration of industry and academic researchers had successfully produced the world's first transgenic goats through cloning. The three goats are genetically identical copies of a goat embryo whose genetic material was modified to produce milk containing the human anticlotting protein known as antithrombin III (ATIII).

May 27, 1999

A study was published indicating that Dolly, the cloned sheep, may be susceptible to premature aging. Researchers in Scotland found that 3-year-old Dolly, who was cloned from a 6-year-old ewe, has cells that appear to be at least 9 years old.

June 1, 1999

University of Hawaii researchers reported the first documented cloning of an adult male animal. Fibrio, a mouse, was cloned from cells clipped from the tip of a male mouse's tail. This was a different method than was used to produce Gene, the bull, who was cloned from fetal stem cells. Up to now, all animals cloned from adult cells had been female.

June 14, 1999

It was reported that two companies, Geron Corporation (Menlo Park, CA) and Advanced Cell Therapeutics (Worcester, MA) had started programs to grow their own embryos (human clones or human-cow hybrids) by cloning. The clones would be used as sources of embryonic stem cells, which may have potential in the treatment of a number of human conditions, including Parkinson's disease and diabetes.

2000

January 12, 2000

Dr. Gerald Schatten and his colleagues at the Oregon Regional Primate Center, Beaverton, OR, announced the first successful example of the cloning of a monkey called Tetra by using a technique called "embryo splitting." This technique splits an eight-cell embryo into four identical two-cell clones and then implants them in surrogate mothers. Clones made from split embryos are genetically identical.

This technique is different than the one used to clone Dolly the sheep, the first clone of an adult female animal. In that case, the clone was created by taking an adult animal's cell nucleus and implanting it in an unfertilized egg cell from which the original nucleus had been removed.

Some scientists hope that embryo splitting can be used to develop genetically identical laboratory animals better suited for testing therapies that may eventually be used to treat humans.

March 5, 2000

Researchers at the Blacksburg, VA, facility of PPL Therapeutics announced the production of the first cloned pigs—Millie, Christa, Alexis, Carrel, and Dotcom. The Scotland-based company, which also has a research facility in New Zealand, is the same firm that cloned the sheep Dolly. Basically, the pigs were created with the same technique used to create Dolly; but the researchers have not yet disclosed the "additional inventive steps" they said were used to create the pigs. Researchers hope that cloned pigs can eventually become a source of organ and cell transplants for humans.

July 21, 2000

Alexander Kind and his colleagues at PPL Therapeutics, the Edinburgh-based company that helped create Dolly, announced that they had successfully cloned Cupid, Diana, and a third unnamed transgenic lamb. The lambs were the first transgenic livestock to carry specifically chosen modifications in their genes.

The lambs were created using the same method applied for Dolly in 1997. DNA was taken from another sheep and transferred into unfertilized eggs. This time, however, the team picked a specific point on one of the sheep's chromosomes and inserted a new DNA gene sequence into it. This technique is called "gene targeting" and had previously only been possible in laboratory mice.

The inserted gene allowed the sheep to produce the human protein alpha 1-antitrypsin in their milk. This protein may someday be used to treat a variety of lung diseases, including cystic fibrosis; but the true significance of the feat lies in the wider application of gene-altering technology. Another potential application of this technology is the development of animals that could supply organs for human patients.

2001

March 28, 2001

The House Committee on Energy and Commerce, Subcommittee on Oversight and Investigations held a hearing on human cloning, entitled *Issues Raised by Human Cloning Research.*

April 13, 2001

PPL Therapeutics in Scotland announced the creation of the world's first five genetically-modified cloned pigs, a critical milestone towards producing spare part transplant organs for human patients. Inserted into the genetic make-up of the five clones was a marker gene, for fluorescence taken from a jellyfish. Tissue removed from the clones glowed under ultraviolet light, showing that the marker gene had been successfully integrated into their genetic blueprints.

Dr. Alan Colman of PPL Therapeutics stated that the same technique could allow scientists to create pigs whose organs would not be rejected by human patients' immune systems, since pigs have a gene which causes humans to reject their organs. The birth of the piglets suggested that the gene can be altered.

July 16, 2001

H.R. 2505, Human Cloning Prohibition Act of 2001, a bill to amend Title 18 of the *United States Code* to prohibit human cloning, was introduced. This bill passed in the House of Representatives by a 256-162 vote (roll call no. 303) on July 31, 2001. On August 1, 2001, it was received in the Senate, where it received no further action.

July 31, 2001

H. Amdt. 284 to H.R. 2505, was passed in the House of Representatives by voice vote. This amendment required the General Accounting Office to conduct a study to assess any need for the amendment of the prohibition on human cloning, as defined in Section 301 of Title 18, *United States Code,* as added by this Act and to transmit such study to Congress within 4 years.

November 25, 2001

Scientists at Advanced Cell Technology (ACT), a private company in Worcester, MA, announced that they had cloned the first human embryos. They said their aim was not to produce cloned human beings, but to create genetically matched stem cells to treat a wide range of diseases. However, the cloned embryos that they produced stopped developing after dividing into just a few cells — not enough to yield medically useful stem cells. ACT used two techniques to produce human embryos — cloning and a second process called parthenogenesis.

In cloning, the researchers obtained egg cells from seven female volunteers. They stripped the DNA from 19 egg cells and replaced it with genetic material from another person of unspecified gender. The new genetic material came from a skin cell or from ovarian material called a cumulus cell. Seven of the eggs began to divide and grow. These early embryos were clones, or offspring that carried genes from only one adult (the person who had donated the skin or cumulus cell). Two embryos divided into four cells each, and one embryo divided into six cells before the growth stopped. The growth occurred over a 3-day period.

In parthenogenesis, an egg cell is treated with chemicals that cause it to start dividing into an embryo without being fertilized by sperm. ACT exposed 22 human eggs to those chemicals. After 5 days, six eggs had matured into larger masses of cells. Scientists believe embryos created this way could mature long enough to be useful in medical treatment but would be unable to grow to term.

Both the cloned and parthenogenetically produced embryos had significant shortcomings. None developed stem cells, which can grow into any type of body cell or tissue.

November 28, 2001

President Bush issued Executive Order 13237, establishing the President's Council on Bioethics. The Council is to advise the President on bioethical issues that may emerge as a consequence of advances in biomedical science and technology.

The executive order establishing the Council may be accessed on the Web at [http://www.whitehouse.gov/news/releases/2001/11/20011128-13.html].

December 31, 2001

The following legislation to prohibit cloning of humans in some manner was introduced in 2001 during the first session of the 107th Congress: H.Res. 214, H.R. 1260, H.R. 1372, H.R. 1608, H.R. 1644, H.R. 2172, H.R. 2608, H.R. 3495, S. 704, S. 790, S. 1758, and S. 1893. H.R. 2505 with H.Amdt. 284 passed in the House of Representatives (see July 16 and 31, 2001). No floor action was taken on the other measures.

2002

January 3, 2002

PPL Therapeutics in Scotland, the company that cloned Dolly, the sheep, announced that it had cloned another five genetically modified pigs—Noel, Angel, Star, Joy, and Mary. The pigs were born on Christmas Day 2001 at the firm's research facility in Blacksburg, VA. As a step toward producing pigs with organs and cells that could be safely transplanted into humans, the cloned pigs (all female) had a gene that was inactivated, or "knocked out." This is the pig gene that attaches sugar molecules to the surfaces of organs. When organs are transplanted into humans, the human immune system attaches to those sugars, recognizes the transplanted organs as foreign, and rejects them. Several other genetic modifications, including the addition of up to three human genes, would be needed to avoid rejection of the pig parts. The work was being funded by a $2 million grant from the National Institute of Standards and Technology.

January 24, 2002

The Senate Appropriations Committee Subcommittee on Labor, Health and Human Services, and Education, held a hearing on human cloning. At this time, no hearing transcript is available.

January 24, 2002

S. 2893, Human Cloning Ban and Stem Cell Research Protection Act of 2002, a bill to ban human cloning while protecting stem sell research, was introduced. This bill was referred to the Senate Health, Education, Labor, and Pensions Committee.

January 28, 2002

S. 1899, Human Cloning Prohibition Act of 2001, a bill to amend Title 18 of the *United States Code* to prohibit human cloning, was introduced. This bill was referred to the Senate Judiciary Committee.

January 30, 2002

Researchers at Advanced Cell Technology in Worcester, MA, announced that they have used cells derived from cloned cow embryos to grow functioning kidney-like organs that were not rejected when implanted into adult cows, marking the first use of cloning technology to grow personalized, genetically-matched organs for transplantation.

This was the first time that cells taken from a cloned embryo had been used to grow an organ which, like kidneys, removes toxins from the body and produces urine. Researchers are still checking to see if it carries out all the functions of a kidney. For this work to be replicated in humans, the kidney would have to be created by using cloned human embryos. Cells taken from the human patient would be used to produce a cloned human embryo genetically identical to the patient. The experimental procedure would then harvest cells from the embryo to grow the organs needed for transplant, which theoretically would not be rejected by the patient because they would be genetically identical. Information on this work has not been published in a scientific journal, nor has the work been confirmed by others.

February 15, 2002

Researchers at Texas A&M's College of Veterinary Medicine announced that they had created the first cloned cat, a shorthaired calico named CC (short for "carbon copy" and "copy cat"). It was cloned with cells from a cat named Rainbow. Delivered by cesarean section on December 22, 2001, in a university laboratory, CC was the first household pet to be cloned.

Working first with an adult male cat, the researchers harvested cells from the animal's mouth and then fused them with cat donor eggs that had been emptied of genetic material. This created 82 cloned embryos that were transferred into the wombs of seven cats. The process yielded only a single fetal clone, and it died in utero.

In a second attempt, researchers used cumulus cells from the ovaries of the female cat named Rainbow and created five cloned embryos. They were implanted in Allie, another female cat. This time, an embryo took hold and grew. Sixty-six days later, CC arrived.

The kitten was anything but an exact copy of Rainbow. Although tests indicated that CC was a genetic duplicate of the cat that donated the original ovary cell, CC's markings were quite different from those of Rainbow. Calico markings such as those possessed by CC are the result of random molecular changes that occur during fetal development.

Mark Westhusin, the project's lead scientist, stated, "This is reproduction, not resurrection." He warned pet owners that cloning will never return their old pets, although the clones will probably resemble their predecessors in looks and temperament.

The work was funded by Arizona millionaire John Sperling, who gave Texas A&M about $3.7 million to develop technology to clone his pet dog Missy, an aging border collie-Siberian husky mix. Although several pregnancies have been achieved by using cells from Missy, none of the clones have survived to term. Parallel work on cats went faster, Westhusin said, in part because cat eggs grow and mature in culture dishes better than dog eggs.

To commercialize the work, 2 years ago, Sperling created a Texas company called Genetic Savings and Clone, which holds the licensing rights to any proprietary pet cloning techniques developed by Texas A&M's "Missyplicity Project."

March 29, 2002

French researchers reported that they had cloned rabbits by using genetic materials from adult cells. The four cloned rabbits, all females, were born a year earlier at the National Institute for Agronomical Research outside Paris. A team of developmental biologists led by Jean-Paul Renard did not report the achievement until the animals were mature and in good health. Neither these rabbits nor their parents (the adult females that provided the genes and eggs from which the cloned rabbits were made) were given names.

The French team is collaborating with other scientists to try to clone rabbits with the gene defect responsible for cystic fibrosis in human beings. The gene in the two species is very similar, Renard said, and the scientists hope cloned rabbits will provide a better model for studying the disease than other animals.

Robert Lanza, medical director at Advanced Cell Technology, said that since rabbits reach sexual maturity and breed very quickly, they provide offspring quickly as well, which is very important in cloning research. In contrast, a cow fetus spends 9 months in the womb, and it is expensive to produce large numbers of them.

May 15, 2002

The House Government Reform Committee's Subcommittee on Criminal Justice, Drug Policy, and Human Resources held a hearing, entitled *Medical Science and Bioethics: Attack of the Clones.*

December 10, 2002

Stanford University announced plans to create a $120 million institute to study the overlapping biology of cancer and stem cells, including a plan to start cloning new stem cells from human embryos. An anonymous donor gave $12 million to seed a fund-raising campaign for the new center. The new Institute for Cancer/StemCell Biology and Medicine will be directed by Dr. Irving Weissman of Stanford. Researchers at the institute hope to use cloning techniques and stem cells to make genetically tailored material for transplants and to study the course of disease in specific types of human cells.

December 27, 2002

Brigitte Boiselier, a former university chemist who is the director of Clonaid, a company founded by the Raelians (a religious sect that believes humans are descended from space aliens) claimed that Clonaid had created the world's first human clone, a baby named Eve, who is allegedly a genetic carbon copy of her mother. Boiselier said the baby was born by cesarean section, but did not specify where the birth took place or identify the parents. No evidence was offered to back up the claim.

December 31, 2002

The following legislation to prohibit cloning of humans in some manner was introduced in 2002 during the second session of the 107th Congress: S. 2893, S. 1899, S. 2076, and S. 2439. No floor action was taken on these measures, and they died at the end of the 107th Congress.

2003

January 4, 2003

Clonaid announced the birth of a second human cloned baby born to a Dutch couple on January 3, 2003. No evidence was offered to back up the claim.

January 8, 2003

H.R. 234, Human Cloning Prohibition Act of 2003, a bill to amend Title 18 of the *United States Code* to prohibit human cloning, was introduced.

H.R. 246, Departments of Labor, Health and Human Services, and Education, and Related Agencies Appropriations Act, 2003, was introduced. Section 510 includes human clones in its prohibition of funding research using human embryos.

January 29, 2003

S. 245, Human Cloning Prohibition Act of 2002, a bill to amend the Public Health Service Act to prohibit human cloning, was introduced.

February 5, 2003

H.R. 534, Human Cloning Prohibition Act of 2003, a bill to amend Title 18, *United States Code* to prohibit human cloning, was introduced. On February 25, 2003, the bill was reported to the Committee on Judiciary, H.Rept. 108-18. This bill passed in the House by a 241-155 vote (roll call no. 39) on February 27, 2003.

February 5, 2003

S. 303, Human Cloning Ban and Stem Cell Research Protection Act 2003, a bill to prohibit human cloning and protect stem cell research, was introduced.

February 13, 2003

H.R. 801, a bill to amend the Federal Food, Drug, and Cosmetic Act with respect to the cloning of humans, and for other purposes, was introduced.

February 14, 2003

The Roslin Institute, Edinburgh, Scotland, announced that Dolly the sheep was euthanized, short of her normal lifespan after being diagnosed with a progressive lung disease. The decision to end the life of the 6-year-old sheep was made after a veterinary examination confirmed the lung disease.

February 25, 2003

H.R. 916, a bill to prohibit the expenditure of federal funds to conduct or support research on the cloning of humans, and to express the sense of Congress that other countries should establish substantially equivalent restrictions, was introduced.

February 26, 2003

H.R. 938, a bill to prohibit federal payments to any individual, business, institution, or organization that engages in human cloning, was introduced.

February 27, 2003

H.Amdt. 4 to H.R. 534 amendment, as modified, requiring the General Accounting Office, after consultation with the National Academy of Sciences, to conduct a study to assess the need (if any) for amendment to the prohibition on human cloning contained in the bill, was introduced and agreed to by voice vote.

H.Amdt. 5 to H.R. 534, an amendment in the nature of substitute sought to prohibit human somatic cell nuclear transfer technology to initiate a pregnancy and allow its use for medical research, was introduced and failed by recorded vote: 174-231, 1 Present (roll call no. 37).

May 29, 2003

The combined research team led by Gordon Woods and Dirk Vanderwall of the Northwest Equine Reproduction Laboratory at the University of Idaho and by Ken White of Utah State University announced the May 4, 2003, birth of a live cloned mule, the first successful cloning of an equid. Named Idaho Gem, after the state in which he was born, the foal is a sibling of Taz, a world champion racing mule. The researchers did not want to clone an adult animal because they "wanted to take the aging component out of the equation," said Woods. Some researchers suspect that the first cloned mammal, Dolly the sheep, aged prematurely because her DNA was derived from an adult cell. So the team rebred Taz's parents, took a somatic cell from the 45-day-old fetus, and fused it with an enucleated horse oocyte that was then implanted into a mare. Idaho Gem, born after a normal 346-day gestation, is not only the first member of the horse family but also the first sterile animal to be cloned. Mules, sired by donkeys and borne by horses, cannot reproduce.

June 11, 2003

Two weeks after announcing the birth of the first cloned mule, the same combined research team announced the June 9, 2003, birth of a second cloned mule, named Utah Pioneer.

June 26, 2003

S.1356, Department of Health and Human Services Appropriations Act 2004, Title V, General Provisions, Sec. 510, prohibits the use of funds made available in this Act for (1) the creation of a human embryo for research purposes; or (2) research in which a human embryo is destroyed, discarded, or knowingly subjected to risk of injury or death greater than that allowed for research on fetuses in utero under specified federal regulations and the Public Health Service Act. Defines "human embryo or embryos" to include any organism, not protected as a human subject under specified federal regulations as of the date of enactment that is derived by fertilization, parthenogenesis, cloning, or any other means from one or more human gametes or human diploid cells.

WEB SITES

President Clinton established the National Bioethics Advisory Commission (NBAC) by Executive Order 12975 on October 3, 1995. This executive order is in the *Federal Register,* October 5, 1995, p. 52063, and may be accessed at the Web site [http://frwebgate.access.gpo.gov/cgi-bin/getdoc.cgi?dbname=1995_register&docid= fr05oc95-126].

Information on the NBAC is at [http://bioethics.georgetown.edu/nbac/]. The NBAC charter expired on October 3, 2001.

Roslin Institute
[http://www.roslin.ac.uk]

PPL Therapeutics
[http://www.ppl-therapeutics.com/]

Oregon Regional Primate Research Center
[http://www.ohsu.edu/orprc/]

ABS Global, Inc.
[http://www.absglobal.com/]

Advanced Cell Technology
[http://www.advancedcell.com/]

Northwest Equine Reproduction Laboratory, University of Idaho
[http://www.avs.uidaho.edu/nerl/]

Utah State University, Department of Animal, Dairy, and Veterinary Science
[http://www.advs.usu.edu/]

Chapter 2

STATE LAWS ON HUMAN CLONING*

Jody Feder

SUMMARY

Although numerous human cloning bills have been introduced in the 108th Congress and one — H.R. 534 — has passed in the House of Representatives, cloning legislation has yet to be enacted at the federal level. In the absence of federal legislation, state laws provide the only existing framework for regulating human cloning activity. Currently, nine states have enacted laws that expressly regulate cloning. This chapter provides a summary and comparison of existing state laws on human cloning, as well as a brief discussion of similar legislation currently under consideration in Congress.

BACKGROUND

In 1997, scientists announced that they had successfully cloned a sheep, whom they named Dolly. Fearing that scientists would next seek to clone a human being, President Clinton quickly issued an executive memorandum banning the use of federal funds for purposes of cloning.[1] This ban continues to remain in force today. Meanwhile, legislators at the state and federal level have sought to enact legislation that would regulate human cloning specifically, regardless of funding source.[2] In 2002, Clonaid, an organization affiliated with the Raelian sect, added urgency to this pursuit when it announced, in an unsubstantiated claim, that the world's first cloned baby had been born.

---

* Excerpted from CRS Report RS21517. Updated May 14, 2003.

1 Memorandum on the Prohibition on Federal Funding for Cloning of Human Beings, 33 Weekly Comp. Pres. Doc. 281 (March 4, 1997).

2 Although many states have laws that regulate research or experimentation on embryos and fetuses and although these laws may be broad enough to include embryos in the early stages of development, it is unlikely that courts would construe these laws to apply to cloned embryos. Lori B. Andrews, *The Current and Future Legal Status of Cloning, in* Cloning Human Beings, Vol. II at F-5 (National Bioethics Advisory Commission, June 1997). As a result, this report focuses exclusively on state laws that regulate cloning specifically.

Since these events, numerous human cloning bills have been introduced in Congress. Although one bill — H.R. 534 — has passed in the House of Representatives, cloning legislation has yet to be enacted at the federal level. In the absence of federal legislation, state laws provide the only existing framework for regulating human cloning activity. Currently, nine states have enacted laws that expressly regulate cloning. This chapter provides a summary and comparison of existing state laws on human cloning, as well as a brief discussion of similar legislation currently under consideration in Congress.

What is Cloning?

Cloning has several meanings. As used in this chapter, it refers to a process in which cellular material from a DNA donor is transferred to an egg whose own DNA has been removed. When stimulated, the egg begins to divide. The resulting embryo is genetically identical to the DNA of the original cell donor.[3]

Human cloning may be divided into two categories: reproductive cloning and therapeutic cloning. In reproductive cloning, the cloned embryo is implanted in a woman's uterus, where it potentially results in pregnancy and the birth of a cloned human being. Therapeutic cloning, on the other hand, allows scientists to create an abundant source of stem-cells for research purposes.[4]

In general, there is a consensus among legislators and scientists that reproductive cloning, which poses a large number of safety and ethical concerns, should be banned. Far less agreement, however, exists where therapeutic cloning is concerned. Proponents of therapeutic cloning contend that the process of extracting stem cells from cloned human embryos is essential for researching new therapies and developing cures for debilitating or life-threatening diseases like Alzheimer's and Parkinson's disease. Opponents, however, argue that creating cloned human embryos for research purposes unethically treats human life as a commodity and contend that destroying embryos in order to extract stem cells is tantamount to murder.[5]

Partly as a result of this debate, the manner in which states have chosen to regulate human cloning varies widely, but state laws generally fall into one or more of several categories, including laws that ban reproductive cloning, therapeutic cloning, and/or the use of state funds for cloning activities. In addition, state laws may also be categorized according to the type of penalties — civil, criminal, or administrative — that they impose for human cloning violations. In addition to the nine states that have enacted human cloning legislation, another twenty-two states have cloning bills pending before their state legislatures.[6] Each of the state cloning laws currently in force is detailed below.

---

[3] Alissa Johnson, National Conference of State Legislatures, *Human Cloning* (June 2002), [http://www.ncsl.org/programs/health/genetics/Humancloning.pdf].

[4] *Id.*

[5] Mary Agnes Carey, *Cloning Debate Undiminished Despite House-Passed Ban*, CQ Weekly, March 1, 2003, at 508.

[6] For detailed information about proposed cloning legislation in the states, see National Conference of State Legislatures, *2003 Legislative Activity Human Cloning* (April 24, 2003), [http://www.ncsl.org/programs/health/Genetics/03clone.htm].

State Human Cloning Laws

Many of the states that have enacted legislation prohibiting human cloning have distinguished between reproductive and therapeutic cloning. Although some state laws prohibit both types of cloning, other states have opted to ban reproductive cloning but to allow therapeutic cloning. Of the nine states that currently have human cloning legislation in place, four states — Arkansas, Iowa, Michigan, and North Dakota — prohibit both reproductive and therapeutic cloning, while three states — California, Louisiana, and Rhode Island — ban reproductive cloning but permit therapeutic cloning. Of the remaining two states, Missouri forbids the use of state funds to conduct reproductive cloning but allows therapeutic cloning, while Virginia law clearly bans reproductive cloning but is unclear with regard to therapeutic cloning. Beginning with the states that prohibit both types of human cloning, each of the state laws is discussed in detail below.

In Arkansas, both reproductive and therapeutic cloning are banned, as is the shipment or transfer of any cloned human embryo. Violations of the ban on cloning are a felony, while violations of the shipment provision are a misdemeanor. In addition to imposing criminal liability for violations of the statute, the Arkansas statute levies a fine of the greater of $250,000 or twice the amount of any profit gained through illegal cloning activity.[7] Similarly, both reproductive and therapeutic cloning are banned in Iowa, where the transfer or receipt of a cloned human embryo is also banned. Violations of the prohibition against cloning are punishable as a felony, while violations of the transfer prohibition are considered to be an aggravated misdemeanor. As in Arkansas, any person who gains financially from cloning activity is subject to a civil penalty in an amount that is twice the financial gain. Finally, violations of Iowa's human cloning statute are also punishable with the denial or revocation of any license, permit, or certification required to practice a trade, occupation, or profession regulated by the state.[8]

Like Arkansas and Iowa, Michigan forbids reproductive and therapeutic cloning. Violations of this prohibition give rise to administrative sanctions imposed by the health department, civil penalties of $10 million, or felony criminal charges punishable by up to 10 years of imprisonment. In addition, Michigan bars the use of state funds for purposes of human cloning and prohibits health facilities or agencies from permitting individuals to engage in cloning activity.[9] Finally, North Dakota state law also prohibits both types of human cloning, as well as the transfer or receipt of any product of human cloning. Under state law, criminal sanctions are imposed for violations of the statute, with violations of the cloning provisions punishable as a felony and violations of the transfer provision punishable as a misdemeanor.[10]

In contrast to the four states described above, California, Louisiana, and Rhode Island prohibit reproductive cloning but permit therapeutic cloning. In California, violations of the ban on reproductive cloning give rise to civil but not criminal charges. A violator who is a corporation, firm, clinic, hospital, laboratory, or research facility is liable for a fine of up to $1 million, while a violator who is an individual is liable for a fine of up to $250,000.

---

7 2003 Ark. SB 185.

8 Iowa Code § 707B.1-4.

9 MCLS §§ 333.16274-16275, 333.20197, 333.26401-26406, 750.430a.

10 2003 N.D. HB 1424.

However, any violator who profits from his illegal activity is liable for twice the amount of his financial gain.[11] In addition to these civil penalties, a violation of the cloning statute constitutes unprofessional conduct and is grounds for the revocation of a business license.[12]

Like California, Louisiana prohibits reproductive cloning but allows therapeutic cloning. Violations of the ban on reproductive cloning give rise to criminal penalties leading to up to ten years in prison or civil penalties of up to $5 million for individuals and up to $10 million for corporations, firms, clinics, hospitals, laboratories, or research facilities. However, violators who profit from their illegal activities are liable for twice the amount of their financial gain. In addition, violations of Louisiana's cloning law constitute unprofessional conduct and are grounds for denying or revoking any stateissued license, certification, or permit required to practice a trade, occupation, or profession.[13] Finally, Rhode Island also prohibits reproductive cloning but permits therapeutic cloning. Like California, Rhode Island imposes civil but not criminal penalties for violations of the statute; these penalties are identical in amount and structure to the fines imposed in California. Currently, Rhode Island's ban on reproductive cloning is set to expire in 2010.[14]

Of the remaining two states with laws that regulate human cloning, Missouri's law is the most straightforward. Under Missouri law, no state funds may be used for research with respect to the cloning of a human being. The statute is silent with respect to cloning activities conducted using private funds.[15] Virginia law is somewhat more complicated, for although it is clear that state law prohibits reproductive cloning, it is unclear whether or not the ban extends to therapeutic cloning. Despite this confusion, Virginia law clearly imposes a civil penalty of up to $50,000 for each violation of the cloning statute.[16]

As noted above, another 22 states are currently considering various bills that would regulate human cloning. In addition, federal legislators are considering certain cloning bills pending before the United States Congress. These congressional bills are described in the following section.

HUMAN CLONING BILLS IN CONGRESS

In February, the House of Representatives passed H.R. 534, the Human Cloning Prohibition Act of 2003. The bill, which is awaiting consideration by the Senate, would forbid both reproductive and therapeutic cloning, as well as prohibit the shipment, receipt, or importation of any embryo produced by human cloning or and product of such embryo. The bill would impose both civil and criminal penalties for violations of the cloning prohibition. Under the legislation, criminal penalties could result in up to ten years in prison, while any violator who profits from illegal cloning activity would be fined the greater of $1 million or twice the amount of financial gain. In enacting H.R. 534, the House of Representatives rejected an alternative bill — H.R. 801 — that would have prohibited reproductive cloning but permitted therapeutic cloning.

---

[11] Cal. Health & Saf. Code §§ 24185-87, 24189.
[12] Cal. Bus. & Prof. Code §§ 2260.5, 16004, 16105.
[13] La. R.S. 40:1299.36-36.6.
[14] R.I. Gen. Laws §§ 23-16.4-1-4.
[15] § 1.217 R.S.Mo.
[16] Va. Code Ann. §§ 32.1-162.21-22.

Despite winning substantial support in the House, H.R. 534 faces an uncertain future in the Senate. In the 107th Congress, a similar measure passed the House but failed to reach a floor vote in the Senate. Currently, the Senate version of the bill — S. 245, sponsored by Senator Brownback — does not appear to have enough backers to withstand a filibuster.[17] Furthermore, Brownback's bill faces competition from another human cloning bill — S. 303, sponsored by Senator Hatch — that would prohibit reproductive cloning but allow therapeutic cloning for the purpose of conducting stem cell research. Although President Bush has expressed his support for a comprehensive ban on human cloning that would bar both reproductive and therapeutic cloning, prospects for such legislation currently remain unclear due to the dispute in the Senate. As a result, the states are likely to continue their role as groundbreakers in enacting legislation to regulate human cloning.

[17] Mary Agnes Carey, *Cloning Debate Undiminished Despite House-Passed Ban*, CQ Weekly, March 1, 2003, at 508.

Chapter 3

HUMAN CLONING[†]

Judith A. Johnson

INTRODUCTION

On December 27, 2002, a representative of Clonaid announced the overseas birth of the first cloned human to a 31-year old American woman. Although the company said genetic tests would show that the baby is a clone of the birth mother, tests results have not been released. Claims of a second birth of a cloned baby to a Dutch couple on January 3, 2003 have also not been proven. These announcements are likely to rekindle debate in the 108th Congress on the moral and ethical implications of human cloning as the disclosure by Advanced Cell Technology (ACT) did in the 107th Congress. In November 2001 ACT announced the creation of the first cloned human embryos (which survived only for a few hours). ACT intended to use the embryos to derive stem cells to produce therapies for diseases like diabetes and Parkinson's disease.

President Bush announced in August 2001 that for the first time federal funds will be used to support research on human embryonic stem cells, but funding will be limited to "existing stem cell lines." Federal funds will not be used for the cloning of human embryos for any purpose, including stem cell research. The President's Council on Bioethics was established in November 2001 to consider all of the medical and ethical ramifications of biomedical innovation. In July 2002 the Council released its report on human cloning which unanimously recommended a ban on reproductive cloning and, by a vote of 10 to 7, a 4-year moratorium on cloning for medical research purposes.

In January 2002, the National Academies released its report entitled *Scientific and Medical Aspects of Human Reproductive Cloning*. The panel recommended that the U.S. ban human reproductive cloning that is aimed at creating a child. It suggested the ban should be legally enforceable and carry substantial penalties rather than rely simply on voluntary actions. The panel noted that the ban should be reconsidered within 5 years, but only if compelling new data on safety and efficacy are presented and a national dialogue on the

---

[†] Excerpted from CRS Report RL31358. Updated March 10, 2003.

social and ethical issues suggests that a review is warranted. However, the panel concluded that research using cloning procedures to produce stem cells should be permitted because of the considerable potential for developing new therapies and advancing biomedical knowledge.

The House passed H.R. 534 (Weldon), the Human Cloning Prohibition Act of 2003, on February 27, 2003. H.R. 534 would ban the process of human cloning as well as the importation of any product derived from an embryo created via cloning. Cloning could not be used for reproductive purposes or for research on therapeutic purposes, which has implications for stem cell research. The House defeated a substitute amendment, H.Amdt. 5, that would have banned *only* human reproductive cloning; the ban would have sunset after 10 years. H.Amdt 5 is similar to H.R. 801 (Greenwood). Supporters of H.R. 534 argue that a partial ban on human cloning, such as H.R. 801, would be impossible to enforce. Critics of H.R. 534 argue that the measure would curtail medical research and prevent Americans from receiving lifesaving treatments created overseas. In the Senate, S. 234 (Brownback)would ban reproductive cloning and research on therapeutic cloning; S. 303 (Hatch) would ban only reproductive cloning.

BACKGROUND

The term "cloning" is used by scientists to describe many different processes that involve making copies of biological material, such as a gene, a cell, a plant or an animal. The cloning of genes, for example, has led to new treatments developed by the biotechnology industry for diseases such as diabetes and hemophilia. In the context of this chapter, a human embryo produced via cloning involves the process called somatic[1] cell nuclear transfer (SCNT). In SCNT, the nucleus of an egg is removed and replaced by the nucleus from a mature body cell, such as a skin cell. In cloning, the embryo is created without sexual reproduction.

Concern over the possibility of producing a human clone increased with the announcement on February 24, 1997, that scientists in Scotland had used SCNT in 1996 to produce the first cloned adult mammal, Dolly, the sheep. Scientists at the Roslin Institute in Edinburgh removed the nucleus from a sheep egg and replaced it with the nucleus of a mammary gland cell from an adult sheep. The resulting embryo was then transferred to the uterus of a surrogate sheep. A total of 277 such embryos were transferred, but only one lamb was born.[2] Analyses of Dolly's genetic material confirmed that she was derived from the sheep mammary cell. Dolly was euthanized on February 14, 2003, after developing a lung infection. Although some claim that her somewhat early death at 6 years was related to being a clone, scientists at the Roslin Institute believe her ailment may be due to the fact that she was raised indoors (for security reasons) rather than as a pastured sheep, which can live to 11 or 12 years of age.[3]

Although scientists have been successful in using SCNT to produce other animals (such as a cat, goat, cow, pig and mouse), the efficiency of the procedure is still very low and

---

[1] A somatic cell is a body cell, as opposed to a germ cell, which is an egg or sperm cell.

[2] Wilmut, I., et al. Viable Offspring Derived From Fetal and Adult Mammalian Cells. *Nature*, v. 385, February 27, 1997. p. 810-813.

[3] Kolata, G. First Mammal Clone Dies; Dolly Made Science History. The New York Times, February 15, 2003, p. A4.

frequently results in abnormal development. Proponents maintain that one day cloning may be very useful for a number of agriculture applications, including the improvement of livestock.

Clonaid

On December 27, 2002, a representative of Clonaid announced the birth of the first cloned human, a seven pound baby girl nicknamed Eve. The baby was born at 11:55 AM on December 26, 2002, at an undisclosed location outside the United States. Although the company offered no proof of its claim, Dr. Brigette Boisselier, Managing Director of Clonaid, stated that genetic tests would show that the baby is the clone of the 31 year old American woman who is the birth mother. To date tests have not been performed; the company claims that the parents fear the test results could lead to legal actions and loss of custody of the child.[4]

Dr. Boisselier stated that as of December 27, 2002, four other women were pregnant with clones created by Clonaid scientists, and two of the women were carrying clones of children who have died. On January 4, 2003, the company announced the birth of a second cloned baby to a Dutch couple. The baby was born on January 3, 2003; no additional information or evidence was provided by Clonaid. The company plans to implant cloned embryos in 20 additional women in January 2003. Clonaid was founded in 1997 by the leader of the Raelians, an international sect of 55,000 people in 84 countries, which claims that life on Earth was created via genetic engineering by a human extraterrestrial race.[5]

The Food and Drug Administration (FDA) is investigating the company's actions; the agency would consider any human cloning activity to be illegal if performed in the United States.[6] In April 2001 FDA investigated a Clonaid laboratory in Nitro, West Virginia; the laboratory closed shortly thereafter.[7]

Advanced Cell Technology

On November 25, 2001, Advanced Cell Technology (ACT) of Massachusetts announced that it had created the world's first human embryos produced via cloning; the results were published the following day in an electronic journal.[8] ACT used two techniques to produce human embryos — SCNT and a second process called parthenogenesis. ACT researchers obtained eggs from seven women, ages 24 to 32, who were paid $3000 to $5000. In the SCNT approach, scientists removed the nucleus from 19 eggs and replaced it with a nucleus from another adult cell. For 11 of the eggs, the nucleus came from a skin cell, for the remaining eight eggs, from cells which cling to the egg and are called cumulus cells. None of the eggs that received a skin cell nucleus divided; seven of the eggs with the cumulus cell nucleus began to divide. Two embryos divided into four cells each, and one embryo divided

4 Chang, K. Scientist in Clone Tests Says Hoax is Possible. *The New York Times*, January 7, 2003, p. A12.

5 For further information, see: [http://www.clonaid.com] and [http://www.rael.org].

6 Greenhouse, L. FDA Exploring Human Cloning Claim. *The New York Times*, December 30, 2002, p.

7 Kolata, G. and K. Chang. For Clonaid, a Trail of Unproven Claims. *The New York Times*, January 1, 2003, p. A13.

8 Cibelli, J.B., et al. Somatic Cell Nuclear Transfer in Humans: Pronuclear and Early Embryonic Development. *Journal of Regenerative Medicine*, v. 2, November 26, 2001, p. 25-31.

into six cells before division stopped. In parthenogenesis, an egg cell is treated with chemicals causing it to divide without being fertilized by a sperm. ACT exposed 22 human eggs to the chemicals. After 5 days, six eggs had matured into a larger mass of cells before division stopped. None of the embryos developed by ACT through cloning divided sufficiently to produce stem cells, which occur at the blastocyst stage in 1-week-old embryos.

The stated goal of ACT's work is not to produce a cloned human baby (which requires implantation of the cloned embryo into a woman's uterus), but human embryonic stem cells.[9] Other research groups have derived stem cells from mice and cattle using SCNT. ACT intends to derive stem cells from human embryos to develop new therapies for diseases such as diabetes and Parkinson's disease. Some scientists believe that stem cells transplanted into a patient could treat disease or injury by replacing damaged tissue. If the cell nucleus used in SCNT is from the patient, the stem cells would be genetically identical to the patient, recognized by the patient's immune system, and avoid any tissue rejection problems that could occur in other stem cell therapeutic approaches. Because of this, many scientists believe the SCNT technique may provide the best hope of eventually treating patients using stem cells for tissue transplantation. A California biotechnology company, Geron Corporation, has also explored creating stem cells via SCNT.[10]

Dr. Richard Seed, Dr. Panos Zavos, Dr. Severino Antinori

Within a year of the Dolly announcement, concerns over human cloning were heightened when Dr. Richard Seed, a Chicago scientist, announced on January 7, 1998, his intention to clone a human being. In response, bills were introduced in the 105th Congress that would have banned human cloning indefinitely or imposed a moratorium. The legislation was opposed by a number of medical organizations, the biotechnology industry and many scientists and was not enacted. Others who have expressed an interest in reproductive cloning include Dr. Panos Zavos, of the University of Kentucky, and Dr. Severino Antinori, director of a fertility clinic in Rome. At one time, Dr. Zavos and Dr. Antinori were working together to help infertile couples have children via cloning. In April 2002, there were unconfirmed reports in the media that Dr. Antinori had implanted cloned human embryos in women. Dr. Antinori claimed there were 3 such pregnancies of 6 to 9 weeks duration, 2 in Russia and 1 in an Islamic state. His claim was disputed by his former partner Dr. Zavos.

ETHICAL AND SOCIAL ISSUES

The possibility of using cloning technology not just for therapeutic purposes but also for reproducing human beings raises profound moral and ethical questions. In response to the creation of Dolly and concerns about the potential application of cloning humans, on February 24, 1997, President Clinton asked the National Bioethics Advisory Commission[11]

[9] For more information about stem cells, see CRS Report RL31015, *Stem Cell Research*, by Judith A. Johnson.

[10] Weiss, R. Embryo Work Raises Spector of Human Harvesting. *Washington Post*, June 14, 1999. p. A01.

[11] NBAC was established by Presidential Executive Order 12975 on October 3, 1995, to provide guidance to federal agencies on the ethical conduct of current and future human biological and behavioral research. A September 16, 1999 executive order extended the NBAC charter until October 2001. NBAC has been replaced by the

(NBAC) to review the ethical and legal issues associated with the use of cloning technology; NBAC reported its findings and recommendations on June 9, 1997.[12] NBAC recommended a continuation of the moratorium on the use of federal funding in the support of any attempt to create a child by SCNT, and an immediate request to all non-federally funded investigators to comply voluntarily with the intent of the federal moratorium. NBAC also recommended that federal legislation be enacted, with a 3- to 5- year sunset clause, to prohibit anyone from attempting to create a child through the use of SCNT in a research or clinical setting. The NBAC found it morally unacceptable to attempt to clone humans for the purpose of human reproduction because scientific data from animal experiments indicate the method is not safe for mother or baby. In addition to concerns about physical safety, the NBAC report pointed out that SCNT raises issues about the individuality, autonomy, objectification and kinship of the resulting children.

Cloning, if allowed for human reproduction, could affect society's perception of what it means to be a human being. Uncertainties over a cloned individual's personal uniqueness or freedom to create one's own identity may haunt him or her. Relatives or friends could have specific expectations regarding the cloned individual's talents and abilities. Others might ill treat or discriminate against a cloned individual. Some worry that cloning would lead to diminished respect for human life in general, and for cloned individuals in particular, since the cloned person might simply be replaced with another clone. Others point out, however, that this altered perception does not occur today with identical twins, who are naturally produced clones. Cloning human embryos also raises difficult questions about the rights of parents to control their own embryos and other issues concerning reproductive rights and privacy. Some observers believe that it would be ethical to clone human embryos to help infertile couples conceive. Lastly, the possibility of human cloning is offensive to the religious and other deeply held beliefs of many people.

BRIEF HISTORY OF FEDERAL POLICY INVOLVING HUMAN EMBRYO RESEARCH

Currently no U.S. laws or regulations would prohibit all cloning research. However, federal funding of *any* type of research involving human embryos, starting with *in vitro* fertilization (IVF) then later cloning and stem cells, has been blocked by various policy decisions dating back 25 years.

Ethics Advisory Board

Following the birth of the first IVF baby, Louise Brown, in July 1978, the Ethics Advisory Board (EAB) was tasked with considering the scientific, ethical, legal, and social

President's Council on Bioethics, which was described by the Bush Administration in its August 9, 2001 policy decision on human embryonic stem cell research. The President's remarks on embryonic stem cell research are available at: [http://www.whitehouse.gov/news/releases/2001/08/20010809-2.html].

[12] National Bioethics Advisory Commission. *Cloning Human Beings*. June 1997

issues surrounding human IVF.[13] The EAB released its report on May 4, 1979, which found that IVF research was acceptable from an ethical standpoint and could be supported with federal funds. The EAB's recommendations were never adopted by HHS, the EAB was dissolved in 1980, and no other EAB was ever chartered. Because federal regulations that govern human subject research (45 CFR 46) stipulated that, at the time, federally supported research involving human IVF must be reviewed by an EAB, a so-called "de facto moratorium" on human IVF research resulted. Other types of embryo research ensuing from the development and use of IVF, such as cloning and stem cells, were therefore also blocked. The de facto moratorium was lifted with the enactment of the National Institutes of Health (NIH) Revitalization Act of 1993 (P.L. 103-43, Section 121(c)) which nullified the regulatory provision (45 CFR 46.204(d)) requiring EAB review of IVF proposals.

NIH Human Embryo Research Panel

In response, the NIH established the Human Embryo Research Panel to assess the moral and ethical issues raised by this research and develop recommendations for NIH review and conduct of human embryo research. The NIH Panel released a report providing guidelines and recommendations on human embryo research in September 1994. It recommended that some areas of human embryo research be considered for federal funding, including SCNT, stem cells and (under certain limited conditions) embryos created solely for the purpose of research.[14] The NIH Panel also identified areas of human embryo research it considered to be unacceptable, or to warrant additional review. It determined that certain types of cloning[15] without transfer to the uterus warranted additional review before the Panel could recommend whether the research should be federally funded. However, the Panel concluded that federal funding for such cloning techniques followed by transfer to the uterus should be unacceptable into the foreseeable future. The Panel's report was unanimously accepted by the NIH Advisory Committee to the Director (ACD) on December 2, 1994.

After the ACD meeting on December 2, 1994, President Clinton directed NIH not to allocate resources to "support the creation of human embryos for research purposes." The President's directive did not apply to research involving so-called "spare" embryos, those that sometimes remain from clinical IVF procedures performed to assist infertile couples to become parents. Nor did it apply to human parthenotes, eggs that begin development through artificial activation, not through fertilization. Following the Clinton December 2, 1994 directive to NIH, the agency proceeded with plans to develop guidelines to support research using spare embryos.

---

[13] The EAB was created in 1978 by the Department of Health Education and Welfare (HEW), the forerunner of the Department of Health and Human Services (HHS). The EAB was formed at the recommendation of the National Commission for the Protection of Human Subjects of Biomedical and Behavioral Research. The National Commission operated from 1974 to 1978 and issued 10 reports, many of which formed the basis of federal regulations for research involving human subjects (45 CFR 46).

[14] National Institutes of Health. *Report of the Human Embryo Research Panel*, September 27, 1994.

[15] These were **blastomere separation**, where a two- to eight-cell embryo is treated causing the cells (blastomeres) to separate; and, **blastocyst division**, in which an embryo at the more advanced blastocyst stage is split into two.

Dickey Amendment

NIH plans to develop guidelines on embryo research were halted on January 26, 1996, with the enactment of P.L. 104-99 which contained a rider that affected FY1996 funding for NIH. The rider prohibited HHS from using appropriated funds for the creation of human embryos for research purposes or for research in which human embryos are destroyed. This same rider, often referred to as the Dickey amendment, has been attached to the Labor, HHS and Education Appropriations Acts for FY1997 through FY2002.[16] For FY2003, the provision is found in Section 510 of Division G in H.J.Res. 2, which is the Labor, HHS and Education division of the Omnibus FY2003 appropriation bill. It prohibits HHS from using FY2003 appropriated funds for:

1. the creation of a human embryo or embryos for research purposes; or,
2. research in which a human embryo or embryos are destroyed, discarded, or knowingly subjected to risk of injury or death greater than that allowed for research on fetuses in utero under 45 CFR 46.208(a)(2) and Section 498(b) of the Public Health Service Act (42 U.S.C. 289g(b)). For purposes of this section, the term "human embryo or embryos" includes any organism, not protected as a human subject under 45 CFR 46 ... that is derived by fertilization, parthenogenesis, cloning, or any other means from one or more human gametes [sperm or egg] or human diploid cells.

One month after the Dolly announcement, on March 4, 1997, President Clinton sent a memorandum to the heads of all executive departments and agencies making it "absolutely clear that no federal funds will be used for human cloning." This action extended the congressional ban beyond HHS to all federally supported research. Clinton also urged the private sector to adopt a voluntary ban on the cloning of human beings. The *NIH Guidelines on Stem Cell Research*, published by the Clinton Administration in August 2000, would not have funded research in which: human stem cells are used for reproductive cloning of a human; human stem cells are *derived* using SCNT; or, human stem cells that were derived using SCNT are *utilized* in a research project.

Bush Administration Policy Regarding Human Embryo Research

On August 9, 2001 President Bush announced that for the first time federal funds will be used to support research on human embryonic stem cells, but funding will be limited to "existing stem cell lines." In the speech, President Bush stated that he strongly opposes human cloning. Although not mentioned specifically in the August 9 speech, a fact sheet on

[16] The original rider, introduced by Representative Jay Dickey, is in Section 128 of P.L. 104-99; it affected NIH funding for FY1996 contained in P.L. 104-91. For subsequent fiscal years, the rider is found in Title V, General Provisions, of the Labor, HHS and Education Appropriations Acts in the following public laws: FY1997, P.L. 104-208; FY1998, P.L. 105-78; FY1999, P.L. 105-277; FY2000, P.L. 106-113; FY2001, P.L. 106-554; and FY2002, P.L. 107-116.

the White House website states that federal funds will not be used for "the cloning of human embryos for any purpose."[17] In his speech, President Bush announced his intention to name a President's council, chaired by Dr. Leon Kass of the University of Chicago, "to consider all of the medical and ethical ramifications of biomedical innovation." The President's Council on Bioethics, was established for a period of up to 2 years by Executive Order 13237 on November 28, 2001. The White House announced the other 17 members of the council on January 16, 2002.

The first meeting of the President's Council on Bioethics was held on January 17-18, 2002, in Washington, D.C.[18] Dr. Kass announced that the first topic to be addressed by the Council would be human cloning. At the Council's second meeting, the terminology of cloning was discussed in order to reach a consensus on the terms used to describe the two types of cloning: reproductive vs. therapeutic or research cloning. All Council members voted in opposition to reproductive cloning; however, they could not come to an agreement on articulating the precise nature of their objection, whether solely on safety grounds or which of the various moral objections were most important. On the issue of therapeutic cloning, what the Council prefers to call research cloning, the Council also could not come to agreement. Dr. Kass proposed that the Council's final report should reflect both the arguments supporting cloning for the purpose of medical treatment and those against. He asserted that the report should also provide the soundest arguments for each position and indicate how many Council members supported each viewpoint.

The third meeting of the Council was held on April 25 and 26, 2002. The Council heard presentations on the scientific and therapeutic promise of embryonic stem cells from John Gearhart of Johns Hopkins University and the potential of adult stem cells from Catherine Verfaillie of the University of Minnesota. In an informal vote, about half of the 18 members of the Council voiced their support for the therapeutic use of human cloning. The May 2002 meeting was cancelled.

At the June 20, 2002, meeting, nine Council members voted to support cloning for medical research purposes, without a moratorium, provided a regulatory mechanism was established.[19] Because one member of the Council had not attended the meetings and was not voting, the vote seemed to be 9 to 8 in favor of research cloning. However, draft versions of the Council report sent to Council members on June 28, 2002, indicated that two of the group of nine members had changed their votes in favor of a moratorium. Both made it clear that they have no ethical problem with cloning for biomedical research, but felt that a moratorium would provide time for additional discussion.[20] The changed vote took many Council members by surprise, and some on the Council believe that the moratorium option, as opposed to a ban, was thrown in at the last minute and did not receive adequate discussion. In addition, some on the Council believe that the widely reported final vote of 10 to 7 in favor of a moratorium does not accurately reflect the fact "that the majority of the council has no problem with the ethics of biomedical cloning."[21] The final report, *Human Cloning and*

---

[17] The White House Fact Sheet on embryonic stem cell research is available at: [http://www.whitehouse.gov/news/releases/2001/08/20010809-1.html].

[18] A transcript of the first meeting and papers developed by staff for discussion during the meeting can be found at [http://www.bioethics.gov].

[19] Hall, S.S. President's Bioethics Council Delivers, *Science*, v. 297, July 19, 2002, p. 322-324.

[20] Ibid., p. 324.

[21] Ibid., p. 322.

Human Dignity: An Ethical Inquiry, was released at the July 11, 2002, meeting of the Council.

In March 2001, the FDA sent letters to the research community stating that the creation of a human being using cloning is subject to FDA regulation under the Public Health Service Act and the Food, Drug and Cosmetic Act.[22] FDA stated that such research could only occur when an investigational new drug application (IND) is in effect. Some legal scholars believe that there is no legal basis for the regulation of cloning by FDA.[23] They find little evidence to support FDA's position that cloned human embryos are "drugs." However, the biotechnology industry and the American Society for Reproductive Medicine believe FDA has the authority to regulate cloning and legislation is unnecessary because FDA regulation is preferred to any new action by Congress.[24]

On January 18, 2002, the National Academies released its report entitled *Scientific and Medical Aspects of Human Reproductive Cloning.*[25] The panel recommended that the U.S. ban human reproductive cloning that is aimed at creating a child. Based on the results of animal cloning experiments, the panel was concerned for the safety of both the woman and the fetus and judged the procedure to be too dangerous for use in humans at the present time. It recommended that the ban should be legally enforceable and carry substantial penalties rather than be based simply on voluntary actions. The panel stated that the ban should be reconsidered within 5 years, but only if compelling new data on safety and efficacy are presented and a national dialogue on the social and ethical issues suggests that a review is warranted. However, the panel concluded that research using SCNT to produce stem cells should be permitted because of the considerable potential for developing new therapies and advancing biomedical knowledge. This position is in agreement with a previous National Academies' report entitled *Stem Cells and the Future of Regenerative Medicine* which was released on September 11, 2001.[26]

LEGISLATION

On February 27, 2003, the House passed H.R. 534 (Weldon), the Human Cloning Prohibition Act of 2003 by a vote of 241-155. H.R. 534 amends Title 18 of the United States Code and would ban the process of human cloning as well as the importation of any product derived from an embryo created via cloning. Under this measure, cloning could not be used for reproductive purposes or for research on therapeutic purposes, which would have implications for stem cell research. H.R. 534 includes a criminal penalty of imprisonment of not more than 10 years and a civil penalty of not less than $1 million.

H.R. 534 is essentially identical to the measure which passed the House in the 107th Congress (H.R. 2505). During floor debate on H.R. 534, an amendment, H.Amdt. 4 (Scott),

---

[22] The FDA position statement and letters to the research community are available at [http://www.fda.gov/cber/genetherapy/clone.htm].

[23] Weiss, R. Legal Barriers to Human Cloning May Not Hold Up. *Washington Post*, May 23, 2001. p. A1.

[24] Ibid.

[25] The National Academies are the National Academy of Sciences, the National Academy of Engineering, the Institute of Medicine, and the National Research Council. The report on human cloning is available at: [http://www.nap.edu/catalog/10285.html?onpi_topnews_011802].

[26] The National Academies' report on stem cell research is available at: [http://www.nap.edu/catalog/10195.html?onpi_topnews_091101].

was agreed to by voice vote. H.Amdt. 4 requires that the General Accounting Office (GAO), in consultation with the National Academy of Sciences, conduct a study on the impact of the cloning ban on medical technology and assess the need (if any) for modification of the cloning ban contained in the bill. A report to Congress with findings and recommendations would be required within 2 years of enactment. An amendment in the nature of a substitute, H.Amdt 5 (Greenwood), was not adopted by a vote of 174 to 231. The amendment would have prohibited human SCNT technology when used to initiate a pregnancy but allowed SCNT to be used in medical research. H.Amdt 5 is similar to H.R. 801 (Greenwood) (see below).

H.R. 534 was introduced on February 5, 2003, and reported (19-12 vote) by the House Judiciary Committee on February 12, 2003 (H.Rept. 108-18). During markup, four amendments were defeated by 12-19 or by voice vote. The amendments attempted to either limit the ban to 3 years, exempt the importation of medical treatments, exempt the use of cloning in research, or in the creation of additional stem cell lines. A fifth amendment that would add the GAO study was withdrawn when Chairman Sensenbrenner assured his support if it was added to the bill during floor debate.

A companion bill, S. 245 (Brownback), was introduced on January 29, 2003. It is similar to H.R. 534, except that: (1) it does not contain the ban on importation of products derived from therapeutic cloning; and (2) it amends Title 4 of the Public Health Service Act (42 U.S.C. 289 et seq.) instead of Title 18 of the United States Code. S. 245 includes a criminal penalty of imprisonment of not more than 10 years and a civil penalty of not less than $1 million. It requires the General Accounting Office to conduct a study to assess the need (if any) for any changes of the prohibition on cloning in light of new developments in medical technology, the need for SCNT to produce medical advances, current public attitudes and prevailing ethical views on the use of SCNT and potential legal implications of research in SCNT. The study is to be completed within 4 years of enactment. S. 245 has been referred to the Senate Health, Education, Labor, and Pensions Committee.

H.R. 801 (Greenwood), the Cloning Prohibition Act of 2003, was introduced on February 13, 2003. H.R. 801 would prohibit human reproductive cloning while allowing cloning for medical research purposes, including stem cell research. The bill includes a civil penalty of up to $10 million and a criminal penalty of up to 10 years in prison for those convicted of using SCNT for human reproductive purposes, or for importing the products of human cloning if the products would be used to initiate a pregnancy. The bill amends the Food, Drug and Cosmetic Act (21 U.S.C. 301 et seq.) and requires that all researchers performing SCNT on human cells must register their research activity with the Secretary; such registration would most likely be submitted to the FDA.

H.R. 801 stipulates that all research involving human SCNT shall be conducted in accordance with Part 50 (Protection of Huma Subjects) and Part 56 (Institutional Review Boards) of Title 21 of the Code of Federal Regulations (CFR). Under the bill, individuals whose cells are used for such research (presumably the donor of the unfertilized egg and the donor of the somatic cell) would be considered human subjects for the purposes of Parts 50 and 56 of 21CFR. In addition to the requirements under Parts 50 and 56 of 21CFR, the human cell donors must sign an informed consent statement declaring that: (1) the cells are donated for research purposes; (2) the donor understands that Federal law regulates SCNT and use of SCNT to initiate a pregnancy is a criminal act; and, (3) the individual does not intend for the donated cells to be used to initiate a pregnancy. A sunset provision states that the prohibition would expire 10 years after enactment.

H.R. 801 requires the Secretary of HHS to request a study reviewing the current state of knowledge on: (1) the biological properties of stem cells obtained from embryos, fetal tissue, and adult tissue; (2) any biological differences of such stem cells and the consequences for research and medicine; and (3) the ability of stem cells to generate different types of tissue and their potential clinical uses. The study must be conducted by the Institute of Medicine or another appropriate public or nonprofit private entity.

S. 303 (Hatch), the Human Cloning Ban and Stem Cell Research Protection Act of 2003, was introduced on February 5, 2003. Although S. 303 would amend Title 18 of the United States Code while H.R. 801 would amend Title 21, both bills have the same intent: human reproductive cloning would be banned but cloning for medical research purposes would be allowed, including stem cell research. S. 303 includes a criminal penalty of imprisonment of not more than 10 years and a civil penalty of not less than $1 million.

S. 303 requires the Comptroller General to prepare a report within 1 year of enactment that describes the actions taken by the Attorney General to enforce the prohibition on human reproductive cloning, the personnel and resources used to enforce the prohibition, and a list of any violations of the prohibition. The Comptroller General must also prepare a report within 1 year of enactment on similar state laws that prohibit human cloning and actions taken by the States' attorney general to enforce the provisions of any similar state law along with a list of violations. A report on the coordination of enforcement actions among the federal, state and local governments must also be prepared by the Comptroller General within 1 year of enactment, as well as a report on laws adopted by foreign countries related to human cloning.

S. 303 also would amend the Public Health Service Act by requiring that nuclear transplantation research be conducted in accordance with the ethical requirements (such as informed consent, examination by an Institutional Review Board, and protections for safety and privacy) contained in subpart A of 45CFR46, or Parts 50 and 56 of 21CFR. In contrast, H.R. 801 requires that all such research shall be conducted in accordance with Part 50 and 56 of 21CFR and does not refer to subpart A of 45CFR46.[27]

S. 303 contains a prohibition on conducting SCNT on fertilized human eggs (oocytes), and states that "unfertilized blastocysts" shall not be maintained after more than 14 days from its first cell division, aside from storage at temperatures less that zero degrees centigrade. S. 303 stipulates that a human egg may not be used in SCNT research unless the egg is donated voluntarily with the informed consent of the woman donating the egg; H.R. 801 contains a similar egg donation and informed consent provision. S. 303 also specifies that human eggs or unfertilized blastocysts may not be acquired, received or otherwise transferred for valuable consideration if the transfer affects interstate commerce. Under S. 303, SCNT may not be conducted in a laboratory in which human eggs are subject to assisted reproductive

---

[27] Subpart A 45CFR46, often referred to as the Common Rule, "applies to all research involving human subjects conducted, supported or otherwise subject to regulation by any federal department or agency which takes appropriate administrative action to make this policy applicable to such research." The Common Rule covers 18 federal agencies by force of law or Executive Order. FDA has regulatory authority over research on the products the agency regulates (food, drugs, medical devices) and applies its own set of regulations on the protection of human subjects (21CFR 50, 56) that are generally but not entirely the same as subpart A of 45CFR46. For further information, see National Bioethics Advisory Commission, *Ethical and Policy Issues in Research Involving Human Participants*, Appendix C: The Current Oversight System: History and Description, August 2001.

technology treatments or procedures, such as in vitro fertilization, for the treatment of infertility. Violation of these provisions in S. 303 regarding ethical requirements would result in a civil penalty of not more than $250,000. S. 303 has been referred to the Senate Judiciary Committee.

During floor debate in the 107th Congress, supporters of a ban on human cloning (such as that contained in H.R. 534 introduced in the 108th Congress) argued that a partial ban on human cloning (such as that contained in S. 303 introduced in the 108th Congress) would be impossible to enforce. Critics of the ban on human cloning argued that SCNT creates a "clump of cells" rather than an embryo, and that the measure would curtail medical research and prevent Americans from receiving lifesaving treatments created overseas. President Bush has stated his support for a prohibition on all forms of human cloning and has endorsed the cloning ban legislation introduced in the 107th Congress (H.R. 2505) and the 108th Congress (H.R. 534). However, 40 Nobel Laureates, who are in favor of nuclear transplantation technology (SCNT) for research and therapeutic purposes, announced their strong opposition to the legislation.[28] The statement asserted that the legislation "would impede progress against some of the most debilitating diseases known to man."

Former President Gerald Ford stated his strong opposition to the legislation in a April 25, 2002, letter to President Bush.[29] In the letter, Ford indicated that during his administration, the controversy over recombinant DNA research was "successfully addressed with 'careful thought' and the implementation of safety regulations."[30] Former President Ford expressed his "full support for therapeutic cloning, arguing a prohibition of this technology 'would adversely impact scientific research and should not become law.'"[31]

Former First Lady Nancy Reagan has indicated she also is opposed to legislation that would limit embryonic stem cell research and its promise in aiding patients afflicted with serious diseases which have no treatment, such as Alzheimer's disease. In 1994, it was disclosed that former President Ronald Reagan was suffering from the effects of Alzheimer's disease. In a recent letter to Senator Orrin Hatch, Mrs. Reagan states her support for stem cell research and S. 303 which will allow the use of therapeutic cloning.[32]

The U.S. Supreme Court has recognized in past cases certain personal rights as being fundamental and protected from government interference.[33] Some legal scholars believe a ban on human cloning may be struck down by the Supreme Court because it would infringe upon the right to make reproductive decisions which is "protected under the constitutional right to privacy and the constitutional right to liberty."[34] Other scholars do not believe that noncoital, asexual reproduction, such as cloning, would be considered a fundamental right by the Supreme Court. A ban on human cloning research may raise other constitutional issues: scientists' right to personal liberty and free speech. In the opinion of some legal scholars, any

---

[28] The American Society for Cell Biology statement by the 40 Nobel Laureates is available at: [http://www.ascb.org /publicpolicy/Nobelletter.html].

[29] Hafner, L. Revised Feinstein/Kennedy Cloning Bill Has Criminal and Civil Penalties, Requires Research Review. *Washington Fax*, May 2, 2002.

[30] Ibid.

[31] Ibid.

[32] Complete text of the Reagan letter can be found at: [www.senate.gov/~hatch/].

[33] For further discussion of these issues and their relationship to human cloning, see CRS Report RL31422, *Substantive Due Process and a Right to Clone*, by Jon O. Shimabukuro.

[34] Andrews, L. B. Is There a Right to Clone? Constitutional Challenges to Bans on Human Cloning. *Harvard Journal of Law and Technology*, summer 1998. p. 643-680.

government limits on the use of cloning in scientific inquiry or human reproduction would have to be "narrowly tailored to further a compelling state interest."[35] However, no case involving these issues is scheduled to come before the Supreme Court this term.

[35] Ibid., p. 667.

Chapter 4

BIBLIOGRAPHY WITH ABSTRACTS

A brief introduction to the methods for novel gene cloning Abstract: There are a lot of methods for novel gene cloning, but how to clone candidate gene(s) quickly and correctly? This is a brief introduction to methods of novel gene cloning, these methods includes: differential display reverse transcriptase polymerase chain reaction(DD RT-PCR), suppression subtractive hybridization (SSH), RNA arbitrarily primed PCR(RAP-PCR), representational difference analysis(RDA), yeast two-hybrid system, cDNA capturation, et al. We not only introduced these methods, but also discussed the advantages and disadvantages of them. However, no single method is omnipotent, one should pick up the method most suitable for a special purpose. Author(s): Sun, C X; Yu, A C Language: chi Source Sheng Li Ke Xue Jin Zhan 2000 Jan;31(1):35-42.

A broad spectrum PCR method for the detection of polyomaviruses and avoidance of contamination by cloning vectors. Abstract: Polyomaviruses induce tumours of different histological types when inoculated into experimental animals, but their aetiological role in the development of malignant tumours in humans remains questionable. We developed a degenerate PCR assay in an attempt to identify additional, presently unknown human polyomavirus types which may be involved in the malignant transformation of human tissues. Degenerate oligonucleotide primers were deduced from four different conserved amino acid motifs in the highly conserved viral capsid protein, VP1. Three different sets of primers were included for the each test. Bladder carcinomas, Hodgkin's lymphomas, meningiomas, Kaposi-tumours and -cell lines were analysed. No polyomavirus DNA sequences could be detected. A comparative analysis led to the recognition of the presence of SV40 DNA sequences in more than 200 vectors available in the EMBL and Genbank Databanks and commonly used in laboratories worldwide. The majority of primers used to detect polyomavirus sequences in human tumours are distributed throughout these regions also present in the vectors. Only a small stretch of 286bp in the overlapping region of the VP1, VP2 and VP3 genes is not present in the vector sequences. We propose to use this region for the design of additional non-contamination primers. Author(s): Volter, C; zur Hausen, H; Alber, D; de Villiers, E M Language: eng Source Dev Biol Stand 1998;94:137-42.

A century of (epithelial) transport physiology: from vitalism to molecular cloning. Abstract: During the past century, the generally accepted view of the function of biological membranes has evolved from that of a static, inert lipoprotein envelope that simply separates the intracellular machinery from the outside milieu, to that of a dynamic structure that is, in fact, an integral part of that machinery actively involved in homeostasis and bioenergetics. This brief review traces that paradigm shift with special emphasis on the evolution of central concepts in epithelial transport physiology. Author(s): Schultz, S G Language: eng Source Am J Physiol 1998 Jan;274(1 Pt 1):C13-23.

A differential cloning procedure for rearranged or altered genomic DNA based on in-gel competitive reassociation. Abstract: We have developed a substantially improved differential cloning procedure designed for cloning anonymous altered restriction DNA fragments from higher organisms. The improvements include (i) in-gel dissociation and reassociation of biotinylated restriction digests of target DNA fragments, (ii) replacement of agarose gel by a synthetic gel material for electrophoresis, (iii) use of a reassociation enhancing reagent (CTAB) for in-gel reassociation, and (iv) introduction of PCR. After several cycles of IGCR, we attained considerable enrichment of altered or rearranged DNA fragments which were originally present at one copy or less per complex eukaryotic genome. Examples of enrichment include those of an exogenously added DNA fragment, a chromosomal DNA sequence that has undergone a deletion, and DNA fragments containing a recombination junction. Author(s): Kiyama, R; Inoue, S; Ohki, R; Kikuya, E; Yokota, H; Oishi, M Language: eng Source Adv Biophys 1995;31:151-61.

A high-resolution physical map of human chromosome 11q and its application for positional cloning; Hosoda, F; Ohki, M Language: jpn Subject(s): Chromosomes, Human, Pair 11/genetics; Cloning, Molecular; Genome, Human; Leukemia; Oncogenes; Restriction Mapping methods; Translocation (Genetics) Source Tanpakushitsu Kakusan Koso 1997 Dec;42(17 Suppl):2721-8.

A positional cloning approach to inherited renal disease. Author(s): Germino, G G; Somlo, S Language: eng Subject(s): Amino Acid Sequence; Base Sequence; Chromosome Mapping; Cloning, Molecular; DNA; Female; Human; Kidney Diseases; Male; Molecular Sequence Data; Pedigree; Polycystic Kidney, Autosomal Dominant genetics Source Semin Nephrol 1992 Nov;12(6):541-53.

A reappraisal of aldosterone effects on the kidney: new insights provided by epithelial sodium channel cloning. Abstract: Sodium homeostasis is crucial for the control of extra-cellular volume and blood pressure. Regulation of sodium reabsorption is mainly achieved in the distal nephron by the mineralocorticoid aldosterone, but the molecular pathway of aldosterone action has largely remained unclear. Molecular genetic analysis of inherited diseases disturbing sodium homeostasis has now demonstrated that the amiloride-sensitive epithelial sodium channel is a major effector of aldosterone action. Mechanisms by which aldosterone regulates the epithelial sodium channel activity are beginning to emerge and will be of great importance for a better understanding of salt-sensitive hypertension. Author(s): Grunder, S; Rossier, B C Language: eng Source Curr Opin Nephrol Hypertens 1997 Jan;6(1):35-9.

A simple system of cloning in phage M13 and DNA sequencing with terminators Abstract: A compilation of techniques for DNA cloning in filamentous phage M13 based vectors for a novice in cloning is presented. It does not require either specialized microbiological facilities, or any specific knowledge in Escherichia coli genetics. The cloning strategy uses only blunt-end ligation into a vector that has been prepared once for several hundred experiments. The first part describes the isolation, preparation and checking of a blunt-ended M13 vector (with M13 mp series vectors as an example), and also the isolation of clonable fragments, transformation of competent cells and preliminary analysis of recombinants. The second part describes procedures and equipment, which enable to sequence recombinant M13 clones by the chain termination procedure of Sanger et al. It includes simplified procedures for the preparation of sequencing gels, and the rules of interpretation of the sequencing ladders. Reference material is added, which includes trouble-shooting guide, E. coli K12 strain list and polylinker sequences for use of mp-series vectors as well as a fully documented cloning and sequencing experiment. Author(s): Kraev, A S Language: rus Source Mol Biol (Mosk) 1988 Sep-Oct;22(5):1164-97.

A spare or an individual? Cloning and the implications of monozygotic twinning. Abstract: The creation of Dolly, the cloned sheep, raises the scenario of cloning in humans. Neither the case for, nor against, the ethics of cloning in humans is discussed in this paper. Instead, it considers the neglected issue of the likely happiness or otherwise of the resulting children if they are born as monozygotic twins or triplets. The advantages and disadvantages of twinship are discussed in detail, and it is concluded that recognized medical risks, and incompletely understood psychological effects, should be given serious consideration. Author(s): Bryan, E M Language: eng Source Hum Reprod Update 1998 Nov-Dec;4(6):812-5.

Achievements and problems in cloning and making transgenic mammals Abstract: The current use of nuclear transfer for cloning of mammals providing its application in agriculture, for biomedical aims and restoration of biodiversity was analyzed. Main problems associated with the method complexity and its low efficiency as well as ethic problems were discussed. Future application prospects were evaluated. Author(s): Triapitsyna, N V; Glazko, V I Language: rus Source Tsitol Genet 2002 Jul-Aug;36(4):57-71.

Actinobacillus pleuropneumoniae: virulence and gene cloning. Abstract: Actinobacillus pleuropneumoniae is the causal agent of porcine contagious pleuropneumonia (PCP). The infection produces important economic losses in porciculture due to its high morbidity and mortality. Survivors are asymptomatic carriers infectious to other pigs and have low alimentary conversion. The causative agent possesses several virulence factors: adhesion fimbriae, lipopolysaccharide of the outer membrane, capsule, and cytolysins. In addition, our group has reported secretion proteases of a wide pH range of activity. These proteases degrade different substrates such as porcine gelatin, hemoglobin and IgA, and bovine or human hemoglobin. To control PCP dissemination, farmers require serodiagnostic tests which detect carriers and discriminate between vaccinated and infected animals. Bacterines used as immunogens are serotype specific and do not prevent the infection. Genes have been cloned that codify a cohemolysin, cytolysins, and an iron-binding protein. We have cloned A. pleuropneumoniae genes using the expression plasmids pUC19 and Bluescript, in Escherichia coli Q358 and DH5 alpha; the screening for antigen production was made in four groups of pigs (vaccinated, experimentally infected, naturally infected, and from slaughterhouses); two E. coli clones expressed polypeptides recognized by sera from all the groups. Author(s): Negrete-Abascal, E; Tenorio, V; Garcia, C; Godinez, D; Serrano, J J, de la Cuadra, J A; de la Garza, M A Language: eng Source Arch Med Res 1994 Summer;25(2):229-33.

Active urea transport in the rat inner medullary collecting duct: functional characterization and initial expression cloning. Abstract: Active transport of urea has been proposed to exist in the inner medullary collecting duct (IMCD) of low-protein fed mammals for over 30 years. We perfused IMCD subsegments from rats fed a standard (18%) or a low (8%) protein diet and tested for the presence of active urea transport. We found no active urea transport in terminal IMCDs, regardless of diet. In initial IMCDs from rats fed 18% protein or fed 8% protein for one to two weeks, we again found no active urea transport. However, in rats fed 8% protein for three to four weeks, we found significant net urea reabsorption. This active urea reabsorption was inhibited when Na+,

K(+)-ATPase activity was inhibited by adding 1 mM ouabain or removing bath potassium, suggesting a secondary active transport process. Removing sodium from the perfusate completely inhibited net urea reabsorption, demonstrating that this active urea transport is dependent upon the presence of sodium in the tubule lumen. Unlike the facilitated urea transporter, the active urea transporter was not inhibited by phloretin nor stimulated by vasopressin, suggesting that it is a distinct transport protein. To test this hypothesis, we size-separated poly(A)(+)-RNA prepared from inner medullae of rats fed 8% protein for three weeks and injected it into Xenopus laevis oocytes. RNA from a 4.4 to 8.4 kb size fraction increased urea permeability fourfold compared to water-injected oocytes or injecting RNA from other size-fractions. We conclude that feeding rats a low-protein diet for three weeks induces the expression of an unique, secondary active, sodium-dependent urea transporter whose cDNA is between 4.4 and 8.4 kb in size. In addition, our results suggest that it will be possible to clone the cDNA for this sodium-urea cotransporter by expression in Xenopus laevis oocytes. Author(s): Sands, J M; Martial, S; Isozaki, T Language: eng Source Kidney Int 1996 Jun;49(6):1611-4.

Adler, Robert E., 1946- Science firsts: from the creation of science to the science of creation / Robert Adler. Published/Created: New York: John Wiley, c2002. Description: vii, 232 p.: ill.; 25 cm. ISBN: 0471401749 Summary: "Tells the engaging and inspiring stories of thirty-five landmark scientific discoveries. From the first accurate prediction of an eclipse in 585 B.C. to the cloning of Dolly the sheep...Adler clearly explains the context and importance of these discoveries." Contents: Thales and natural causation -- Anaximander orders the cosmos -- Pythagoras numbers the cosmos -- Atoms and the void -- Aristotle and the birth of biology -- Aristarchus, the forgotten Copernicus -- Archimedes's physics -- Ibn al-Haitham illuminates vision -- Copernicus moves the earth -- Galileo discovers the skies -- Kepler solves the planetary puzzle -- Van Leeuwenhoek explores the microcosm -- Newton: gravity and light -- A breath of fresh air -- Humphry Davy, intoxicated with discovery -- Visionaries of the computer -- Darwin's great truth -- A genius in the garden -- Mendeleev charts the elements -- In the realm of radioactivity -- Planck's quantum leap -- Wired on wireless -- Rutherford dissects the atom -- Einstein: matter, energy, space, and time -- Wegener sets the continents adrift -- Hubble's expanding universe -- Out of Africa -- Fermi and the fire of the gods -- McClintock's chromosomes -- A bit of genius -- The dynamic duo of DNA -- Echoes of creation -- We are not what we seem -- Planetary pioneers -- After Dolly, life will never be the same. Notes: Includes bibliographical references (p. 207-219) and index. Subjects: Science--History. Scientists. LC Classification: Q125 .A32 2002 Dewey Class No.: 509 21

Affinity labelling and cloning of steroid and thyroid hormone receptors / edited by H. Gronemeyer. Published/Created: Weinheim, Federal Republic of Germany: VCH; Chichester, England: E. Horwood, 1988. Related Names: Gronemeyer, H. (Hinrich) Description: 322 p.; 25 cm. ISBN: 3527265570 Notes: Includes bibliographies and index. Subjects: Steroid hormones--Receptors--Affinity labeling. Thyroid hormones--Receptors--Affinity labeling. Molecular cloning. Series: Ellis Horwood series in biomedicine, 0930-3367 LC Classification: QP572.S7 A343 1988 Dewey Class No.: 599/.01927 19

Allelic chromosomal deletion using microsatellite polymorphism--the clue to the positional cloning of the tumor suppressor genes responsible for gastric carcinoma Abstract: Cancers are caused by an accumulation of multiple genetic lesions that we are now beginning to understand. Gastric cancer has a different morphological appearance, ability to invade or metastasize, and response to treatment. The pattern of loss of heterozygosity is similar, but not necessarily the same among the different pathological lesions. It is likely that several unknown tumor suppressor genes could be participate in tumor behavior. Positional cloning using microsatellite genes would facilitate the identification of such additional tumor suppressor genes in the genome. Author(s): Sato, J; Fukutomi, H Language: jpn Source Nippon Rinsho 1996 Apr;54(4):992-5.

Alonso, Kenneth. Shall we clone a man?: genetic engineering and the issues of life: a view from a Catholic physician scientist / Kenneth Alonso. Published/Created: Atlanta, Ga.: Allegro Press, c1999. Description: 177 p.; 21 cm. ISBN: 1890731056 (pbk.) Subjects: Catholic Church--Doctrines. Human cloning--Moral and ethical aspects. Genetic engineering--Moral and ethical aspects. Genetic engineering--Religious aspects--Catholic Church. Bioethics--Religious aspects--Christianity.

Ames, Mildred. Anna to the infinite power / Mildred Ames. Published/Created: New York: Scribner, c1981. Description: 198 p.; 22 cm. ISBN: 0684168553 : Summary: A 12-year-old math whiz accidentally learns the startling facts about her true identity and her role in an important secret experiment. Subjects: Cloning--Fiction. Science fiction. LC Classification: PZ7.A5143 An Dewey Class No.: [Fic] 19

Amino acid neurotransmitter reuptake: mechanistics, biochemistry and molecular cloning. Author(s): Kanner, B I Language: eng Subject(s): Animal; Biological Transport; Molecular; Gene Library; Membrane Proteins; Nerve Tissue gamma-Aminobutyric; Metabolism Source Biochem Soc Trans 1991 Feb;19(1):92-5.

An answer to a complex problem: cloning the mouse t-complex responder. Abstract: The t-complex is maintained in wild mouse populations by its high transmission (up to 99%) from heterozygous males and provides an example of "meiotic drive". Its molecular basis has remained obscure despite long and intensive study. In a major advance, the t complex responder gene, thought to be the key gene on which several distorters act, has now been cloned. Author(s): Lyon, M F Language: eng Source Mamm Genome 2000 Oct;11(10):817-9.

An aspartic proteinase from flowers of Centaurea calcitrapa. Purification, characterization, molecular cloning, and modelling of its three-dimensional structure. Author(s): Domingos, A; Xue, Z T; Guruprasad, K; Clemente, A; Blundell, T; Pais, M S; Brodelius, P E Language: eng Source Adv Exp Med Biol 1998;436:465-72.

Analysis of differential gene expression in the kidney by differential cDNA screening, subtractive cloning, and mRNA differential display. Abstract: It is becoming increasingly evident that significant changes in gene expression occur during the course of disease progression in both genetic and nongenetic kidney diseases. Knowledge of the differentially expressed genes may yield information about the abnormal biochemical events that occur in the initiation and pathogenesis of these diseases. The purpose of this review is to provide an overview of some of the current approaches for identifying and analyzing differentially expressed genes. The power of these techniques lies in their their ability to detect differences in the levels of specific mRNAs in the diseased compared to the nondiseased kidney without prior knowledge of their identity. The three basic techniques considered are differential cDNA library screening, subtracted cDNA libraries, and PCR-based differential display. Emphasis is placed on cDNA library construction and differential screening. Also reviewed are the analysis of differentially expressed cDNAs by Southern and Northern blot hybridization, S1-protection, RT-PCR, DNA sequencing, and DNA sequence analysis. Author(s): Maser, R L; Calvet, J P Language: eng Source Semin Nephrol 1995 Jan;15(1):29-42.

Andrews, Lori B., 1952- The clone age: adventures in the new world of reproductive technology / Lori B. Andrews. Edition Information: 1st ed. Published/Created: New York: Henry Holt, c1999. Description: 264 p.; 22 cm. ISBN: 0805060804 (acid-free paper) Notes: Includes bibliographical references (p. 263-264). Subjects: Human reproductive technology--Moral and ethical aspects. Human cloning--Moral and ethical aspects. Human reproductive technology--Law and legislation. Human cloning--Research--Law and legislation. LC Classification: RG133.5 .A57 1999 Dewey Class No.: 176 21

Androgen and estrogen sulfotransferases of the rat liver: physiological function, molecular cloning, and in vitro expression. Abstract: Cloning of the androgen and estrogen sulfotransferases has allowed studies on their spatiotemporal regulation and physiological function. These two enzymes appear to be involved in regulating target cell sensitivity for the androgenic and estrogenic steroids. Recombinant androgen sulfotransferase produced in insect Sf9 cells through a baculoviral expression vector is capable of transferring the sulfate from PAPS (phosphoadenosine phosphosulfate) to 3 alpha, 3 beta, 17 alpha, and 17 beta hydroxyl groups of a number of steroid substrates indicating a broad range of substrate specificity. Author(s): Chatterjee, B; Song, C S; Kim, J M; Roy, A K Language: eng. Source Chem Biol Interact 1994 Jun;92(1-3):273-9.

Angiotensin II receptors: cloning and regulation. Abstract: Angiotensin II subtype 1 (AT1) receptor cDNA was cloned by expression cloning from bovine adrenal cortical cells. Human AT1 receptor was also cloned. These receptors were found to have a seven transmembrane structure. The receptor seems to interact with more than one type of G-proteins. AT1 consists of subtypes. cDNA for AT1A was cloned from rat kidney and that for AT1B was cloned from rat adrenal by plaque hybridization. They have similar base sequences in the coding region but are different in non-coding regions. Their functional implication is not clear. The regulation of the receptors occur at many stages. Expression of mRNA is studied in cultured rat mesangial cells. It was down regulated by angiotensin II and cAMP. On the other hand in whole body experiments, chronic infusion of angiotensin II was shown to upregulate adrenal AT1, and bilateral nephrectomy or losartan (CAS 124750-99-8) administration reduced AT1 mRNA expression. In addition to AT1 and AT2 the presence of a new subtype AT3 has been shown. Author(s): Inagami, T; Iwai, N; Sasaki, K; Guo, D F; Furuta, H; Yamano, Y; Bardhan, S; Chaki, S; Makito, N; Badr, K Language: eng Source Arzneimittelforschung 1993 Feb;43(2A):226-8.

Angiotensin II receptors: cloning, structure, function and regulation. Author(s): Inagami, T; Yamano, Y; Bardhan, S; Iwai, N; Sasaki, K; Chaki, S; Ohyama, K; Makita, N; Badr, K F Language: eng Subject(s): Amino Acid Sequence; Cloning, Molecular; Gene Expression Regulation; Human; Receptors, Angiotensin/chemistry/genetics/physiology Source Nippon Jinzo Gakkai Shi 1993 May;35(5):421-3.

Animal cloning and nuclear reprogramming. Author(s):Tsunoda, Yukio; Kato, Yoko Language: jpn Subject(s): Abnormalities/genetics; Animal; Cattle/genetics; Cell Nucleus/genetics/physiology/transplantation; Cloning, Organism/adverse effects/methods; Meiosis; Mitosis; Species Specificity Source Tanpakushitsu Kakusan Koso 2002 Oct;47(13):1797-803.

Animal cloning by somatic nuclear transfer; Tsunoda, Y; Kato, Y Language: jpn Subject(s): Animal; Cell Cycle; Cell Nucleus/transplantation; Cloning, Organism/methods; Oocytes Source Tanpakushitsu Kakusan Koso 2000 Sep;45(13 Suppl):2015-20.

Animal cloning--the route to new genomics in agriculture and medicine. Abstract: This paper reviews the origin and development of animal cloning in metazoans starting with primitive experiments performed during the late 1880's and early 1900's, followed by nuclear transplantation in amphibians in 1952, then extended to fish and insects in the 1960's, and finally to mammals in the 1980's. Emphasis is placed on the applications of mammalian cloning to agriculture, medicine, and the conservation of endangered species. In addition, the introduction of genes via random insertion or gene targeting into the genome of donor cells to be used for cloning has opened up another route for new genomics in agriculture and medicine. The production of transgenic clones starting in 1997 has indeed contributed a milestone to scientific research. Although cloning efficiency is still low, certain kinds of experiments are quite feasible, and we anticipate improvements in the future. Author(s): Di Berardino, M A Language: eng Source Differentiation 2001 Sep;68(2-3):67-83.

Another jigsaw piece towards solving the iron metabolism puzzle: the cloning of the iron exporter, ferroportin. Author(s): Richardson, D R Language: eng Subject(s): Biological Transport, ; Carrier Proteins/genetics; Cloning, Molecular; Iron/metabolism; Membrane Proteins Source Redox Rep 2000;5(1):7-9.

Antibiotics--cloning of biosynthetic pathways. Abstract: Biosynthetic pathways leading to antibiotics have often been found to be clustered, and new organizational forms of multifunctional enzymes have been discovered. Such polyenzymes accomplish the synthesis of complex metabolites such as peptides or polyketides by a sequence of enzymatic reactions. So, reactions leading to the tripeptide precursor of beta-lactam antibiotics, ACV, or to the cycloundecapeptide cyclosporine have been fused into single polypeptide chain synthetases, respectively. In certain isofunctional sites restricted similarities have been detected. Author(s): Kleinkauf, H; von Dohren, H Language: eng Peptide Source FEBS Lett 1990 Aug 1;268(2):405-7.

Application of molecular cloning to studies on the complement system. Abstract: The isolation of cDNA and, in certain cases, genomic clones has been reported for the following complement proteins: C1q, C2, C3, C4, C5, C9, and factor B, C4b-binding protein and C1-inhibitor. The availability of cloned DNA has allowed rapid advances to be made in the understanding of the structure (from the derived amino acid sequences), function, biosynthesis and genetics of these proteins. This is most strongly illustrated from recent studies on the C2, factor B and C4 genes, which code for the class III molecules of the major histocompatibility complex, especially as certain allelic forms of these genes may be associated with disease susceptibility. Author(s): Reid, K B Language: eng Source Immunology 1985 Jun;55(2):185-96.

Arabidopsis map-based cloning in the post-genome era. Abstract: Map-based cloning is an iterative approach that identifies the underlying genetic cause of a mutant phenotype. The major strength of this approach is the ability to tap into a nearly unlimited resource of natural and induced genetic variation without prior assumptions or knowledge of specific genes. One begins with an interesting mutant and allows plant biology to reveal what gene or genes are involved. Three major advances in the past 2 years have made map-based cloning in Arabidopsis fairly routine: sequencing of the Arabidopsis genome, the availability of more than 50,000 markers in the Cereon Arabidopsis Polymorphism Collection, and improvements in the methods used for detecting DNA polymorphisms. Here, we describe the Cereon Collection and show how it can be used in a generic approach to mutation mapping in Arabidopsis. We present the map-based cloning of the VTC2 gene as a specific example of this approach. Author(s): Jander, Georg; Norris, Susan R; Rounsley, Steven D; Bush, David F; Levin, Irena M; Last, Robert L Language: eng Source Plant Physiol 2002 Jun;129(2):440-50.

Ardley, Neil. Fact or fantasy? / Neil Ardley. Published/Created: London; New York: F. Watts, 1982. Description: 36 p.: col. ill.; 29 cm. ISBN: 0531044734 (U.S.) Summary: Examines space travel, colonies in space, cloning, time travel, and other futuristic possibilities and explores the likelihood of their eventual existence. Notes: Includes index. Subjects: Voyages, Imaginary--Juvenile literature. Forecasting. Outer space--Exploration--Juvenile literature. Outer space--Exploration. Series: World of tomorrow LC Classification: TL788.7 .A73 1982 Dewey Class No.: 003/.2 19

Are we ready for cloning?; Hernandez, Jose J; Wilbois, Ronald P Language: eng Subject(s): Animal; Cloning, Organism/legislation & jurisprudence/standards; Ethics, Medical; Human; United States Source J Androl 2002 Jan-Feb;23(1):23-4.

At the crossroads of developmental genetics: the cloning of the classical mouse T locus. Abstract: The discovery, more than 60 years ago, of a mutant mouse with a short tail led to the birth of the new field of developmental genetics. Over the years since, numerous investigators have probed the biology of the original short-tail mutation at the T locus, as well the naturally-occurring t haplotypes that were uncovered as a result of their interaction with this mutation. Although the T locus ranks among the best characterized developmental loci in the mouse, it was not among the first to be cloned. This situation has now been rectified with two recent reports from Herrmann, Lehrach and their colleagues. While the T locus is expressed uniquely in the embryonic tissues predicted from the mutant phenotype, the gene itself, as well as the predicted amino acid sequence of the T product, show no strong homology to any known sequence. For the moment, at least, the mystery behind the function of the T locus still awaits definitive resolution. Author(s): Silver, L M Language: eng Source Bioessays 1990 Aug;12(8):377-80.

Bacteriophage lambda as a cloning vector. Abstract: Extensive research has been directed toward the development of multipurpose lambda vectors for cloning ever since the potential of using coliphage lambda as a cloning vector was recognized in the late 1970s. An understanding of the intrinsic molecular organization and of the genetic events which determine lysis or lysogeny in lambda has allowed investigators to modify it to suit the specific requirements of gene manipulations. Unwanted restriction sites have been altered and arranged together into suitable polylinkers. The development of a highly efficient in vitro packaging system has permitted the introduction of chimeric molecules into hosts. Biological containment of recombinants has been achieved by introducing amber mutations into the lambda genome and by using specific amber suppressor hosts. Taking advantage of the limited range of genome size (78 to 105% of the wild-type size) for its efficient packaging, an array of vectors has been devised to accommodate inserts of a wide size range, the limit being 24 kbp in Charon 40. The central dispensable fragment of the lambda genome can be replaced by a fragment of heterologous DNA, leading to the construction of replacement vectors such as Charon and EMBL. Alternatively, small DNA fragments can be inserted without removing the dispensable region of the lambda genome, as in lambda gt10 and lambda gt11 vectors. In addition, the introduction of many other desirable properties, such as NotI and SfiI sites in polylinkers (e.g., lambda gt22), T7 and T3 promoters for the in vitro transcription (e.g., lambda DASH), and the mechanism for in vivo excision of the intact insert (e.g., lambda ZAP), has facilitated both cloning and subsequent analysis. In most cases, the recombinants can be differentiated from the parental phages by their altered phenotype. Libraries constructed in lambda vectors are screened easily with antibody or nucleic acid probes since several thousand clones can be plated on a single petri dish. Besides the availability of a wide range of lambda vectors, many related techniques such as rapid isolation of lambda DNA, a high efficiency of commercially available in vitro packaging extracts, and in vitro amplification of DNA via the polymerase chain reaction have collectively contributed to lambda's becoming one of the most powerful and popular tools for molecular cloning. Author(s): Chauthaiwale, V M; Therwath, A; Deshpande, V V Language: eng Source Microbiol Rev 1992 Dec;56(4):577-91.

Bacteriophage lambda cloning vectors. Author(s): Sorge, J A Language: eng Subject(s): Bacteriophage lambda/genetics; Cloning, Molecular; DNA/genetics; DNA Replication; DNA Restriction Enzymes; Genetic Vectors; Virus Replication Source Biotechnology 1988;10:43-60.

Balkwill, Frances R. Gene machines / Fran Balkwill & Mic Rolph. Published: Cold Spring Harbor, NY: Cold Spring Harbor Laboratory Press, c2002. Related Names: Rolph, Mic. Description: 31 p.: col. ill.; 21 cm. ISBN: 0879696168 (alk. paper) 0879696117 (pbk.: alk. paper) Summary: An introduction to how genes work, including basic information about cloning and gene therapy. Subjects: Genetics--Cloning--Juvenile literature. Series: Enjoy your cells; 4 LC Classification: QH437.5 .B35 2002 Dewey Class No.: 572.8 21

Ban human cloning: report of the American Bioethics Advisory Commission. Published/Created: Stafford, VA: American Life League, c1997. Description: 77 p.; 28 cm. ISBN: 1890712264 Notes: "A project of American Life League"--Letter. Includes bibliographical references. Subjects: Human cloning--Moral and ethical aspects. LC Classification: QH442.2 .B36 1997 Dewey Class No.: 174/.25 21

Basic cloning procedures / Valdis Berzins (ed.). Published/Created: Berlin; New York: Springer, c1998. Related Names: Berzins, Valdis, 1943- Description: 169 p.: ill.; 24 cm. ISBN: 3540635394 (wire-o-bind: alk. paper) Notes: Includes bibliographical references and index. Subjects: Molecular cloning--Laboratory manuals. Series: Springer lab manual LC Classification: QH442.2 .B37 1998 Dewey Class No.: 572.8/645 21

Basic factors of pathogenicity and cloning of pseudomonads; Anishchenko, M A Language: rus; Bacterial Proteins/metabolism; Bacterial Toxins/toxicity; Cloning, Molecular; English Abstract; Escherichia coli/genetics; Gene Expression; Genes, Bacterial; Source Mol Gen Mikrobiol Virusol 1991 May;(5):3-8.

Baudrillard, Jean. The vital illusion / Jean Baudrillard; edited by Julia Witwer. Published/Created: New York: Columbia University Press, c2000. Related Names: Witwer, Julia. Description: 102 p.; 19 cm. ISBN: 0231121008 (alk. paper) Notes: Includes bibliographical references and index. Subjects: Reality. Virtual reality. Cloning. Series: Wellek Library lecture series at the University of California, Irvine. LC Classification: BD331 .B287 2000 Dewey Class No.: 194 21

BeauSeigneur, James. Acts of God / James BeauSeigneur. Published/Created: New York: Warner Books, 2003. Description: p. cm. ISBN: 0446531278 Subjects: End of the world--Fiction. Second Advent--Fiction. Armageddon--Fiction. Cloning--Fiction. Series: BeauSeigneur, James. Christ clone trilogy; bk. 3. LC Classification: PS3552.E2319 A65 2003 Dewey Class No.: 813/.54 21

BeauSeigneur, James. Birth of an age / by James BeauSeigneur. Edition Information: 1st ed. Published/Created: Rockville, Md.: SelectiveHouse Publishers, c1997. Description: 241 p.; 22 cm. ISBN: 0965694860 (pbk.) Subjects: Asteroids--Collisions with Earth--Fiction. End of the world--Fiction. Second Advent--Fiction. Cloning--Fiction. Series: BeauSeigneur, James. Christ clone trilogy; bk. 2. LC Classification: PS3552.E2319 B57 1997 Dewey Class No.: 813/.54 21

BeauSeigneur, James. In his image / by James BeauSeigneur. Edition Information: 1st ed. Published: Rockville, Md.: Selective House Publishers, 1997. Description: 366 p.; 22 cm. ISBN: 0965694852 Notes: Includes bibliographical references. Subjects: Second Advent--Fiction. Holy Shroud--Fiction. Cloning--Fiction. Series: BeauSeigneur, James. Christ clone trilogy; bk. 1. LC Classification: PS3552.E2319 I5 1997 Dewey Class No.: 813/.54 21

BeauSeigneur, James. In his image / James BeauSeigneur. Edition Information: Warner Books ed. Published/Created: New York: Warner Books, c2003. Description: 416 p.; 24 cm. ISBN: 0446531251 Subjects: Second Advent--Fiction. Holy Shroud--Fiction. Cloning--Fiction. Series: BeauSeigneur, James. Christ clone trilogy; bk. 1. LC Classification: PS3552.E2319 I5 2003 Dewey Class No.: 813/.54 21

Beecroft, Simon. Superhumans: a beginner's guide to bionics / written by Simon Beecroft; illustrated by Stephen Sweet & Stuart Squires. Published/Created: Brookfield, Conn.: Copper Beech Books, 1998. Related Names: Squires, Stuart, ill. Sweet, Stephen, 1965- ill. Description: 32 p.: col. ill.; 28 cm. ISBN: 0761306218 (lib. bdg.) 0761306366 (pbk.) Summary: Looks at possible future methods for prolonging life, such as genetic engineering and the use of artificial body parts. Also discusses robots, artificial intelligence, cybernetics, cloning, and other technologies. Notes: Includes index. Subjects: Longevity--Juvenile literature. Bionics--Juvenile literature. Genetic engineering--Juvenile literature. Longevity. Bionics. Genetic engineering. Series: Future files LC Classification: RA776.75 .B44 1998 Dewey Class No.: 610/.28 21

Bendick, Jeanne. Super people: who will they be? / By Jeanne Bendick. Published/Created: New York: McGraw-Hill, c1980. Description: 128 p.: ill.; 24 cm. ISBN: 0070045038 : Summary: Considers the existence of human beings endowed with super senses, such as ESP, the implications of creating human-like androids with super powers, and the possibility that intelligent machines could take over the world. Subjects: Androids--Conscious automata--Cloning--Juvenile literature. Robots. LC Classification: TJ211 .B46 Dewey Class No.: 629.8/92

Benefits and problems with cloning animals. Abstract: Animal cloning is becoming a useful technique for producing transgenic farm animals and is likely to be used to produce clones from valuable adults. Other applications will also undoubtedly be discovered in the near future, such as for preserving endangered breeds and species. Although cloning promises great advantages for commerce and research alike, its outcome is not always certain due to high pregnancy losses and high morbidity and mortality during the neonatal period. Research into the mechanisms involved in the reprogramming of the nucleus is being conducted throughout the world in an attempt to better understand the molecular and cellular mechanisms involved in correcting these problems. Although the cause of these anomalies remains mostly unknown, similar phenotypes have been observed in calves derived through in vitro fertilization, suggesting that culture conditions are involved in these phenomena. In the meantime, veterinarians and theriogenologists have an important role to play in improving the efficiency of cloning by finding treatments to assure normal gestation to term and to develop preventative and curative care for cloned neonates. Author(s): Smith, L C; Bordignon, V; Babkine, M; Fecteau, G; Keefer, C Language: eng Source Can Vet J 2000 Dec;41(12):919-24.

Between sanctity and value of human life: in perspective of human cloning Abstract: The more we know, the more duties and greater responsibility we have. The dynamic development of biology carries a lot of hope for the freeing of mankind from genetic diseases. But the introduction of scientific thought necessarily has to be bound with the development of technology. It is wrong when technology dictates science a direction of development; when technique comes before ethics; and when technology does not respect the essence of a human being. The uncritical introduction of eugenics, especially cloning of people and rejecting all moral arguments, recalls inglorious acts of science when the ideology of progress determined the range of problems that researchers focused on. The same ideology of progress, although originating from other sources, pushes science toward a second extreme, into utilitarianism. The author wishes to substantiate the necessity for researchers to respect ethical norms. Recognition of natural laws alone doesn't provide science with full development if the rights of conscience are violated. Author(s): Dyk, W Language: pol Source Med Wieku Rozwoj 2001;5(1 Suppl 1):63-73.

Bill Clinton statement re: cloning. [1997-03-04] / CNN. Published/Created: United States: CNN, 1997-03-04. Description: 1 videocassette of 1 (19 min.): sd., col.; 3/4 in. master. Summary: Live CNN coverage and analysis of Bill Clinton statement re: his directive on human cloning research, and Clinton's responses to reporters' questions. Notes: Sources: Vanderbilt Television News Archive website; label on container. Cast: Principal reporter: Leon Harris. Source of Acquisition: Received: 7-18-2000; master; purchase; Vanderbilt Television News Archive Collection. Genre/Form: News--Television special LC Classification: VSD 4167 (master)

Bill Clinton statement re: cloning. [1997-06-09] / CNN. Published/Created: United States: CNN, 1997-06-09. Description: 1 videocassette of 1 (17 min.): sd., col.; 3/4 in. master. Summary: Live CNN coverage and analysis of Bill Clinton statement re: cloning, preceded by National Bioethics Advisory Commission Chairman Dr. Harold Shapiro statement. Analysis includes an interview with University of Chicago Law Professor Cass Sunstein. Notes: Sources: Vanderbilt Television News Archive website; label on container. Cast: Principal reporter: Sonia Ruseler. Source of Acquisition: Received: 7-18-2000; master; purchase; Vanderbilt Television News Archive Collection. Genre/Form: News--Television special LC Classification: VSD 4191 (master)

Biochemical genetics: examples of life after cloning. Abstract: The blend of biochemistry and molecular biology required to understand the pathogenesis of genetic disease is assuming an increasing role in research. We review three example of this inevitable post-cloning trend: first, the surprising relationship between mice with albino deletions and human hereditary tyrosinemia type I; second, the discovery that choroideremia is due to defect in prenylation; and third, fibrillin mutations in the Marfan syndrome. Author(s): McInnes, R R; Byers, P H Language: eng Source Curr Opin Genet Dev 1993 Jun;3(3):475-83.

Biotechnology and the ethics of cloning: how far should we go?: hearing before the Committee on Science, Subcommittee on Technology, U.S. House of Representatives, One Hundred Fifth Congress, first session, March 5, 1997. Published/Created: Washington: U.S. G.P.O.: For sale by the U.S. G.P.O., Supt. of Docs., Congressional Sales Office, 1997. Description: iii, 59 p.; 24 cm. ISBN: 0160552672 Notes: Distributed to some depository libraries in microfiche. Shipping list no.: 97-0339-P. "No. 3." Subjects: Cloning--Moral and ethical aspects--Genetic engineering--Research--Government policy--United States. Bioethics--United States. LC Classification: KF27 .S39957 1997b Dewey Class No.: 174/.957 21 Govt. Doc. No.: Y 4.SCI 2:105/3

Bisson, Terry. The 6th day / Terry Bisson; based on the screenplay by Cormac Wibberley & Marianne Wibberley. Edition Information: 1st mass market ed. Published/Created: New York: Tor, c2000. Description: 231 p.; 18 cm. ISBN: 081257947X Subjects: Cloning--Fiction. LC Classification: CPB Box no. 2003 vol. 19

Bonnicksen, Andrea L. Crafting a cloning policy: from Dolly to stem cells / Andrea L. Bonnicksen. Published/Created: Washington, D.C.: Georgetown University Press, c2002. Description: xii, 220 p.; 24 cm. ISBN: 0878403701 (cloth: alk. paper) 087840371X (pbk.: alk. paper) Notes: Includes bibliographical references (p. 194-213) and index. Subjects: Cloning--Government policy--United States. LC Classification: QH442.2 .B66 2002 Dewey Class No.: 174/.966065 21

Bothwell, Al. Methods for cloning and analysis of eukaryotic genes / Al Bothwell, George D. Yancopolous, Frederick W. Alt. Published: Boston: Jones and Bartlett Publishers, c1990. Related Names: Yancopoulos, George Damis. Alt, Frederick W. Description: xii, 324 p.; 24 cm. ISBN: 0867201037 Notes: Includes bibliographies and index. Subjects: Molecular cloning--Recombinant DNA--analysis--Genetic Engineering--laboratory manuals. LC Classification: QH442.2 .B67 1990 Dewey Class No.: 574.87/328 20

Brain melanocortin receptors: from cloning to function. Abstract: The cloning of brain melanocortin (MC) receptors, the mapping of their expression pattern and the identification of MC receptor selective ligands have opened a new avenue towards elucidating the role of the melanocortin system in the brain. MC receptors have now been implicated in melanocortin-induced grooming behavior in rats, in the melanocortin-induced lowering of blood pressure and in the control of weight homeostasis. Functional opioid antagonism and the anti-pyretic and anti-inflammatory effects of melanocortins are probably also mediated via MC receptors. However, the effects of melanocortins on avoidance behavior and the effect of gamma 2-MSH on increasing blood pressure are not mediated via one of the cloned brain MC receptors. Structure of brain MC receptors, their expression pattern, the selective ligands and the function of MC receptors are briefly reviewed. Author(s): Adan, R A; Gispen, W H Source Peptides 1997;18(8):1279-87.

Brass, Perry. The harvest: a novel / by Perry Brass. Edition Information: 1st ed. Published/Created: Bronx, NY: Belhue Press, c1997. Description: 216 p.; 22 cm. ISBN: 096271237X Subjects: Gay men--Happiness--Cloning--Fiction. LC Classification: PS3552.R336 H37 1997 Dewey Class No.: 813/.54 22

Brin, David. Kiln people / David Brin. Edition Information: 1st ed. Published/Created: New York: Tor, 2002. Description: 459 p.; 25 cm. ISBN: 0765303558 (alk. paper) Notes: "A Tom Doherty Associates book." Subjects: Cloning--Fiction. LC Classification: PS3552.R4825 K55 2002 Dewey Class No.: 813/.54 21

Broad-host-range plasmid cloning vectors for gram-negative bacteria. Author(s): Schmidhauser, T J; Ditta, G; Helinski, D R Language: eng Subject(s): Cloning, Molecular; Cosmids; DNA Restriction Enzymes; Genetic Vectors; Source Biotechnology 1988;10:287-332.

Brown, T. A. (Terence A.) Gene cloning and DNA analysis: an introduction / T.A. Brown. Edition Information: 4th ed. Published/Created: Oxford; Malden, Mass.: Blackwell Science, c2001. Related Names: Brown, T. A. (Terence A.). Gene cloning. Description: xii, 363 p.: ill.; 25 cm. ISBN: 063205901X Notes: Prev. ed. published with title: Gene cloning. Includes bibliographical references and index. Subjects: Molecular cloning. Nucleotide sequence. DNA--Analysis. LC Classification: QH442.2 .B76 2001 Dewey Class No.: 572.8/633 21

Brown, T. A. (Terence A.) Gene cloning: an introduction / T.A. Brown. Edition Information: 3rd ed. Published/Created: London; New York: Chapman and Hall, 1995. Description: vii, 334 p.: ill. (some col.); 25 cm. ISBN: 0412622408 Notes: Includes bibliographical references and index. Subjects: Molecular cloning. LC Classification: QH442.2 .B76 1995 Dewey Class No.: 574.87/322 20

Brown, T. A. (Terence A.) Gene cloning: an introduction / T.A. Brown. Edition Information: 2nd ed. Published: London; New York: Chapman and Hall, University and Professional Division, 1990. Description: xii, 286 p.: ill.; 24 cm. ISBN: 0412342103 Notes: Includes bibliographical references and index. LC Classification: QH442.2 .B76 1990 Dewey Class No.: 574.87/322 20

Bujold, Lois McMaster. Mirror dance / Lois McMaster Bujold. Published: Riverdale, NY: Baen; New York: Distributed by Simon & Schuster, c1994. Description: 392 p.; 25 cm. ISBN: 0671722107 : Notes: "A Baen Books original"--T.p. verso. Subjects: Vorkosigan, Miles (Fictitious character)--Cloning--Inheritance and succession--Fiction. Series: Bujold, Lois McMaster. Vorkosigan adventure. LC Classification: PS3552.U397 M57 1994 Dewey Class No.: 813/.54 20

Burden, David W. (David Wilson), 1958- Biotechnology: proteins to PCR: a course in strategies and lab techniques / David W. Burden, Donald B. Whitney. Published/Created: Boston: Birkhäuser, c1995. Related Names: Whitney, Donald B., 1955- Description: xv, 317 p.: ill.; 25 cm. ISBN: 0817637567 (hardcover: alk. paper) 0817638431 (pbk.: alk. paper) 3764337567 (hardcover: alk. paper) 3764338431 (pbk.: alk. paper) Notes: Includes bibliographical references and index. Subjects: Proteins--Analysis--Laboratory manuals--Purification--Molecular cloning-- Polymerase chain reaction--Biotechnology--Laboratory Manuals. Genetic Engineering--laboratory manuals. LC Classification: QP551 .B95 1995 Dewey Class No.: 660/.63 20

Burton, Robert Alan, 1941- Cellmates / Robert A. Burton. Published/Created: San Francisco: Russian Hill Press, 1997. Description: 271 p.; 24 cm. ISBN: 0965352447 Subjects: Genetics--Research--Cloning--San Francisco (Calif.)--Fiction. LC Classification: PS3552.U7824 C45 1997 Dewey Class No.: 813/.54 21

Ca2+-activated Cl- channels: focus on "molecular cloning and transmembrane structure of hCLCA2 from human lung, trachea, and mammary gland". Author(s): Cuppoletti, J; Malinowska, D H Language: eng Source Am J Physiol 1999 Jun;276(6 Pt 1):C1259-60.

California Advisory Committee on Human Cloning presents its findings and recommendations on cloning and stem cell research: January 15, 2002 informational hearing / California Legislature, Senate Select Committee on Genetics, Genetic Technologies and Public Policy. Published/Created: Sacramento, Calif.: Senate Publications, [2002] Related Names: California Advisory Committee on Human Cloning. Cloning Californians. Description: 1 v. (various pagings); 28 cm. Contents: Agenda -- Background paper -- Transcript text -- Related 2002-03 legislation -- Appendix: Report of the California Advisory Committee on Human Cloning. Notes: Cover title. Senate Publications stock no.: 1149-S. Includes bibliographical references. Subjects: Human cloning--Stem cells--Research--Moral and ethical aspects. Human cloning--Cloning--Government policy California. Genetic engineering. Bioethics. LC Classification: QH442.2 .C35 2002

Can mammalian cloning combined with embryonic stem cell technologies be used to treat human diseases? Abstract: Cloning is commonly perceived as a means of generating genetically identical individuals, but it can also be used to obtain genetically matched embryo-derived stem cells, which could potentially be used in the treatment of patients. A recent report offers the first 'proof of principle' of such cloning for therapeutic purposes, referred to as nuclear transplantation to produce stem cells for autologous transplantation. Author(s): Hadjantonakis, Anna-Katerina; Papaioannou, Virginia E Language: eng Source Genome Biol 2002 Jul 30;3(8):REVIEWS1023.

Carmen, Ira H. Cloning and the Constitution: an inquiry into governmental policymaking and genetic experimentation / Ira H. Carmen. Published/Created: Madison: University of Wisconsin Press, 1985. Description: xv, 223 p.; 24 cm. ISBN: 0299103404 : Notes: Includes index. Bibliography: p. 209-215. Subjects: Genetic engineering--Recombinant DNA--Research--Constitutional law--Molecular cloning--Government policy--United States. LC Classification: KF3827.G4 C37 1985 Dewey Class No.: 344.73/095 347.30495 19

CCR2: from cloning to the creation of knockout mice. Author(s):Charo, I F Language: eng Subject(s): Arteriosclerosis/etiology; Chemotaxis; Cloning, Molecular; Mice; Mice, Knockout; Receptors, Cytokine; HIV/physiology Source Chem Immunol 1999;72:30-41.

CDNA cloning and analysis of tissue specific genes by differential hybridization screening procedures: analysis of OSF-1 gene specifically expressed in brain and osteoblasts; Hashimoto, T; Tsujimura, A; Kuriyama, K Language: jpn Source Nippon Yakurigaku Zasshi 1993 Jul;102(1):47-52.

cDNA cloning of galactosyltransferase associated with tumor (GAT); Uejima, T; Uemura, M; Narimatsu, H Language: jpn Source Tanpakushitsu Kakusan Koso 1992 Aug;37(11 Suppl):1683-8.

cDNA cloning of possible mammalian homologs of the yeast secretory pathway Ca(2+)-transporting ATPase. Author(s): Shull, G E; Clarke, D M; Gunteski-Hamblin, A M Language: eng Subject(s): Amino Acid Sequence; Animal; Base Sequence; Ca(2+)-Transporting ATPase/genetics; Cloning, Molecular; Comparative Study; DNA/genetics; Endoplasmic Reticulum/enzymology; Mammals; Molecular Sequence Data; Sarcoplasmic Reticulum/enzymology; Amino Acid; Sequence Homology, Nucleic Acid Source Ann N Y Acad Sci 1992 Nov 30;671:70-80; discussion 81.

cDNA cloning of small nuclear and cytoplasmic RNA-associated proteins. Author(s): van Dam, A; Winkel, I; Smeenk, R; Cuypers, T Language: eng Subject(s): Autoantigens/immunology; Cloning, DNA/genetics; Epitopes; Ribonucleoproteins/genetics/immunology; Source Clin Rheumatol 1990 Mar;9(1 Suppl 1):20-8.

cDNA library protocols / edited by Ian G. Cowell and Caroline A. Austin. Published/Created: Totowa, N.J.: Humana Press, c1997. Related Names: Cowell, Ian G. Austin, Caroline A. Description: x, 321 p.: ill.; 24 cm. ISBN: 089603383X (alk. paper) Notes: Includes bibliographical references and index. Subjects: Gene libraries--Laboratory manuals. DNA--Laboratory manuals. Antisense DNA--Laboratory manuals. Molecular cloning--Laboratory manuals. Gene Library. DNA, Complementary. Genetic Techniques. Series: Methods in molecular biology (Clifton, N.J.); v. 69. Variant Series: Methods in molecular biology; v. 69 LC Classification: QH442.4 .C36 1997 Dewey Class No.: 574.87/3282 20

cDNA selection and other approaches in positional cloning. Author(s): Parimoo, S; Patanjali, S R; Kolluri, R; Xu, H; Wei, H; Weissman, S M Language: eng Subject(s): Animal; Base Sequence; Cloning, Molecular; DNA/analysis; DNA, Complementary/genetics/isolation & purification; Human; Molecular Probe Techniques; Molecular Sequence Data Source Anal Biochem 1995 Jun 10;228(1):1-17.

Cefrey, Holly. Cloning and genetic engineering / Holly Cefrey. Published/Created: Danbury, Conn.: Children's Press, c2002. Description: 48 p.: col. ill.; 23 cm. ISBN: 0516239163 (lib. bdg.) 0516240064 (pbk.) Summary: Introduces cloning and genetic engineering, exploring the technology and social issues involved and looking toward what the future might bring as it becomes possible to duplicate even human DNA. Contents: Understanding the terms -- Learning through history -- Gene genies and clone arrangers -- What the future holds. Notes: Includes bibliographical references (p. 44) and index. Subjects: Cloning--Juvenile literature. Genetic engineering--Juvenile literature. Cloning. Genetic engineering. Genetics. Series: Life in the future High interest books LC Classification: QH442.2 .C445 2002 Dewey Class No.: 660.6/5 21

Cell cycle co-ordination in embryo cloning by nuclear transfer. Abstract: Exciting new opportunities in embryo cloning have been made possible by recent studies on the interaction of the donor nucleus with the recipient cytoplasm after embryo reconstruction. This article reviews information regarding the co-ordination of nuclear and cytoplasmic events during embryo reconstruction, in particular the direct and indirect effects of maturation/ meiosis/mitosis-promoting factor (MPF), upon the transferred nucleus. This will be discussed in relation to DNA replication, the maintenance of correct ploidy, the occurrence of chromosomal abnormalities and development of reconstructed embryos. Although this review is primarily concerned with the reconstruction of mammalian embryos, specific examples from amphibians will also be cited. Author(s): Campbell, K H; Loi, P; Otaegui, P J; Wilmut, I Language: eng Subject(s): Cell Cycle/physiology; Cell Nucleus/transplantation; Cloning, Organism; Cytoplasm/physiology; Cytoskeleton/physiology; DNA Replication; Maturation-Promoting Factor/physiology Source Rev Reprod 1996 Jan;1(1):40-6.

Cell-free system for RNA replication, expression and cloning; Chetverin, A B Language: rus; Subject(s): Cloning, Molecular; Gene Expression, Molecular Sequence Data; Q beta Replicase; RNA, Viral biosynthesis Source Mol Biol (Mosk) 1996 Sep-Oct;30(5):981-90.

Characterization, cloning and expression of the Tage4 gene, a member of the immunoglobulin superfamily. Abstract: pE4 is a rat carcinoma-associated antigen identified by monoclonal antibody E4, which was raised against a rat colon carcinoma cell line. This glycoprotein is expressed at the surface of all the rat colon carcinoma cell lines tested, as determined by immunofluorescence analysis. In contrast, a barely detectable level was found on normal adult rat colon and lung and no expression could be detected on the other normal rat tissues tested. The corresponding Tage4 gene (tumor-associated glycoprotein E4) is also expressed in rat colon tumors induced by 1,2-dimethylhydrazine and in Min mouse intestinal adenomas. The Tage4 gene product is closely related to the hepatocellular carcinoma antigen TuAg.1. The Tage4 cDNA has been isolated and sequenced. Analysis of the deduced aminoacid sequence indicated that Tage4 is a member of the immunoglobulin supergene family. This family contains cell adhesion molecules which have wide-ranging functions and mediate a variety of homotypic and heterotypic cellular interactions playing a general role in cell surface recognition. The Tage4 gene has been mapped to rat chromosome 1q22 and mouse 7A2-B1, regions that are homologous to the long arm of human chromosome 19. Summary review of our work is presented. Author(s): Denis, M G Language: eng Source Int J Oncol 1998 May;12(5):997-1005.

Chimeric single-stranded DNA phage-plasmid cloning vectors. Abstract: A variety of ssDNA phagemid cloning vectors have been constructed that combine the advantages of the filamentous coliphages with a number of plasmid and other bacteriophage-encoded functions. The practical and biological advantages of a chimeric phage-plasmid vector are considerable, and the trend toward converting existing plasmids into ssDNA phagemids and consolidating any number of useful features into one or a few vectors will undoubtedly accelerate. A new helper phage specifically designed for the isolation of large amounts of single-stranded plasmid DNA simplifies the use of these cloning vehicles. Additional refinements in phagemid-helper phage systems should extend the potential of these vectors even further. Author(s): Mead, D A; Kemper, B Language: eng Source Biotechnology 1988;10:85-102.

Choi, Mieyoung. Sequencing and cloning of the N4-coded single-stranded DNA binding protein gene [microform]: identification and functional analysis of active domains / by Mieyoung Choi. Published: 1992. Description: xi, 194 leaves: ill.; 28 cm. Notes: Thesis (Ph. D.)--University of Chicago, 1992. Microfilm. Chicago, Ill.: University of Chicago, Joseph Regenstein Library, Dept. of Photoduplication, 1993. 1 microfilm reel; 35 mm. Subjects: Cloning. Escherichia coli--Genetics. LC Classification: Microfilm 94/2883 (Q)

Cholecystokinin receptor family. Molecular cloning, structure, and functional expression in rat, guinea pig, and human. Abstract: A review of the literature encompassing numerous pharmacological, physiological, and biochemical studies indicates the presence of at least four CCK receptor types, CCKA, CCKB, gastrin, and CG-4 receptors. Multiple subtypes of the CCKAR have been postulated to account for the differences in pharmacology or affinity cross-linking of CCKARs between pancreas and gallbladder and the presence of high and low affinity CCKARs on pancreatic acini. Multiple subtypes of the CCKBR have been postulated to explain the differences in pharmacology and physiology between gastric and gallbladder smooth muscle CCKBRs. We recently cloned and functionally expressed both the CCKAR and the CCKBR from rat, guinea pig, and human. The CCKAR and CCKBR are 48% homologous and constitute a family of receptors within the guanine nucleotide-binding regulatory protein-coupled superfamily of receptors. Each receptor is highly conserved between species. A single cDNA encoding a single protein is present in both pancreas and gallbladder and can account for both high and low affinity CCKARs found on pancreatic acini when transfected into COS-7 cells. A single cDNA encoding a single CCKBR protein is present in both the central nervous system and the periphery including the gastrointestinal system. Therefore, the gastrin receptor is actually a CCKBR present on parietal cells. Genomic and cDNA library hybridization as well as Northern and Southern hybridization studies among rat, guinea pig, and human species identifies only two members of the CCK receptor family, CCKAR and CCKBR. Although these studies do not identify other closely related members of the CCK receptor family, they do not rule out the existence of other less closely related members. Furthermore, differences in tissue and species-specific posttranslational processing, receptor coupling, and associated membrane protein and lipid heterogeneity may be among some of the other factors that may account for the phenotypic expression of more receptor subtypes than molecular studies would predict. Author(s): Wank, S A; Pisegna, J R; de Weerth, A Language: eng Source Source Ann N Y Acad Sci 1994 Mar 23;713:49-66.

Chromatin remodeling in nuclear cloning. Abstract: Nuclear cloning is a procedure to create new animals by injecting somatic nuclei into unfertilized oocytes. Recent successes in mammalian cloning with differentiated adult nuclei strongly indicate that oocyte cytoplasm contains unidentified remarkable reprogramming activities with the capacity to erase the previous memory of cell differentiation. At the heart of this nuclear reprogramming lies chromatin remodeling as chromatin structure and function define cell differentiation through regulation of the transcriptional activities of the cells. Studies involving modification of chromatin elements such as selective uptake or release of binding proteins, covalent histone modifications ..., and DNA methylation should provide significant insight into molecular mechanisms of nuclear dedifferentiation and redifferentiation in oocyte cytoplasm. Author(s): Wade, Paul A; Kikyo, Nobuaki Language: eng Source Eur J Biochem 2002 May;269(9):2284-7.

Chromosome landing: a paradigm for map-based gene cloning in plants with large genomes. Abstract: The original concept behind map-based or positional cloning was to find a DNA marker linked to a gene of interest, and then to 'walk' to the gene via overlapping clones (e.g. cosmids or YACs). While chromosome walking is straightforward in organisms with small genomes, it is difficult to apply in most plant species, which typically have large, complex genomes. The strategy of chromosome walking is based on the assumption that it is difficult and time consuming to find DNA markers that are physically close to a gene of interest. Recent technological developments invalidate this assumption for many species. As a result, the mapping paradigm has now changed such that one first isolates one or more DNA marker(s) at a physical distance from the targeted gene that is less than the average insert size of the genomic library being used for clone isolation. The DNA marker is then used to screen the library and isolate (or 'land' on) the clone containing the gene, without any need for chromosome walking and its associated problems. Chromosome landing, together with the technology that has made it possible, is likely to become the main strategy by which map-based cloning is applied to isolate both major genes and genes underlying quantitative traits in plant species. Author(s): Tanksley, S D; Ganal, M W; Martin, G B Language: eng Source Trends Genet 1995 Feb;11(2):63-8.

Chromosome microdissection and cloning: a practical guide / [edited by] Nabil G. Hagig and Michael V. Viola. Published/Created: San Diego: Academic Press, 1993. Related Names: Hagag, Nabil G. Viola, Michael V. Description: x, 160 p.: ill.; 24 cm. ISBN: 0123133203 Notes: Includes bibliographical references and index. Subjects: Molecular cloning--Technique. Chromosomes--ultrastructure--Dissection--methods--laboratory manuals. LC Classification: QH442.2 .C46 1993 Dewey Class No.: 575.1/0724 20

Clinical correlations with the human tumor cloning assay. Author(s): Hanauske, A R; Von Hoff, D D Language: eng Subject(s): Aminoacridines/therapeutic use; Amsacrine; Anthracenes/therapeutic use; Breast Neoplasms; Clinical Trials; Clone Cells; Dactinomycin/therapeutic use; Drug Resistance; Interferon Type I/therapeutic use; Leukemia ; Lung Neoplasms; Melanoma; Multiple Myeloma/drug therapy; Neoplasms/drug therapy; Ovarian Neoplasms/drug therapy; Prospective Studies; Retrospective Studies; Tissue Culture/methods; Tumor Stem Cell Assay Source Cancer Invest 1985;3(6):541-51.

Cloned mice raise possibility of human cloning. Author(s):Nasir, J Language: eng Subject(s): Animal; Cell Nucleus/transplantation; Cloning, Organism; G0 Phase/physiology; Human; Mice; Microinjections Source Clin Genet 1999 Jan;55(1):11-2.

Clones / edited by Jack Dann & Gardner Dozois. Edition Information: Ace ed. Published/Created: New York: Ace Books, 1998. Related Names: Dann, Jack. Dozios, Gardner R. Description: xi, 254 p.; 17 cm. ISBN: 0441005225 Contents: Nine lives / Ursula K. Le Guin -- Mary / Damon Knight -- The extra / Greg Egan -- Out of copyright / Charles Sheffield -- The Phantom of Kansas / John Varley -- Blood sisters / Joe Haldeman -- Past magic / Ian R. MacLeod -- Clone sister / Pamela Sargent -- Where late the sweet birds sang / Kate Wilhelm. Notes: 9 stories. Subjects: Science fiction, American. Genetic engineering--Fiction. Cloning--Fiction. Variant Series: Ace science fiction LC Classification: PS648.S3 C57 1998 Dewey Class No.: 813/.0876208356 21

Clones and clones: facts and fantasies about human cloning / edited by Martha C. Nussbaum and Cass R. Sunstein. Edition Information: 1st ed. Published/Created: New York: Norton, c1998. Related Names: Nussbaum, Martha Craven, 1947- Sunstein, Cass R. Description: 351 p.; 22 cm. ISBN: 0393046486 Notes: Includes bibliographical references. Subjects:

Human cloning--Social aspects. Human cloning--Moral and ethical aspects. LC Classification: QH442.2 .C55 1998 Dewey Class No.: 174/.25 21

Clones, genes, and reproductive autonomy. The ethics of human cloning. Author(s):Harris, J Language: eng Subject(s): Cloning, Organism/legislation & jurisprudence; Ethics, Medical; Freedom; Human; Human Rights/legislation & jurisprudence; Philosophy, Medical; Power (Psychology); Reproduction; Variation (Genetics)/genetics; World Health Organization Source Ann N Y Acad Sci 2000 Sep;913:209-17.

Cloning - controversies Abstract: Cloning of the human being is not only highly controversial; in the opinion of the authors it is impossible - we are not able to reproduce human behaviour and character traits. Reproduction through cloning is limited to personal genome resources. The more important is protection of genomic characteristics as private property and taking advantage of cloning for production of the human organs directly or through xenotransplants. In this paper we present the legislation related to cloning in Poland, in the European Union and other countries. We also indicate who and why is interested in cloning. Author(s): Twardowski, T; Michalska, A Language: pol Source Med Wieku Rozwoj 2001;5(1 Suppl 1):181-94.

Cloning / Bruno Leone, book editor. Published/Created: Farmington Hills, MI: Greenhaven Press; San Diego: Thomson/Gale , c2003. Related Names: Leone, Bruno, 1939- Description: 93 p.; 23 cm. ISBN: 0737713399 (pbk.: alk. paper) 0737713380 (lib.: alk. paper) Notes: Includes bibliographical references (p. 88-89) and index. Subjects: Cloning--Social aspects. Human cloning--Social aspects. Cloning--Moral and ethical aspects. Cloning--Religious aspects. Series: Opposing viewpoints series At issue (San Diego, Calif.) LC Classification: QH442.2 .C5642 2003 Dewey Class No.: 174/.966 21

Cloning / edited by M.L. Rantala and Arthur J. Milgram. Published: Chicago, Ill.: Open Court, c1999. Related Names: Rantala, M. L. Milgram, Arthur J., 1964- Description: xi, 277 p.; 24 cm. ISBN: 0812693744 (hbk.: alk. paper) 0812693752 (pbk.: alk. paper) Notes: Includes bibliographical references and index. Subjects: Human cloning--Moral and ethical aspects--Philosophy--Miscellanea. Series: For and against; [v. 3] LC Classification: QH442.2 .C564 1999 Dewey Class No.: 176 21

Cloning / Lisa Yount, book editor. Published: San Diego, Ca.: Greenhaven Press, 2000. Related Names: Yount, Lisa. Description: 176 p.; 24 cm. ISBN: 073770330X Notes: Includes bibliographical references (p. 168-170) and index. Subjects: Cloning--Social aspects--Moral and ethical aspects. Series: Contemporary issues companion LC Classification: QH442.2 .C56 2000 Dewey Class No.: 571.8/9 21

Cloning / Paul A. Winters, book editor. Published/Created: San Diego, Calif.: Greenhaven Press, c1998. Related Names: Winters, Paul A., 1965- Description: 89 p.; 24 cm. ISBN: 1565107535 (alk. paper) 1565107527 (pbk.: alk. paper) Notes: Includes bibliographical references (p. 84-86) and index. Subjects: Cloning--Moral and ethical aspects. Series: At issue (San Diego, Calif.) LC Classification: QH442.2 .C56 1998 Dewey Class No.: 174/.957 21

Cloning agricultural plants via in vitro techniques / editor, B. V. Conger. Published/Created: Boca Raton, Fla.: CRC Press, c1981. Description: 273 p.; 26 cm. ISBN: 0849357977 Notes: Includes bibliographical references and index. Subjects: Plant cell culture. Cloning. Plants, Plant micropropagation. LC Classification: SB123.6 .P74 Dewey Class No.: 631.5/3 19

Cloning allergen-specific antibody fragments (Fabs); tools for allergen standardization and therapy of type I allergy. Abstract: Recombinant allergens have made it possible to dissect the mechanisms of allergen-antibody interactions at a molecular level. It becomes clear that monoclonal human IgG antibodies as well as animal derived antibodies can block the interaction of specific IgE antibodies as well as the allergen induced allergic effector reaction. Using PCR technology and combinatorial plasmid vectors, recombinant antibody fragments can be produced and it has even become possible to isolate allergen-specific IgE Fabs out of combinatorial IgE libraries constructed from allergic patients lymphocytes. Recombinant Fabs will represent useful tools to study the IgE-allergen interaction as well as for the standardization of allergen extracts and quantitative allergen measurements. Moreover, allergen-specific recombinant Fabs which block the allergen-IgE interaction have to be considered as tools for local therapy in effector organs of allergic patients. Author(s): Valenta, R; Steinberger, P; Laffer, S; Dolecek, C; Wiedemann, P; Flicker, S; Kraft, D Language: eng Source Arb Paul Ehrlich Inst Bundesamt Sera Impfstoffe Frankf A M 1997;(91):222-9.

Cloning and assembly strategies in microbial genome projects. Author(s):Frangeul, L; Nelson, K E; Buchrieser, C; Danchin, A; Glaser, P; Kunst, F Language: eng Subject(s): Cloning, Molecular; Genome, Bacterial; Genome, Fungal; Source Microbiology 1999 Oct;145 (Pt 10):2625-34.

Cloning and associated physiopathology of gestation Abstract: Normal fertile offspring can be produced with nuclear transfer (NT) of somatic cells. This technique is associated with important gestational losses in early pregnancy during the first two to three months and in the late fetal and perinatal periods in cows. In cows and sheep, recent studies suggest that early losses may be associated with placental vascularisation deficiencies. In late gestation and at term, a syndrome, commonly called the large offspring syndrome (LOS), causes important perinatal deaths. This syndrome is associated with increased fetal and placental growth, disturbed placental function and fetal abnormalities. Moreover, prolonged gestations are common. Live offspring occasionally exhibit a respiratory distress syndrome and several types of abnormalities that may hinder their survival. This paper reviews the current knowledge about these pathologies and their incidence in somatic and embryonic cloning. Author(s): Chavatte-Palmer, P; Heyman, Y; Renard, J P Language: fre Source Gynecol Obstet Fertil 2000 Sep;28(9):633-42.

Cloning and characterisation of amphibian ClC-3 and ClC-5 chloride channels. Abstract: Amphibians have provided important model systems to study transepithelial transport, acid-base balance and cell volume regulation. Several families of chloride channels and transporters are involved in these functions. The purpose of this review is to report briefly on some of the characteristics of the chloride channels so far reported in amphibian epithelia, and to focus on recently cloned members of the ClC family and their possible physiological roles. The electrophysiological characterisation, distribution, localisation and possible functions are reviewed and compared to their mammalian orthologs. Author(s): Schmieder, S; Lindenthal, S; Ehrenfeld, J Language: eng Source Biochim Biophys Acta 2002 Nov 13;1566(1-2):55-66.

Cloning and characterization of an extracellular Ca(2+)-sensing receptor from parathyroid and kidney: new insights into the physiology and pathophysiology of calcium metabolism. Author(s): Brown, E M; Pollak, M; Riccardi, D; Hebert, S C Language: eng Subject(s): Metabolism; Cloning, Molecular; Human; Hypercalcemia; Hyperparathyroidism ; Kidney metabolism; Parathyroid Glands metabolism Source Nephrol Dial Transplant 1994;9(12):1703-6.

Cloning and characterization of beta-lactam biosynthetic genes. Abstract: Seven genes coding for two different enzymes of the penicillin/cephalosporin biosynthetic pathway have been cloned from fungal and bacterial sources. Using amino acid sequences derived from the purified enzymes, oligonucleotide probes were designed to hybridize to their cognate genes in a genomic library. The high degree of similarity (57%) between enzymes of bacterial and fungal origin suggests a horizontal transfer of a primordial beta-lactam pathway, probably from a bacterial cell to a fungal cell. Overproduction of the proteins in Escherichia coli has allowed further study of the mechanism of action of these important enzymes. Author(s): Miller, J R; Ingolia, T D Language: eng Source Mol Microbiol 1989 May;3(5):689-95.

Cloning and characterization of cDNA for DRPLA interacting protein Abstract: Dentatorubral-pallidoluysian atrophy (DRPLA) is associated with CAG repeat expansion. While the DRPLA gene is ubiquitously expressed, neuron death occurs in specific area of the brain, predicting that the DRPLA protein interacts with other proteins, which may play a role in the pathogenesis. We isolated a DRPLA binding protein with a yeast two-hybrid system, and identified it to be a human homologue of insulin receptor substrate protein of 53 kDa (IRSp53). The binding of the DRPLA protein with IRSp53 was ascertained by co-immunoprecipitation, and colocalization. A proline-rich region near polyglutamine of the DRPLA protein and the SH3 domain of IRSp53 were involved in the binding. Extended polyglutamine significantly reduced the binding ability in yeast cells. Author(s): Okamumoho, Y; Yamada, M Language: jpn Source Nippon Rinsho 1999 Apr;57(4):856-61.

Cloning and characterization of DNAs with palindromic sequences. Author(s): Leach, D R Language: eng Subject(s): Bacteria/genetics; Base Sequence; Cloning, Molecular; DNA/biosynthesis/chemistry; DNA Replication; Escherichia coli; Fungi; Nucleic Acid Conformation Source Genet Eng (N Y) 1996;18:1-11.

Cloning and characterization of G protein-coupled receptors. Author(s): Parmentier, M; Libert, F; Perret, J; Eggerickx, D; Ledent, C; Schurmans, S; Raspe, E; Dumont, J E; Vassart, G Language: eng Subject(s): Chemoreceptors/physiology; Cloning, Molecular; Comparative Study; Evolution; GTP-Binding Proteins; Multigene Family/genetics; Oncogene Proteins/genetics; Receptors, Cell Surface/genetics; Signal Transduction Source Adv Second Messenger Phosphoprotein Res 1993;28:11-8.

Cloning and characterization of human chemokine receptors. Author(s): Power, C A; Wells, T N Language: eng Subject(s): Amino Acid Sequence; Chemokines metabolism; Human; Molecular Sequence Data; Receptors, Cytokine genetics metabolism Source Trends Pharmacol Sci 1996 Jun;17(6):209-13.

Cloning and characterization of multiple opioid receptors. Abstract: Over the course of 1 to 2 years, the field has moved swiftly to investigate the functional and structural properties of the newly cloned opioid receptors. Achieving a better understanding of these macromolecules is likely to have profound implications for drug design aimed at the production of better analgesic drugs, for a more fundamental understanding of mechanisms of action of drugs of abuse, and for a more comprehensive knowledge base regarding the biology of opioid peptides in particular, and neuroactive peptides in general. Author(s): Akil, H; Meng, F; Mansour, A; Thompson, R; Xie, G X; Watson, S Language: eng Source NIDA Res Monogr 1996;161:127-40.

Cloning and characterization of primary alloreactive human T-lymphocytes. Author(s): Kornbluth, J; Silver, D M; Dupont, B Language: eng Subject(s): Culture Media; Epitopes; HLA Antigens; Haploidy; Hematopoietic Stem Cells; Killer Cells immunology; Lymphocyte Activation; Lymphocyte Culture Test. Source Immunol Rev 1981;54:111-55.

Cloning and characterization of the bleomycin biosynthetic gene cluster from Streptomyces verticillus ATCC15003. Abstract: Bleomycin (BLM) biosynthesis has been studied as a model for hybrid peptide-polyketide natural product biosynthesis. Cloning, sequencing, and biochemical characterization of the blm biosynthetic gene cluster from Streptomyces verticillus ATCC15003 revealed that (1) the BLM hybrid peptide-polyketide aglycon is assembled by the BLM megasynthetase that consists of both nonribosomal peptide synthetase (NRPS) and polyketide synthase (PKS) modules; (2) BlmIX/BlmVIII/BlmVII constitute a natural hybrid NRPS/PKS/NRPS system, serving as a model for both hybrid NRPS/PKS and PKS/NRPS systems; (3) the catalytic sites appear to be conserved in both hybrid NRPS/PKS and nonhybrid NRPS or PKS systems, with the exception of the KS domains in the hybrid NRPS/PKS systems that are unique; (4) specific interpolypeptide linkers may play a critical role in intermodular communication to facilitate the transfer of the growing intermediates between the interacting NRPS and/or PKS modules; (5) post-translational modification of the BLM megasynthetase has been accomplished by a single PPTase with broad carrier protein specificity; and (6) BlmIV/BlmIII-templated assembly of the BLM bithiazole moiety requires intriguing protein juxtaposition and modular recognition. These results lay the foundation to investigate the molecular basis for intermodular communication between NRPS and PKS in hybrid peptide-polyketide natural product biosynthesis and set the stage for engineering novel BLM analogues by genetic manipulation of genes governing BLM biosynthesis. Author(s): Shen, Ben; Du, Liangcheng; Sanchez, Cesar; Edwards, Daniel J; Chen, Mei; Murrell, Jeffrey M Language: eng Source J Nat Prod 2002 Mar;65(3):422-31.

Cloning and characterization of the human thyrotropin-releasing hormone receptor: regulation of thyrotropin-releasing hormone receptor mRNA levels by thyroid hormone Abstract: 1) We successfully cloned a human thyrotropin-releasing hormone receptor (TRH-R) gene. The human TRH-R consists of 398 amino acids and is a member of the G protein-coupled receptor family. The gene has a single intron in the coding sequence. Although highly homologous to the rat and mouse TRH-Rs, the human TRH-R has several differences in the carboxyl-terminal structure. 2) We next studied the effect of thyroid status on TRH-R mRNA levels both in vivo and in vitro. TRH-R mRNA levels in hypothyroid rats were increased significantly, while no significant changes were observed in vitro. The data indicate that thyroid hormone may regulate TRH-R mRNA levels in different manners in vitro and in vivo. Author(s): Yamada, M; Mori, M Language: jpn Source Nippon Rinsho 1994 Apr;52(4):1001-6.

Cloning and culturing of fetal mouse natural killer cells. Author(s):Brooks, C G Language: eng Subject(s): Animal; Cell Culture/methods; Clone Cells; Human; Killer Cells, Natural/cytology; Mice Source Methods Mol Biol 2000;121:13-24.

Cloning and DNA polymorphisms of the HLA class II antigen genes; regulation of their expression by differential splicing; Inoko, H Language: jpn Subject(s): Cloning, Molecular; H-2 Antigens; Isoantigens; Major Histocompatibility Complex; Polymorphism (Genetics); RNA Splicing Source Tanpakushitsu Kakusan Koso 1985 Dec;30(14 Suppl):1749-67.

Cloning and embryonic stem cells: a new era in human biology and medicine. Abstract: The cloning of mammals using adult cells as nuclear donors has been achieved and the same procedure can be, at least theoretically, used to clone humans. Another recent technological advance, the derivation of human embryonic stem cells, opens up new possibilities in cell and tissue replacement therapy and heralds significant improvements in gene therapy. Besides suggesting new and potentially valuable medical applications, the insights gained through the use of these techniques could significantly enrich our understanding of basic mechanisms regulating human development. On the other hand, these preliminary results are viewed by many as the opening of the Pandora's box and there are loud voices clamoring that research in these areas be forbidden in perpetuity. I suggest in the following article that at present we do not know enough to make anything but an entirely emotional decision about future applications of these techniques. I try to summarize the current state of the knowledge in the field & indicate how much further research is necessary if benefits & drawbacks are to be understood. Author(s): Solter, D Language: eng Source Croat Med J 1999 Sep;40(3):309-18.

Cloning and expression of cDNA for the opioid receptor; Fukuda, K Language: jpn Subject(s): Animal; Cloning, Molecular; DNA/isolation & purification; Receptors, Opioid/classification/genetics Source Masui 1995;44 Suppl:S119-20.

Cloning and expression of genes for the small subunit of ribulosebisphosphate carboxylase. Author(s): Wimpee, C F; Tobin, E M Language: eng Subject(s): Bacteriophage lambda genetics; Cloning, Molecular methods; DNA genetics; DNA Restriction Enzymes; DNA, Recombinant; Gene Expression Regulation/radiation effects; Light; Nucleic Acid Hybridization; Plants/enzymology; Poly A/isolation & purification; RNA/isolation & purification; RNA, Messenger/genetics; Ribulose-Bisphosphate Carboxylase/genetics Source Methods Enzymol 1986;118:396-409.

Cloning and expression of large mammalian cDNAs: lessons from ATM. Author(s):Shiloh, Y; Bar-Shira, A; Galanty, Y; Ziv, Y Language: eng Subject(s): Animal; Base Sequence; Cloning, Molecular/methods; DNA, Complementary/biosynthesis/metabolism; Human; Leucine Zippers; Mammals; Molecular Sequence Data; Open Reading Frames; Proteins/biosynthesis/genetics Source Genet Eng (N Y) 1998;20:239-48.

Cloning and expression of pneumococcal genes in Streptococcus pneumoniae. Abstract: An overview of gene cloning in Streptococcus pneumoniae is presented. The advantages of such cloning, especially for pneumococcal genes, are enumerated. The molecular fate of DNA in transformation of S. pneumoniae, in particular, the conversion of DNA to single-strand segments on entry, determines the mechanisms for plasmid establishment and interaction with the chromosome. One of these mechanisms, the chromosomal facilitation of plasmid establishment, is useful for obtaining recombinant plasmids and for introducing an allele from the chromosome into a plasmid. The difference between linear and circular synapsis of donor DNA strands with the chromosome is illustrated. Circular synapsis can give rise to circular integration, which is useful for insertional mutagenesis of chromosomal genes, for coupled cloning in Escherichia coli, and for sequential cloning of DNA along the pneumococcal chromosome. Cloning in S. pneumoniae is not notably affected by DNA mismatch repair or restriction systems in the host cell. Unusual features of gene expression in S. pneumoniae are discussed. Transcription begins most often at promoters with extended -10 sequences, and in a small but significant number of cases, translation does not require a ribosome-binding site with a Shine-Dalgarno sequence. Author(s): Lacks, S A Language: eng Source Microb Drug Resist 1997 Winter;3(4):327-37.

Cloning and expression of the hEGF gene in Saccharomyces cerevisiae. Author(s): Topczewska, J; Bolewska, K Language: eng Subject(s): Amino Acid Sequence; Cloning, Molecular; Epidermal Growth Factor; Genes, Fungal; Molecular Sequence Data; Recombinant Proteins/biosynthesis/genetics; Source Acta Biochim Pol 1993;40(1):4-7.

Cloning and expression of Treponema pallidum protein antigens in Escherichia coli. Abstract: Difficulties in culturing the bacterium Treponema pallidum have greatly hindered syphilis research. Recently, several laboratories have accomplished the expression of T. pallidum antigens in Escherichia coli. It is anticipated that treponemal antigens produced by recombinant DNA technology will provide new tools for investigating the pathogenesis and immunobiology of T. pallidum infection. Author(s): Stamm, L V; Bassford, P J Jr Language: eng Source DNA 1982;1(4):329-33.

Cloning and expression of trichosanthin and alpha-momorcharin cDNA. Author(s): Shaw, P C; Zhu, R H; Yung, M H; Yeung, H W; Ho, W K Language: eng Subject(s): Amino Acid Sequence; Base Sequence; Cloning, Molecular; DNA; Escherichia coli; Genetic Vectors; Molecular Sequence Data; Plant Proteins/genetics; Sequence Homology, Nucleic Acid; Trichosanthin/genetics Source Targeted Diagn Ther 1992;7:213-21.

Cloning and expression vectors for gene function analysis / edited by Quinn Lu, Michael P. Weiner. Published/Created: Westborough, MA: Biotechniques Press, Eaton Pub., c2001. Related Names: Lu, Quinn, 1962- Weiner, Michael P., 1958- Description: xiii, 284 p.: ill.; 28 cm. ISBN: 188129921X Notes: Includes bibliographical references and index. Subjects: Molecular cloning. Genetic vectors. Gene expression. Series: BioFocus (Westborough, Mass.) LC Classification: QH442.2 .C565 2001 Dewey Class No.: 572.8/6 21

Cloning and functional characterization of a family of receptors for the melanotropic peptides. Author(s): Cone, R D; Mountjoy, K G; Robbins, L S; Nadeau, J H; Johnson, K R; Roselli-Rehfuss, L; Mortrud, M T Language: eng Source Ann N Y Acad Sci 1993 May 31;680:342-63.

Cloning and functional characterization of extracellular Ca(2+)-sensing receptors from parathyroid and kidney. Abstract: Parathyroid cells recognize and respond to (i.e., "sense") minute perturbations in the extracellular ionized calcium concentration (Ca2+o), but the mechanisms underlying this process have remained obscure. Recently, we employed expression cloning in Xenopus laevis oocytes to isolate a cDNA coding for a Ca2+o-sensing receptor from bovine parathyroid. Like the native receptor, the cloned Ca2+o-sensing receptor stimulates phospholipase C (PLC) in a G-protein-dependent manner with a nearly identical pharmacological profile. Its deduced amino acid sequence confirms that it is a member of the superfamily of G-protein-coupled receptors (GPR). Transcripts for the receptor are expressed in parathyroid and other tissues that sense Ca2+o (viz., kidney and thyroidal C-cells) as well as those that have no known role in extracellular Ca2+ homeostasis, such as the brain. The availability of the cDNA clone for the Ca2+o-sensing receptor made it possible to test the hypothesis that mutations in the gene encoding the human homolog of the receptor cause inherited disorders of mineral ion metabolism. Familial hypocalciuric hypercalcemia (FHH) and neonatal severe hyperparathyroidism (NSHPT) are, in fact, caused by mutations that reduce the activity of the receptor when they are present in the heterozygous and homozygous states, respectively. In contrast, we have subsequently discovered a family in which a form of autosomal dominant hypocalcemia results from an activating mutation in the receptor gene. The Ca2+-sensing receptor, therefore, permits Ca2+o to play a "hormonelike" role as an extracellular first messenger in addition to its well described role as an important intracellular second messenger. Author(s): Brown, E M; Pollak, M; Chou, Y H; Seidman, C E; Seidman, J G; Hebert, S C Language: eng Source Bone 1995 Aug;17(2 Suppl):7S-11S.

Cloning and gene expression in lactic acid bacteria Abstract: The possibility of using the genera Lactobacillus and Lactococcus as vector representatives is widely discussed at present. The prospects of the construction of recombinant bacteria are closely connected with the solution of a number of problems: the level of the transcription of cloned genes, the effectiveness of the translation of heterologous mRNA, the stability of protein with respect to bacterial intracellular proteases, the method by protein molecules leave the cell (by secretion or as the result of lysis). To prevent segregation instability, the construction of vector molecules on the basis of stable cryptic plasmids found in wild strains of lactic acid bacteria was proposed. High copying plasmids with low molecular weight were detected in L. plantarum and L. pentosus strains. Several plasmids with molecular weights of 1.7, 1.8 and 2.3 kb were isolated from bacterial cells to be used as the basis for the construction of vector molecules. Genes of chloramphenicol- and erythromycin-resistance from Staphylococcus aureus plasmids were used as marker genes ensuring cell transformation. The vector plasmids thus constructed exhibited high transformation activity in the electroporation of different strains, including L. casei, L. plantarum, L. acidophilus, L. fermentum and L. brevis which could be classified with the replicons of a wide circle of hosts. But the use of these plasmids was limited due to the risk of the uncontrolled dissemination of recombinant plasmids. L. acidophilus were also found to have strictly specific plasmids with good prospects of being used as the basis for the creation of vectors, incapable of dissemination. In addition to the search of strain-specific plasmids, incapable of uncontrolled gene transmission, the use of chromosome-integrated heterologous genes is recommended in cloning to ensure the maximum safety. Author(s): Bondarenko, V M; Beliavskaia, V A Language: rus. Source Zh Mikrobiol Epidemiol Immunobiol 2000 Mar-Apr;(2):67-73.

Cloning and genetic characterization of blood group carrier molecules and antigens. Abstract: Based on the astounding pace of growth in the field of molecular biology over the last 2 decades, we are now able to consider the allogenicity of transfused blood products in genetic terms. In this sense, the allogenic barriers to the transfusion-based transplantation are defined by differences between those portions of the donor and recipient genomes that define the antigenicity and immune response to transfused cells. The genetic basis of the major histocompatibility complex barrier described earlier is because of well-characterized transcription from a polymorphic locus on chromosome 6 in nonerythroid cells. In contrast, the allo-barrier of ABO-matched RBC transfusion is the result of differences in over 2 dozen independent genes expressed during erythropoiesis that encode a wide variety of molecules important in membrane biology. Variability in the structure of those genes among separate donors often provides the molecular basis of many blood group antigens and phenotypes. This article reviews our current knowledge of the human blood group systems in genetic terms, focusing on strategies used to identify the relevant genes and their polymorphisms. Author(s): Logdberg, Lennart; Reid, Marion E; Miller, Jeffery L Language: eng Source Transfus Med Rev 2002 Jan;16(1):1-10.

Cloning and reprogramming of the nucleus in the preimplantation embryo Abstract: In mammals, nuclear transfer (or cloning) consists of introducing the nucleus which is generally transcriptionally active into a new cytoplasmic environment, usually that of a transcriptionally inactive metaphase II oocyte. Following the transfer, intense nucleocytoplasmic exchange takes place, and is responsible for remodeling of the nuclear structure and gene reprogramming. The original development program of the donor nucleus is effaced, and the reconstructed embryo adopts that of the recipient oocyte. Both remodeling and reprogramming are regulated by maternal cytoplasmic factors. Overall transcriptional activity, splicing and translational functions as well as the specific expression of certain genes are fairly similar to corresponding activity in normal embryos. However, the development program seems to be read some hours in advance by the reconstructed embryos, and some basic information is lacking. It is particularly important to study these reprogramming abnormalities during the early stages of development in the context of specific abnormalities in fetal and neonatal development observed after transfer of somatic cells. Author(s): Chastant-Maillard, S; Campion, E; Renard, J P Language: fre Source Gynecol Obstet Fertil 2000 Sep;28(9):649-58.

Cloning and sequencing of cDNA of the hepatitis E virus genome--application to diagnosis Abstract: Hepatitis E is endemic, often provoking epidemics in many developing countries. The hepatitis E virus has been molecularly cloned and sequenced and has a single-stranded, positive-sense RNA genome, 7,194 nucleotides, followed by a poly(A) tail. There are three open reading frames. The non-structural gene, approximately 5 kb, is located at the 5' end, while the structural gene, approximately 2 kb, is located at the 3' end of the genome. There is low level nucleotide variation among hepatitis E virus strains isolated from Myanmar and China. Author(s): Uchida, T Language: jpn Source Nippon Rinsho 1993 Feb;51(2):352-6.

Cloning and stem cells. Published/Created: Larchmont, NY: Mary Ann Liebert, [2001- Description: v.: ill.; 28 cm. Vol. 3, no. 3 (2001)- ISSN: 1536-2302 Notes: Title from cover. Indexed by: Chemical abstracts 0009-2258 Subjects: Cloning--Stem cells--Cloning, Organism--Biotechnology--Cloning, Molecular--Genetic Techniques--Periodicals. LC Classification: QH442.2 .C559 Dewey Class No.: 660 13

Cloning and structure-function of the H2 histamine receptor. Author(s): Birdsall, N J Language: eng Subject(s): Cloning, Molecular; Models, Molecular; Protein Conformation; Receptors, Histamine H2/genetics/physiology Source Trends Pharmacol Sci 1991 Jan;12(1):9-10.

Cloning and the future of human embryo research / edited by Paul Lauritzen. Published/Created: Oxford; New York: Oxford University Press, 2001. Related Names: Lauritzen, Paul. Description: xi, 291 p.: ill.; 25 cm. ISBN: 0195128583 (alk. paper) Notes: Includes bibliographical references and index. Subjects: Human cloning--Moral and ethical aspects. Human embryo--Research--Moral and ethical aspects. Human reproductive technology--Moral and ethical aspects. LC Classification: QH442.2 .C566 2001 Dewey Class No.: 176 21

Cloning and transgenesis in farm animals--an Australian perspective. Author(s):Lewis, I M; McClintock, A E; French, A J; Zuelke, K A; Harford, B A; Trounson, A O Language: eng Subject(s): Animal; Animals, Domestic/genetics; Animals, Genetically Modified; Australia; Cattle/genetics; Cloning, Organism veterinary; Goats/genetics; Sheep/genetics Source Aust Vet J 2000 Oct;78(10):694-7.

Cloning by function: expression cloning in mammalian cells. Author(s): Simonsen, H; Lodish, H F Language: eng Subject(s): Animal; Cells, Cultured; Cloning, Molecular/methods; DNA, Complementary/genetics; Genetic Vectors; Human Source Trends Pharmacol Sci 1994 Dec;15(12):437-41.

Cloning by somatic cell nuclear transfer. Abstract: The birth of the first cloned mammals, produced by the introduction of somatic cell nuclei into enucleated oocytes, was an impressive and surprising development. Although the ethical debate has been intense, the important scientific questions raised by this work have been inadequately discussed and are still unresolved. In this essay we address three questions about nuclear transplantation in the eggs of mice and domestic animals. First, why were the recent experiments on somatic cell cloning successful, when so many others have failed? Second, were these exceptional cases, or is somatic cloning now open to all? Third, what are the future possibilities for increasing the efficiency and wider applicability of the cloning process?; Fulka, J Jr; First, N L; Loi, P; Moor, R M Language: eng Source Bioessays 1998 Oct;20(10):847-51.

Cloning Californians: report of the California Advisory Committee on Human Cloning. Published/Created: Sacramento, Calif.: The Committee, [2002] Description: 1 v. (various pagings); 28 cm. Notes: Cover title. "January 11, 2002." Includes bibliographical references. Subjects: Human cloning--Moral and ethical aspects--Government policy--Cloning--Moral and ethical aspects--California. Bioethics. LC Classification: QH442.2 .C353 2002 Dewey Class No.: 176 21 Govt. Doc. No.: H900.C56 R46 cadocs

Cloning cell death genes. Author(s): Schwartz, L M; Milligan, C E; Bielke, W; Robinson, S J Language: eng Subject(s): Animal; Apoptosis/genetics; Base Sequence; Cloning, Molecular/methods; DNA/genetics; DNA, Complementary/genetics; Gene Expression Regulation; Molecular Sequence Data; Polymerase Chain Reaction; RNA/genetics; Subtraction Technique Source Methods Cell Biol 1995;46:107-38.

Cloning DNA binding proteins from cDNA expression libraries using oligonucleotide binding site probes. Author(s): Cowell, I G; Hurst, H C Language: eng Subject(s): Cloning, Molecular/methods; DNA, Complementary/genetics; DNA-Binding Proteins/genetics/metabolism; Escherichia coli/genetics; Gene Library; Molecular Probe Techniques; Oligonucleotide Probes Source Methods Mol Biol 1994;31:363-70.

Cloning embryos by nuclear transfer. Abstract: Nuclear transfer has been used to study the differentiation process in embryogenesis, as well as a method to produce multiple identical individuals. When nuclei are transferred to activated, enucleated oocytes the nuclei swell in diameter, synthesize DNA, acquire cytoplasmic proteins and release nuclear proteins. This protein exchange is thought to result in specific genomic modifications resulting in the transferred nucleus behaving as a zygotic nucleus. The limitations of development observed with relatively differentiated nuclei are thought to result from asynchronies in the length of the cell cycle between the donor cell and the recipient cytoplasm, as well as insufficient genomic modifications. This results in incomplete DNA synthesis and incomplete reprogramming before the first cell division. These nuclear modifications are discussed with data from amphibians and mammals. Author(s): Prather, R S; First, N L Language: eng Source J Reprod Fertil Suppl 1990;41:125-34.

Cloning for expression of immunoglobulin variable region genes and generation of variable region gene repertoires. Author(s): Gussow, D Language: eng Subject(s): Amino Acid Sequence; Animal; Base Sequence; Cloning, Molecular/methods; Immunoglobulin; Hybridomas /immunology; Immunoglobulin Variable Region/genetics; Molecular Sequence Data; Multigene Family; Oligonucleotide Probes; Polymerase Chain Reaction/methods; Restriction Mapping; Sequence Homology, Nucleic Acid Source Behring Inst Mitt 1990 Dec;(87):25-32.

Cloning genes that encode drug-metabolizing enzymes: developmental pharmacology and teratology. Author(s): Nebert, D W; Chen, Y T; Negishi, M; Tukey, R H Language: eng Subject(s): Animal; Aryl Hydrocarbon Hydroxylases biosynthesis /genetics; Cloning, Molecular; Cytochrome P-450 Enzyme System/biosynthesis genetics immunology; DNA Restriction Female; Fetus enzymology; Gene Expression Regulation; Liver/embryology; Pregnancy; RNA, Messenger/analysis; Teratogens/toxicity Source Prog Clin Biol Res 1983;135:61-79.

Cloning heterologous genes: problems and approaches. Abstract: Model fungi such as Neurospora crassa, Aspergillus niger, and Saccharomyces cerevisiae have provided a wealth of genetic information and are currently the object of cooperative genome sequencing projects. Many agriculturally and medically economic important pathogenic fungi, however, are less well characterized, which makes it difficult to study their genes and gene products. Gene sequences from model fungi offer a unique opportunity to clone cognate genes from pathogenic counterparts. In this review, we propose three basic strategies for cloning such genes: Functional complementation, sequence similarity, and genetic linkage. These strategies involve Southern hybridization, cloning, library screening, genetic complementation, and the polymerase chain reaction. We review the major problems encountered using these strategies and outline useful solutions to these difficulties. Author(s): Agnan, J; Korch, C; Selitrennikoff, C Language: eng Source Fungal Genet Biol 1997 Jun;21(3):292-301.

Cloning high molecular weight DNA fragments by the bacteriophage P1 system. Abstract: The cloning of high molecular weight genomic DNA promises to provide the means of mapping chromosomes, isolating genes, and understanding long-range effects on gene expression. This review describes the background and some recent advances in cloning of high molecular weight DNA using the bacteriophage P1 system. Author(s): Sternberg, N L Language: eng Subject(s): Bacteriophage P1/genetics; Cloning, Molecular/methods; DNA, Recombinant/genetics; Gene Library; Genetic Vectors/genetics; Human Source Trends Genet 1992 Jan;8(1):11-6.

Cloning human beings: report and recommendations of the National Bioethics Advisory Commission. Published/Created: Rockville, MD: The Commission, [1997] Description: 2 v.: ill.; 28 cm. Contents: v. I. [no distinct title] -- v. II. Commissioned papers. Notes: "June 1997." Includes bibliographical references. Subjects: Cloning--Moral and ethical aspects--United States. Genetic engineering. Bioethics. LC Classification: QH442.2 .U56 1997 Dewey Class No.: 174/.957 21 Govt. Doc. No.: PrEx 1.19:B 52/C 62

Cloning in cattle: from embryo splitting to somatic nuclear transfer. Abstract: The ability to obtain genetically identical offspring in cattle (clones) is useful for research and for potential applications to breeding schemes. Experimental possibilities for generating such animals have evolved considerably in the last two decades. Embryo splitting has become a relatively simple technique but is limited to twinning. Embryonic nuclear transfer has improved and is associated with sexing to generate sets of clones despite a great variability of results between parent embryos. The factors of progress are reviewed here. Recently, somatic cells used as a source of nuclei in bovine nuclear transfer has been demonstrated. Here we present the results of the developmental potential of nuclei from skin and muscle cells. Author(s): Heyman, Y; Vignon, X; Chesne, P; Le Bourhis, D; Marchal, J; Renard, J P Language: eng Source Reprod Nutr Dev 1998 Nov-Dec;38(6):595-603.

Cloning in humans--state of the art; Benshushan, A; Schenkar, J G Language: heb; Animal; Cloning, Organism; Ethics, Medical; Human; Religion and Medicine Source Harefuah 1998 Dec 15;135(12):604-7.

Cloning in M13 phage or how to use biology at its best. Abstract: A short historical account of the development of the M13mp phage vectors and of the pUC plasmids is presented. Moreover, a complete list of all the M13mp and pUC vectors is compiled. Author(s): Messing, J Language: eng Subject(s): Cloning, Molecular/methods; Coliphages/genetics; DNA, Single-Stranded/genetics; Escherichia coli/genetics; Genetic Vectors; Lac Operon; Plasmids Source Gene 1991 Apr;100:3-12.

Cloning in reproductive medicine. Abstract: This review article summarizes the historical development of mammalian cloning, presents current advances and presumed risk factors in the field of reproductive cloning, discusses possible clinical applications of therapeutic and diagnostic cloning and outlines prospective commercial trends in pharmaceutical cloning. Predictable progress in biotechnology and stem cell engineering should prove to be advantageous for patients' health and for novel benefits in reproductive and regenerative medicine. Author(s): Illmensee, K Language: eng Source J Assist Reprod Genet 2001 Aug;18(8):451-67.

Cloning in silico. Abstract: Cellulose is a major component of plant cell walls, but identification of the enzymes that synthesize it has proven difficult. Now, however, several candidate proteins with sequence homology to bacterial cellulose synthases have been identified by partial sequencing of anonymous cDNA clones from cotton fibers. Author(s): Cutler, S; Somerville, C Language: eng Source Curr Biol 1997 Feb 1;7(2):R108-11.

Cloning issues in reproduction, science, and medicine: a report from the Human Genetics Advisory Commission and the Human Fertilisation and Embryology Authority. Published/Created: [London?]: The Commission, [1998] Description: 50 p.; 30 cm. Notes: Cover title. "December 1998." "Http://www.dti.gov.uk/"--Cover p. 4. Includes bibliographical references. Subjects: Genetic engineering--Human cloning--Moral and ethical aspects. LC Classification: RB155 .C574 1998

Cloning nematode myosin genes. Author(s): Karn, J; Dibb, N J; Miller, D M Language: eng Subject(s): Actins metabolism; Adenosine Triphosphate metabolism; Amino Acid Sequence; Antibodies, Monoclonal immunology; Caenorhabditis genetics physiology; Cloning, Molecular; DNA genetics; Genes, Structural; Isoenzymes genetics; Mutation; Myosins genetics immunology; Phenotype; Protein Conformation; Translation, Genetic Source Cell Muscle Motil 1985;6:185-237.

Cloning Noah's ark. Author(s):Lanza, R P; Dresser, B L; Damiani, P Language: eng Subject(s): Animal; Cats/genetics; Cattle; Cloning, Organism/veterinary; Conservation of Natural Resources; Dogs/genetics; Ruminants/genetics Source Sci Am 2000 Nov;283(5):84-9.

Cloning of a 'D3' receptor subtype expands dopamine receptor family. Author(s): Sibley, D R Language: eng Subject(s): Amino Acid Sequence; Cloning, Molecular; Molecular Sequence Data; Protein Conformation; Receptors, Dopamine/genetics; Terminology Source Trends Pharmacol Sci 1991 Jan;12(1):7-9.

Cloning of a pH sensitive K+ channel in the kidney. Abstract: The renal collecting tubules are responsible for secretion of K+ into the urine and play a major regulatory role in K+ homeostasis of mammals. There are several factors which affect K+ secretion, including aldosterone, K intakes and acid-base balance. We have cloned a complementary DNA encoding a K+ channel of the rabbit renal cortical collecting tubule cells (RACTK1). RACTK1 is a voltage-independent, pH sensitive K(+)-permeable channel, encoding 284 amino-acids, and putatively having two transmembrane segments. This channel is located not only in the luminal membrane of the collecting duct but also in the vascular smooth muscles of the several arteries. The expression of RACTK1 is regulated by external pH but neither by aldosterone nor by K+ concentration in cultured cells. Thus RACTK1 seems to be a unique and distinct from K+ channels in that it is regulated specifically by changes in ambient pH, contributing to the pH dependent urinary K+ excretion. Author(s): Suzuki, M Language: eng Source Nippon Jinzo Gakkai Shi 1995 Aug;37(8):422-7.

Cloning of a putative GABAA receptor from cyclodiene-resistant Drosophila: a case study in the use of insecticide-resistant mutants to isolate neuroreceptors. Abstract: This chapter uses the isolation and cloning of cyclodiene resistance from Drosophila melanogaster to illustrate how mutants resistant to a toxicant can be used to study neuroreceptors. Isolation of mutants from the field, mapping of the single gene responsible and its subsequent cloning are described. As confirmation of gene cloning a susceptible allele of the gene has been used to genetically transform resistant individuals to susceptibility. The gene product appears to code for a subunit of a receptor highly similar to vertebrate GABAA receptor/chloride ion channels, and functional expression studies are described which will elucidate its pharmacology. Cyclodiene resistance is extremely widespread, occurring in both invertebrates and vertebrates. Thus examination of resistance-associated mutations in this receptor in a range of species will enhance our understanding of both the binding sites of toxic ligands and the genetic basis of pesticide resistance. Author(s): ffrench-Constant, R H Language: eng Source EXS 1993;63:210-23.

Cloning of aged animals: a medical model for tissue and organ regeneration. Abstract: Cloning by nuclear transfer has great potential application in pharmaceutical protein production, xeno-transplantation, and perhaps most excitingly, therapeutic cloning. In therapeutic cloning a patient's own skin cells can be used to generate cloned embryos from which embryonic stem cells are isolated. Through targeted differentiation, embryonic stem cells can be directed to develop into the desired tissues/organs for replacement. The combination of homologous recombination of genes and nuclear transfer also offers the promise of correcting defective genes in humans. Demonstration of the successful cloning of aged animals is important for these future medical applications because degenerative diseases often afflict older adults. Our studies have demonstrated that skin fibroblast cells from aged adults, even after prolonged culture, provide nuclear donors equally as competent for cloning as cells from young adults or fetuses. These findings have paved the way for medically treating degenerative diseases of aged humans by tissue regeneration technologies made possible through cloning and homologous recombination. Author(s): Tian, X C; Kubota, C; Yang, X Language: eng Source Trends Cardiovasc Med 2001 Nov;11(8):313-7.

Cloning of Alport syndrome gene. Author(s): Tryggvason, K Language: eng Subject(s): Cloning, Molecular; Collagen/genetics; Human; Nephritis, Hereditary/genetics Source Ann Med 1991 Aug;23(3):237-9.

Cloning of animal cells in tissue culture. Technic and application in cytophysiology; Cinnerova, O Language: cze; Animal; Clone Cells/physiology; Cytogenetics; In Vitro; Mammals; Tissue Culture Source Cesk Fysiol 1968;17(1):11-5.

Cloning of beta 1, 4N-acetylgalactosaminyltransferase cDNAs and its application to the function analysis of carbohydrates; Furukawa, K Language: jpn Subject(s): Animal; Cloning, Molecular; DNA/isolation & purification; Gangliosides/biosynthesis; Gene Expression Regulation, Enzymologic; Human; Leukemia enzymology genetics; Mice; Molecular Structure; N-Acetylgalactosaminyltransferases genetics/physiology; Substrate Specificity Source Seikagaku 1993 Jul;65(7):566-9.

Cloning of cDNA for C3, the third component of complement (author's transl) Abstract: A collection of bacterial plasmids has been constructed carrying cDNA inserts corresponding to mouse liver C3. Eight overlapping cDNA fragments cover 4 700 out of the 5,100 +/- 200 nucleotides of the mRNA including its 3'-end. By partial DNA sequencing it has been found that the beta-chain is encoded in the 5'-half of the mRNA and must thus be located in the amino-terminal portion of the precursor polypeptide pro C3. From DNA sequences it is predicted that the amino-terminus of mouse C3-beta coincides in 12/15 residues with the known amino-acid sequence of guinea-pig C3 beta and in 9/10 residues with that of human C3 beta. A gene bank has been constructed from mouse DNA, from which four C3 genomic clones have been isolated. Two of them carry direct neighbouring fragments of 14- and 18-kilobase pairs of DNA, and contain one entire C3 gene (24-kilobase pairs). The 3'- and 5'-ends of the gene have been mapped. The DNA sequence at the 5'-end predicts that the initial translation product of the C3-mRNA is a prepro C3 molecule which carries at its amino-terminus a signal peptide of 24 amino-acids of typical composition. Human C3 genomic clones have been isolated and used to map the human C3 gene on a chromosome. Author(s): Fey, G; Domdey, H; Wiebauer, K; Odink, K Language: fre Source Ann Immunol (Paris) 1982 Mar-Apr;133C(2):189-97.

Cloning of cDNA specifically involved in the thyroid cAMP mitogenic pathway. Abstract: The activation of the cyclic AMP cascade in dog and human thyroid cells in primary culture induces the expression of differentiated gene expression, hyperfunction and proliferation. These programs are developed simultaneously in quiescent dedifferentiated cells. In this paper the strategy followed by our group to define the genes involved in the cAMP mitogenic cascade is outlined. Author(s): Miot, F; Wilkin, F; Dremier, S; Uyttersprot, N; Lamy, F; Dumont, J E; Maenhaut, C Language: eng Source Horm Res 1994;42(1-2):27-30.

Cloning of cDNAs or genes encoding enzymes of sterol biosynthesis from plants and other eukaryotes: heterologous expression and complementation analysis of mutations for functional characterization. Author(s): Bach, T J; Benveniste, P Language: eng Source Prog Lipid Res 1997 Sep;36(2-3):197-226.

Cloning of contractile protein genes. Author(s): Schwartz, R J; Stone, E M Language: eng Subject(s): Actins genetics; Bacteriophages genetics; Cell Differentiation; Cell Division; Cell Fusion; Cells, Cultured; Cloning, Molecular; Culture Media; DNA/biosynthesis; DNA Restriction Source Cell Muscle Motil 1983;3:195-257.

Cloning of disease genes by chromosome mapping; Suzuki, M; Fujiwara, T; Takahashi, E Language: jpn Subject(s): Animal; Chromosome Mapping; Cloning, Molecular/methods; Genetic Diseases, Inborn/genetics; Human Source Seikagaku 1999 Jan;71(1):65-9.

Cloning of embryos. Abstract: Nuclear transfer for the study of differentiation in amphibians has been used since the 1950s, but not until recently have the same procedures been applied successfully to some mammals. Nuclear transfer, as developed for the amphibian, is successful in sheep, cattle, rabbit, and pig, but not mouse embryos. This fact is discussed in relation to the species-specific timing of the activation of the zygotic genome. Nuclear transfer to an oocyte presumably results in a genomic reprogramming of the transferred nucleus. The limits of differentiation that can be reprogrammed have yet to be determined. Since the cells of early embryos are thought to have identical nuclear genomes, early embryos can be used as a source of donor nuclei; and, when combined with serial nuclear transfer, can theoretically produce an unlimited number of identical offspring. Cloning by splitting does not result in a reprogramming of the genome and is limited in the number of identical offspring that can result. Here we discuss some of the factors to consider concerning micromanipulation and nuclear reprogramming and how they relate to other embryo technologies. Author(s): Prather, R S; First, N L Language: eng Source J Reprod Fertil Suppl 1990;40:227-34.

Cloning of fibrinogen, prothrombin, and alpha 1-antitrypsin. Author(s): Davie, E W Language: eng Subject(s): Animal; Blood Proteins/genetics; Cattle; Cloning, Molecular; DNA; Fibrinogen/genetics; Plasmids; Prothrombin/genetics; Rabbits; alpha 1-Antitrypsin/genetics Source Harvey Lect 1981-82;77:1-21.

Cloning of fish enzymes and other fish protein genes. Abstract: Fish metabolism needs special enzymes that have maximum activity at very different conditions than their mammalian counterparts. Due to the differences in activity, these enzymes, especially cold-adapted proteases, could be used advantageously for the production of some foods. In addition to the enzymes, this review describes some other unique fish polypeptides such as antifreeze proteins, fluorescent proteins, antitumor peptides, antibiotics, and hormones, that have already been cloned and used in food processing, genetic engineering, medicine, and aquaculture. Recombinant DNA technology, which allows these biological molecules to be cloned and overexpressed in microorganisms is also described, highlighting innovative applications. The expected impact of cloning fish proteins in different fields of technology is discussed. Author(s): Macouzet, M; Simpson, B K; Lee, B H Language: eng Source Crit Rev Biotechnol 1999;19(3):179-96.

Cloning of genes from single neurons. Author(s): Dulac, C Language: eng Subject(s): Animal; Cloning, Molecular/methods; Genes, Structural; Mice; Neurons/chemistry/metabolism; Rats Source Curr Top Dev Biol 1998;36:245-58.

Cloning of genes in yeast; Nombela, C; Santamaria, C Language: spa; Cloning, Molecular; Genes, Fungal; Saccharomyces cerevisiae/genetics Source Revis Biol Celular 1985;4:1-25.

Cloning of genes that have environmental and clinical importance from rhodococci and related bacteria. Abstract: Generalised and specialised transduction systems were developed for Rhodococcus by means of bacteriophage Q4. The latter was used in conjunction with DNA from an unstable genetic element of R. rhodochrous to construct resistance plasmids which replicate in strains of R. equi, R. erythropolis and R. rhodochrous. One of the plasmids, pDA21, was joined with Erythropolis coli suicide vector pEcoR251 to obtain shuttle plasmids maintained in both rhodococci and E. coli. Conjugation between these rhodococcal strains demonstrated all were interfertile with each other and that some of the determinants for this were located on the unstable genetic element. Plasmids derived from this element, such as pDA21, carried the conjugative and self-incompatibility capacities; deletion analysis revealed that DNA necessary for self-incompatibility overlapped with that for arsenic resistance. Rifampicin is one of the principal chemotherapeutic agents used to treat infections by rhodococci and related organisms. The genes responsible for two types of inactivation have been cloned. The sequence of the R. equi DNA responsible for decomposition of the antibiotic strongly resembled those of monooxygenases acting upon phenolic compounds, consistent with the presence of a naphthalenyl moiety in the rifampicin molecule. Antibiotic resistance conferred by the gene was surprisingly specific to the semisynthetic compounds rifampicin (150-fold increase) and rifapentine (70-fold). Similar specificity was observed with the other inactivation gene cloned, which ribosylates rifampicin at the 23-hydroxyl position. A 60-bp sequence upstream of the monooxygenase and ribosylation genes is strikingly similar suggesting a shared pattern of regulation. Rhodococcal arsenic resistance and azo dye degradation genes have been cloned and characterised. Author(s): Dabbs, E R Language: eng Source Antonie Van Leeuwenhoek 1998 Jul-Oct;74(1-3):155-67.

Cloning of guanylyl cyclase isoforms. Abstract: The cloning of particulate and soluble guanylyl cyclases is summarized in Table I. With respect to transmembrane signal transduction systems, guanylyl and adenylyl cyclases can be grouped together with some protein tyrosine kinases and protein tyrosine phosphatases to form a diverse protein family with various structural and functional similarities (Garbers, 1989, 1991, 1992; Koesling et al., 1991; Chinkers and Garbers, 1991; Fig. 1). Particulate guanylyl cyclase contains a single transmembrane domain, and the peptide-binding portion (ligand receptor) is on the exterior surface and the catalytic region on the interior, similar to the protein tyrosine kinase/receptor and the protein tyrosine phosphatase/receptor families (Yarden et al., 1986; Charbonneau et al., 1988; Tonks et al., 1988). Protein tyrosine kinases and phosphatases are also activated by ligand binding to the extracellular domain, which in turn results in phosphorylation or dephosphorylation. On the other hand, soluble guanylyl cyclase exists as a heterodimer with two putative catalytic domains, and both subunits are essential for enzyme activity and activation by nitric oxide. It is thus particularly interesting that adenylyl cyclase also contains two catalytic domains, which are both necessary for catalytic activity (Tang et al., 1991). It is possible that particulate guanylyl cyclase may also dimerize on hormonal stimulation and two catalytic domains from two monomers form a functional catalytic center capable of forming cyclic GMP. The catalytic core of GC-A expressed in bacteria was shown to form a homodimer with positively cooperative kinetics (Thorpe et al., 1991). The physiological significance of the existence of multiple forms of soluble guanylyl cyclase subunits remains unclear. Future studies should reveal the differences in tissue distribution and activation by nitrovasodilators in various heterodimers of soluble guanylyl cyclase. Author(s): Nakane, M; Murad, F Language: eng Source Adv Pharmacol 1994;26:7-18.

Cloning of herpesviral genomes as bacterial artificial chromosomes. Abstract: Herpesviruses, which are important pathogens for both animals and humans, have large and complex genomes with a coding capacity for up to 225 open reading frames (ORFs). Due to the large genome size and the slow replication kinetics in vitro of some herpesviruses, mutagenesis of viral genes in the context of the viral genome by conventional recombination methods in cell culture has been difficult. Given that mutagenesis of viral genes is the basic strategy to investigate function, many of the herpesvirus ORFs could not be defined functionally. Recently, a completely new approach for the construction of herpesvirus mutants has been developed, based on cloning of the virus genome as a bacterial artificial chromosome (BAC) in E. coli. This technique allows the maintenance of viral genomes as a plasmid in E. coli and the reconstitution of viral progeny by transfection of the BAC plasmid into eukaryotic cells. Any genetic modification of the viral genome in E. coli using prokaryotic recombination proteins is possible, thereby allowing the generation of mutant viruses and facilitating the analysis of herpes virus genomes cloned as infectious BACs. In this review, we describe the principle of cloning a viral genome as a BAC using murine gammaherpesvirus 68 (MHV-68), a mouse model for gammaherpesvirus infections, as an example. Author(s): Adler, Heiko; Messerle, Martin; Koszinowski, Ulrich H Language: eng Source Rev Med Virol 2003 Mar-Apr;13(2):111-21.

Cloning of human anti-IgE autoantibodies and their role in the regulation of IgE synthesis. Abstract: In vivo experiments using anti-IgE antibodies have clearly documented that they inhibit IgE production. In vitro experiments showed that not only IgE synthesis but also the effector phase of the allergic response may be influenced, because anti-IgE antibodies can prevent basophil sensitization with IgE or even remove IgE-receptor-bound IgE molecules. However, the question remains whether naturally occurring anti-IgE autoantibodies possess similar biological activity. To generate such antibodies for the necessary in vitro studies, we have cloned human Ig variable genes and selected anti-IgE antibodies using phage display libraries. Most of the human anti-IgE antibodies were anti-idiotypes, but anti-isotypes were also isolated. Author(s): Stadler, B M; Stampfli, M R; Miescher, S; Rudolf, M; Vogel, M Language: eng Source Int Arch Allergy Immunol 1995 May-Jun;107(1-3):48-50.

Cloning of human genes Abstract: The fears concerning potential hazards of man made recombinant DNA have been dismissed and the burden of compelling unnecessary safety regulations has been cleared away.

The deciphering of the entire human genome at the nucleotide level is no longer an impossible task. It is based upon the utilization of cloned fragments of DNA used as specific probes. The present catalogue of cloned human genes is still small, but it will expand rapidly in the near future. There are two categories of probes: those corresponding to know genes, and those consisting of random "anonymous" DNA sequences present in genomic libraries. The increasing number of cloned probes will allow us to saturate progressively the entire genome. Medicine will benefit from this progress, first of all with regard to pre-natal and post-natal diagnosis of some genetic diseases and in the field of cancer research. Author(s): Kaplan, J C Language: fre Subject(s): Burkitt Lymphoma/genetics; Child; Cloning, Molecular/methods; English Abstract; Factor IX/genetics; Female; Genes; Human; Linkage (Genetics); Muscle Proteins/genetics; Muscular Dystrophies/genetics; Oncogenes; Pregnancy; Prenatal Diagnosis; Thalassemia/genetics Source Presse Med 1984 Jun 16;13(25):1565-70.

Cloning of human genomic lactoferrin sequence and expression in the mammary glands of transgenic animals. Author(s):Kim, S J; Yu, D Y; Han, Y M; Lee, C S; Lee, K K Language: eng Source Adv Exp Med Biol 1998;443:79-83.

Cloning of human stem cells: some broad scientific and philosophical issues. Author(s): Trosko, J E Language: eng Subject(s): Clone Cells; Cloning, Organism/trends; Ethics, Medical; Human; Stem Cells/cytology/physiology; Technology, Medical Source J Lab Clin Med 2000 Jun;135(6):432-6.

Cloning of human thyroxine-binding globulin cDNA, isolation of the gene, and its transcriptional regulation Abstract: Cloning of human thyroxine-binding globulin (TBG) cDNA and gene revealed that the primary structure of TBG and its gene organization are homologous to those of serine protease inhibitors (serpin). The transfection study of the TBG promoter linked to chloramphenicol acetyltransferase gene demonstrated that the putative hepatocyte nuclear factor-1 site (located at -77 approximately -65) is required for the transcription of the TBG gene. In addition, the sequence located at -218 approximately -102 is responsible for liver-specific expression of the gene. Estrogen, thyroid hormone and glucocorticoid had little effect on the promoter activity, suggesting that the alteration of serum TBG concentration by these hormones is due to their effect on posttranslational steps in TBG synthesis and secretion. Author(s): Kambe, F; Seo, H Language: jpn Source Nippon Rinsho 1994 Apr;52(4):875-9.

Cloning of human tumor stem cells / editor, Sydney E. Salmon. Published/Created: New York: A.R. Liss, c1980. Related Names: Salmon, S. E. (Sydney E.), 1936- Description: xix, 367 p.: ill.; 24 cm. ISBN: 0845100483 Notes: Includes bibliographical references and index. Subjects: Cancer cells. Cloning. Clone cells--Drug evaluation--Neoplasms, Experimental--Congresses. Cell transformation, Neoplastic--Congresses. Series: Progress in clinical and biological research; 48 LC Classification: RC267 .C58 1980 Dewey Class No.: 616.99/4027 19

Cloning of kappa opioid receptors: functional significance and future directions. Author(s): Akil, H; Watson, S J Language: eng Subject(s): Animal; Cloning, Molecular; Endorphins/physiology; Human; Receptors, Opioid, kappa/biosynthesis/genetics Source Prog Brain Res 1994;100:81-6.

Cloning of kidney membrane transport proteins and its clinical implications. Author(s): Sasaki, S; Marumo, F Language: eng Subject(s): Animal; Carrier Proteins; Cloning, Molecular; Kidney Diseases; Membrane Proteins chemistry genetics; Rats; Research Source Clin Sci (Lond) 1994 Aug;87(2):122-3.

Cloning of mammals Abstract: A review of the development of the method of nuclear transplantation and cloning of animals with the use of nuclei of embryonic and adult cells is presented. Also presented are results of studies of nuclear remodeling & reprogramming in the reconstructed oocyte and of cytoplasmic factors that control these processes. Author(s): Lagutina, I S; Galat, V V Language: rus. Source Ontogenez 2001 May-Jun;32(3):180-95.

Cloning of metabolic regulators and regulation of cell proliferation; Murata, K; Kimura, A Language: jpn Subject(s): Aldehydes metabolism; Animal; Cell Division drug effects; Cloning, Molecular; metabolism physiology. Source Tanpakushitsu Kakusan Koso 1986 Sep;31(11):1010-21.

Cloning of microbial epitopes relevant for T- and B-cells. Abstract: This review summarises, and illustrates, the technology that is available for the molecular cloning and precise identification of T-cell and B-cell epitopes, particularly those of bacteria and parasites. Methods include: selective cloning following subtractive hybridization of nucleic acids; selective screening of expression libraries; analysis of subcloned "epitope libraries" or "deletion constructs"; scanning of multiple synthetic peptides, and computer enhanced prediction. The direct sequencing of polymerase chain reaction products allows the rapid analysis of epitope heterogeneity occurring among natural populations. Multiple epitopes can be assembled either by synthesis or by the expression of polymeric epitope-bearing peptides. Prospects for probing expression libraries with T-cells are bleak due to the complexities of antigen processing, presentation and T-cell recognition in vitro. Elucidation of the enzymic steps involved in processing, resolution of peptide/MHC II co-crystals, & pairing of a large number of known epitopes with their functional restriction elements will significantly improve ability to predict T-cell epitopes. Author(s): Miles, M A; Wallace, G R Language: eng Source Behring Inst Mitt 1991 Feb;(88):133-41.

Cloning of muscle regulatory factors. Author(s): Wright, W E Language: eng Subject(s): Animal; Bromodeoxyuridine/pharmacology; Cell Differentiation/drug effects; Cell Line; Cloning, Molecular; DNA/genetics; Gene Expression Regulation; Hybrid Cells; Nucleic Acid Hybridization; RNA. Source Prog Clin Biol Res 1986;226:455-65.

Cloning of Parkin gene and its function; Shimizu, N; Minoshima, S; Asakawa, S Language: jpn Subject(s): Cloning, Molecular; Human; Parkinson Disease/genetics; Proteins/genetics Source No To Shinkei 1999 Jun;51(6):487-91.

Cloning of prepronociceptin has led to the discovery of other biologically active peptides Abstract: Among the opioid receptors family, the cloning of the mu, kappa and delta receptors was followed by that of another member, named ORL1 (Opiate Receptor Like 1). In spite of obvious homologies with the mu, kappa and delta receptors, ORL1 does not display a relevant affinity for the endogenous ligands of these former receptors (beta endorphin, enkephalins, dynorphin A...). This observation has prompted to search for an endogenous ligand of ORL1. A heptadecapeptide which fulfils this function, with a nanomolar affinity, has been found. It was named either nociceptin or orphanin FQ. It demonstrates, according either to the dose or to the route of administration, hyperalgesic, allodynic, antiopioidergic or even analgesic effects. It displays also many behavioural effects, modifying especially locomotion, exploratory behaviour, motivation, anxiety, memory, food intake. Nociceptin results from the cleavage of a large precursor protein, prepronociceptin (PPNOC). In this latter, nociceptin is flanked on its C-terminal region by another peptide which may be regarded either as a heptadecapeptide (NocII), or a bidecapeptide (NocIII) according to the inclusion or not of a fragment constituted by 3 arginine residues. Investigating the functions modulated by NocII, we observed that it stimulates locomotor activity of mice and shortens the forepaws licking latency in the hot plate test (55 degrees C); these effects are not shared by NocIII. The simultaneous administration of NocII and nociceptin resulted in animals put on the hot plate to the appearance of their respective effects, not modified by the presence of the other. A 41 amino acid peptide flanks nociceptin on its N-terminal region in PPNOC. It may be cleaved to generate a heptadecapeptide, named nocistatin on account of its antagonist effect on the hyperalgesia/allodynia induced by nociceptin. Thus, the discovery of ORL1 has led to that of nociceptin, that of its precursor PPNOC, and thereby to that of NocII/NocIII and nocistatin. The functions modulated by these peptides are being investigated whereas their receptors are yet unknown. These multiple targets allow to expect new strategies to modulate their functions. Author(s): Costentin, J; Florin, S; Suaudeau, C; Meunier, J C Language: fre Source C R Seances Soc Biol Fil 1998;192(6):1099-109.

Cloning of rat Ras guanine nucleotide releasing protein 4, and evaluation of its expression in rat mast cells and their bone marrow progenitors. Abstract: We recently cloned a new mast cell (MC) restricted, Ras guanine nucleotide releasing protein (designated mRasGRP4) from IL-3-developed, mouse bone marrow-derived MCs that can activate varied members of the Ras superfamily of small GTP-binding proteins. We now describe the rat ortholog of this MC-specific guanine exchange factor. Using the mRasGRP4 gene and transcript in a homology-based cloning approach, the relevant transcript was isolated and sequenced from the spleen and lungs of Sprague-Dawley rats. Evidence for differential splicing of the rRasGRP4 transcript was obtained in the spleen. The rat basophilic leukemia 1 MC line was found to express rRasGRP4, as well as the MC-committed progenitors residing in the bone marrow and the mature MCs residing in varied tissues of Sprague-Dawley rats. Based on its deduced amino acid sequence, rRasGRP4 is 93% identical to mRasGRP4. rRasGRP4 contains all of the functional domains present in the RasGRP family of guanine nucleotide exchange factors. Like its mouse ortholog, rRasGRP4 is a MC-restricted guanine exchange factor that contains Ca(2+) and phorbol ester/diacylglycerol-binding domains C-terminal of its CDC25-like catalytic domain. Author(s): Li, Lixin; Yang, Yi; Stevens, Richard L Language: eng Source Mol Immunol 2002 Sep;38(16-18):1283-8.

Cloning of serotonin and substance K receptors by functional expression in frog oocytes. Author(s): Levitan, E S Language: eng Subject(s): Animal; Oocytes; RNA, Messenger; Ranidae; Receptors, Neurokinin-2; Receptors, Neurotransmitter; Receptors, Serotonin metabolism; Translation, Genetic Source Trends Neurosci 1988 Feb;11(2):41-3.

Cloning of sheep and cow embryos. Abstract: The results of experiments aimed at cloning of sheep and cattle embryos are described. Two experimental approaches were used to study the developmental potential of blastomeres from sheep and cow embryos: (i) blastomere separation followed by culture and (ii) fusion of isolated blastomeres with enucleated eggs followed by culture. Both approaches allow embryos to be cloned, but whereas blastomere separation allows only a relatively small number of genetically identical animals to be produced, nuclear transplantation will probably open the way for large-scale cloning of livestock. Author(s): Willadsen, S M Language: eng Source Genome 1989;31(2):956-62.

Cloning of testican/SPOCK in man and mouse. Neuromuscular expression perspectives in pathology Abstract: We have recently cloned a novel proteoglycan initially identified in human testis and hence previously called testican. A close examination of the overall protein structure reveals three main regions: four osteonectin/SPARC-like domains encompassing the amino-terminal and central part of the deduced protein, a Kazal-like motif overlapping the third domain, and the CWCV domain in the carboxyl-terminal end region of the protein core. We propose to call it SPOCK, the acronym of SPARC/Osteonectin CWCV and Kazal-like domains proteoglycan, according to its specific multidomain structure. To get further insight into the function, a Northern blot analysis was performed in order to determine the site of expression in various adult tissues; a 5.2 kb transcript appeared only but strongly in mouse brain. The structure of the murine brain proteoglycan was determined through molecular cloning; human and mouse deduced proteins are highly homologous with 95% overall amino acid identity. Murine brain serial sections hybridized with cDNA and immunological probes revealed identical distribution in discrete cerebral regions, such as CA3 hippocampal region and cerebellum. Immunoelectron microscopy showed the antigen selectively localized in the post-synaptic density of scattered pyramidal neurons and Purkinje cells. Structural analysis, a main expression in nervous system and preliminary assignment of the human gene in a critical region for neuropathologies, suggest that SPOCK may be of importance in neural development and neurodegenerative diseases. Author(s): Charbonnier, F; Perin, J P; Roussel, G; Nussbaum, J L; Alliel, P M Language: fre Source C R Seances Soc Biol Fil 1997;191(1):127-33.

Cloning of the APECED gene provides new insight into human autoimmunity. Abstract: Autoimmune polyendocrinopathy-candidiasis-ectodermal dystrophy (APECED) is the only autoimmune disease characterized so far that is caused by a defect in a single gene. We have recently isolated the defective gene in this disease by positional cloning and have identified several different mutations in APECED patients. This novel gene, AIRE, contains two plant homeodomain (PHD)-type zinc finger motifs and a newly described putative DNA-binding domain SAND. We have further shown that the protein encoded by the AIRE gene is localized to the nuclear body-like structures of cell nuclei. Similar discrete speckles within the nucleus have been suggested to be involved in the regulation of transcription, oncogenesis and differentiation of cells. Together with the predicted structural features of the APECED protein the new data obtained both in vitro and ex vivo suggest that this protein participates in the regulation of gene expression in a restricted set of tissues and cells. Author(s): Aaltonen, J; Bjorses, P Language: eng Source Ann Med 1999 Apr;31(2):111-6.

Cloning of the Drosophila cyclodiene insecticide resistance gene: a novel GABAA receptor subtype? Abstract: 1. gamma-aminobutyric acid (GABA) is a major inhibitory neurotransmitter in both vertebrates and invertebrates. GABAA receptors are composed of a number of different subunits that assemble to form a chloride ionophore. 2. Several subunit types, alpha, beta, gamma, delta and rho have been cloned from vertebrates, but until recently these receptors have remained uncloned from invertebrates. 3. GABAA receptors form the proposed site of action of cyclodiene insecticides. Therefore a Drosophila mutant (Rdl), resistant to cyclodienes and the GABAA receptor ligand picrotoxin (PTX), was used to clone the gene responsible for resistance as a putative invertebrate GABAA receptor. 4. Analysis of the predicted amino acid sequence and gene structure shows that Rdl codes for a receptor subunit similar to vertebrate GABAA receptors, but sufficiently different that it may represent a novel class of GABAA receptor subtype. 5. Cyclodiene insecticide resistance accounts for over 60% of reported cases of insecticide resistance and is also found in vertebrates. 6. Therefore elucidating the molecular basis of cyclodiene resistance is not only important to our understanding of pesticide action on the GABAA receptor but also in examining the conservation of the resistance mechanism between vertebrates and invertebrates. Author(s): Ffrench-Constant, R H Language: eng Source Comp Biochem Physiol C 1993 Jan;104(1):9-12.

Cloning of the Duchenne/Becker muscular dystrophy locus. Author(s): Monaco, A P; Kunkel, L M Language: eng Subject(s): Chromosome Mapping; Cloning, Molecular; Muscular Dystrophies/genetics Source Adv Hum Genet 1988;17:61-98.

Cloning of the familial Mediterranean fever gene: expectations for learning about the pathogenesis of amyloid A amyloidosis. Author(s): Pras, M Language: eng Subject(s): Amyloid Protein Amyloidosis etiology therapy; Cloning, Molecular; Familial Mediterranean Fever; Mutation; Proteins. Source Adv Nephrol Necker Hosp 1998;28:343-54.

Cloning of the gene for the fragile X syndrome: implications for the clinical geneticist. Author(s): Connor, J M Language: eng Subject(s): Chromosome Fragility; Cloning, Molecular; Female; Fragile X Syndrome/diagnosis/genetics; Heterozygote Detection; Human; Male; Pregnancy; Prenatal Diagnosis Source J Med Genet 1991 Dec;28(12):811-3.

Cloning of the genes for excitatory amino acid receptors. Abstract: Glutamate is the major excitatory neurotransmitter in the mammalian brain, with receptors on every neuron in the central nervous system; it has major roles in fast synaptic transmission and in the establishment of certain forms of memory. More than 20 years ago Olney and his colleagues described the 'Excitotoxic Hypothesis' which postulates that, in addition to its normal function in the healthy brain, glutamate can kill neurons by prolonged, receptor-mediated depolarization resulting in irreversible disturbances in ion homeostasis. Therefore, glutamate is a two-edged sword; in certain undefined, adverse conditions it undergoes a transition from neurotransmitter to neurotoxin. Its toxicity has been implicated in the death of neurons in ischemia, epilepsy, and the neurodegenerative disorders such as Alzheimer's, Huntington's, and Parkinson's disease. Recent advances in the molecular cloning of the genes for the glutamate family of receptors has revealed a plethora of receptor subtypes and an unexpected level of complexity in the mechanisms of receptor expression and function. Author(s): Henneberry, R C Language: eng Source Bioessays 1992 Jul;14(7):465-71.

Cloning of the genes underlying neurological diseases--chromosome dissection and SUP-PCR; Ikeda, J Language: jpn Subject(s): Base Sequence; Chromosomes, Human, Pair 4; Cloning, Molecular; DNA/isolation & purification; Gene Library; Genetic Code; Human; Mental Disorders/genetics; Molecular Sequence Data; Nervous System Diseases/genetics; Polymerase Chain Reaction; Restriction Mapping Source Tanpakushitsu Kakusan Koso 1993 Feb;38(3):335-45.

Cloning of the hereditary hemochromatosis gene: implications for pathogenesis, diagnosis, and screening. Abstract: Hereditary hemochromatosis (HH) is one of the most common genetic disorders affecting populations of European ancestry. It is characterized by an inappropriately high iron absorption that leads to iron overload preferentially in the parenchymal organs. Although the severity of the phenotype can be modified by a number of factors, it is clear that most homozygotes will eventually become symptomatic. Clinical manifestations are often nonspecific and easily overlooked. Untreated HH can be associated with substantial morbidity and mortality. Diagnosis in the pre-cirrhotic stage, however, results in normal life expectancy. Early diagnosis and the initiation of phlebotomy therapy are therefore keys to preventing complications related to iron overload. The recent isolation of a strong candidate gene for hemochromatosis has the potential to allow for genetic diagnosis and screening in the near future. This exciting finding is likely not only to change clinical practice but also to yield new insights into the poorly understood pathophysiology of the disease. Author(s): Himmelmann, A; Fehr, J Language: eng Source J Lab Clin Med 1999 Mar;133(3):229-36.

Cloning of the holocarboxylase synthetase cDNA and identification of mutations prevalent in Japanese HCS-deficient patients Abstract: Holocarboxylase synthetase (HCS) plays an essential role in biotin utilization in cells and its deficiency causes biotin-responsive multiple carboxylase deficiency in humans. We have cloned the human HCS cDNA, which maps to chromosome 21q22.1. Two mutations in the HCS genes of Japanese patients with HCS deficiency have been identified: a transition from T to C which causes an amino acid substitution of proline for leucine at position 237 (L237P) and A single guanine base deletion (delta G1067) followed by premature termination. Transient expression in cultured fibroblasts from a patient after site-directed mutagenesis showed that the L237P mutation was responsible for decreased HCS activity. Hybridization analysis using allele-specific oligonucleotide probes demonstrated that the prevalence of the mutations--L237P and delta G1067--was 50% and 30%, respectively, among Japanese patients with HCS deficiency. Author(s): Narisawa, K; Suzuki, Y; Aoki, Y Language: jpn Source Nippon Rinsho 1996 Jan;54(1):259-67.

Cloning of the human phenol sulfotransferase gene family: three genes implicated in the metabolism of catecholamines, thyroid hormones and drugs. Abstract: Phenol sulfotransferases (PST) catalyze the sulfonation of catecholamines, thyroid hormones and phenolic drugs. At least two major forms of human PST enzyme have been characterized biochemically from liver, platelets and other tissues, the phenol-preferring PST (P-PST) and the monoamine neurotransmitter-preferring PST (M-PST). Molecular cloning efforts worldwide over the past 7 years have resulted in the identification of numerous PST cDNA isolates representing alleles of three human PST gene loci termed as STP1, STP2 and STM. All three genes have been mapped precisely to a small region on human chromosome 16p12.1-p11.2 (homologous to mouse chromosome 7), using somatic cell hybrids and cosmid clones. The two most closely related genes, STP1 and STP2, encoding P-PST isozymes have been mapped to a single cosmid clone and are, therefore, in close proximity to one another. STP1 and STP2 are approximately 96% identical at the amino acid sequence level, whereas, the STM gene (encoding M-PST) exhibits a lower level of identity (approximately 93-90.5%) relative to STP1 and STP2. STM is located at a distance of ca. 100 Kb from the STP1 and STP2 doublet. One may speculate that the three genes arose by gene duplication and/or gene conversion in humans. Genomic clones have been sequenced to determine the genomic organization for each of the three highly-related genes. All contain seven coding exons, with conserved intron exon boundaries. Sequencing of individual cDNA isolates of STP1 and STM from various tissues has revealed significant heterogeneity in the 5' nontranslated region, likely due to alternative splicing and/or tissue-specific promoter utilization. DNA polymorphisms have been detected in these genes in the human population and may be useful for molecular genetic studies of the metabolism of endogenous and xenobiotic phenolic molecules. Recent advances in the molecular biology of the human PST gene family are summarized. Author(s): Dooley, T P Language: eng Source Chem Biol Interact 1998 Feb 20;109(1-3):29-41.

Cloning of the late genes in the ergosterol biosynthetic pathway of Saccharomyces cerevisiae--a review. Abstract: Research on the ergosterol biosynthetic pathway in fungi has focused on the identification of the specific sterol structure required for normal membrane structure and function and for completion of the cell cycle. The pathway and its end product are also the targets for a number of antifungal drugs. Identification of essential steps in ergo-sterol biosynthesis could provide new targets for the development of novel therapeutic agents. Nine of the eleven genes in the portion of the pathway committed exclusively to ergosterol biosynthesis have been cloned, and their essentiality for aerobic growth has been determined. The first 3 genes, ERG9 (squalene synthase), ERG1 (squalene epoxidase), & ERG7 (lanosterol synthase), have been cloned and found to be essential for aerobic viability since their absence would result in the cell being unable to synthesize a sterol molecule. The remaining 8 genes encode enzymes which metabolize the first sterol, lanosterol, to ultimately form ergosterol. The two earliest genes, ERG11 (lanosterol demethylase) and ERG24 (C-14 reductase), have been cloned & found to be essential for aerobic growth but are suppressed by mutations in C-5 desaturase (ERG3) gene & fen1 & fen2 mutations, respectively. The remaining cloned genes, ERG6 (C-24 methylase), ERG2 (D8AE7 isomerase), ERG3 (C-5 desaturase), & ERG4 (C-24(28) reductase), have been found to be nonessential. The remaining genes not yet cloned are C-4 demethylase & C-22 desaturase (ERG5). Author(s): Lees, N D; Skaggs, B; Kirsch, D R; Bard, M Language: eng Source Lipids 1995 Mar;30(3):221-6.

Cloning of the Pst I restriction-modification system. Author(s): Walder, R Y; Hartley, J L; Donelson, J E; Walder, J A Language: eng Subject(s): Cloning, Molecular; Deoxyribonucleases, Type II Site-Specific/genetics; Genes, Bacterial Source Gene Amplif Anal 1981;1:217-27.

Cloning of the zebrafish retsina blood mutation: a genetic model for dyserythropoiesis and erythroid cytokinesis. Author(s): Paw, B H Language: eng Subject(s): Chloride-Bicarbonate Antiporters; Cloning, Molecular; Erythropoiesis genetics; Models, Animal; Models, Genetic; Mutation; Zebrafish. Source Blood Cells Mol Dis 2001 Jan-Feb;27(1):62-4.

Cloning of thrombopoietin and its physiologic role in megakaryocytopoiesis and thrombopoiesis; Miyazaki, H Language: jpn Subject(s): Blood Platelets cytology; Cloning, Molecular; Thrombocytopenia pathology; Thrombopoietin/genetics/physiology Source Rinsho Ketsueki 1996 Jan;37(1):1-6.

Cloning of tumor suppressor genes involved in solid tumor development. Abstract: OBJECTIVE: The homozygous inactivation of tumor suppressor genes is a common event in the multistep process of tumorigenesis. The identification of most tumor suppressor genes known to date has come from the study of individuals with a defective copy of such a gene that increases their predisposition to malignancy. Although much is still to be learned about the function of these genes, they all seem to play a role in cell growth regulation. Karyotypic and molecular studies of several solid tumors suggest the presence of at least three tumor suppressor genes on the short arm of human chromosome 3. We present an overview on the progress made in the identification of tumor suppressor genes involved in solid tumor development. Conclusion: The explosive effort to map the human genome fueled by the Human Genome Project will facilitate the identification of tumor suppressor genes on chromosome 3 and on other chromosomes within the next 2 to 3 years. This information will not only provide a better understanding in cell cycle regulation and differentiation but also new insights in the design of new therapies for the treatment of human cancer. Author(s): Smith, D I; del Mar Alonso, M Language: eng Source Arch Otolaryngol Head Neck Surg 1993 Nov;119(11):1210-6.

Cloning of vertebrates: successes and problems Abstract: Cloning of vertebrates, in particular, amphibians and mammals, is discussed. In the last decade, significant progress was made cloning mammals, while cloning of adult amphibians remained problematical. Low-traumatic methods of enucleation of recipient oocytes and transplantation of donor nuclei were worked out. In 1997, an adult sheep was cloned in Great Britain, thus demonstrating the possibility of cloning adult mammals. However, methods of cloning mammals need improvement because of the high lethality of reconstructed embryos (nuclear transplants). The use of in vitro cultured low-differentiated stem cells to obtain donor nuclei seems promising. Works on human cloning are not expedient in the near future because of technical and ethical aspects. Author(s): Koniukhov, B V Language: rus Source Genetika 1997 Dec;33(12):1605-20.

Cloning of Y specific repeated DNA and its use in the study of a structurally abnormal Y chromosome; Nakahori, Y; Mitani, K; Yamada, M; Nakagome, Y Language: jpn Subject(s): Base Sequence; Cloning, Molecular; DNA; Human; Male; Repetitive Sequences, Nucleic Acid; Y Chromosome Source Tanpakushitsu Kakusan Koso 1986 Oct;31(13):1331-9.

Cloning PCR products. An overview.. Author(s):Guo, Baotai; Bi, Yuping Language: eng Subject(s): Base Sequence; Cloning, Molecular/methods; DNA/genetics; DNA Primers; Polymerase Chain Reaction/methods; Restriction Mapping/methods Source Methods Mol Biol 2002;192:111-9.

Cloning pigs: advances and applications. Abstract: Although mouse embryonic stem cells have been used widely for over a decade as an important tool for introducing precise genetic modification into the genome, demonstrating the great value of this technology in a range of biomedical applications, similar technology does not exist for domestic animals. However, the development of somatic cell nuclear transfer has bypassed the need for embryonic stem cells from livestock. The production of offspring from differentiated cell nuclei provides information and opportunities in a number of areas including cellular differentiation, early development and ageing. However, the primary significance of cloning is probably in the opportunities that this technology brings to genetic manipulation. Potential applications of gene targeting in livestock species are described with particular emphasis on the generation of pigs that can be used for xenotransplantation, and the production of improved models for human physiology and disease. The development of techniques for somatic cell nuclear transfer in pigs and the challenges associated with this technology are also reviewed. Author(s): Polejaeva, I A Language: eng Source Reprod Suppl 2001;58:293-300.

Cloning researches and ethics of life; Yonemoto, S Language: jpn Subject(s): Animal; Cloning, Organism; DNA; Ethics; Human; Life; Research Source Tanpakushitsu Kakusan Koso 1997 Nov;42(15):2542-7.

Cloning single-stranded DNA. Abstract: DNA sequence and expression analyses have greatly benefited from using M13 and pUC derived cloning vectors and their polycloning sites. A chronology of the original concepts and experiments is reviewed. Author(s): Messing, J Language: eng Source Mol Biotechnol 1996 Feb;5(1):39-47.

Cloning strategies for peptide hormone receptors. Abstract: With the advancement in molecular biology in the last decade many receptors have been cloned and the understanding of their mechanism of function has improved proportionally. In addition the concurrent discovery of receptor gene families has been used to design elegant cloning strategies which in turn have facilitated the characterization of new receptors and receptor subtypes. This review covers the most currently used cloning strategies and gives some examples. Author(s): Wright, M S; Gautvik, V T; Gautvik, K M Language: eng Subject(s): Animal; Cloning, Molecular/methods; DNA/genetics; Gene Library; Human; Receptors, Cell Surface/genetics Source Acta Endocrinol (Copenh) 1992 Feb;126(2):97-104.

Cloning strategies. Author(s): Intres, R; Crabb, J W Language: eng Subject(s): Biotechnology; Cloning, Molecular/methods; DNA; Polymerase Chain Reaction; Toxins/genetics Source Targeted Diagn Ther 1992;7:43-53.

Cloning the calcium channel. Author(s): Alsobrook, J P 2nd; Stevens, C F Language: eng Subject(s): Amino Acids analysis; Calcium Channels analysis; Cloning, Molecular Source Trends Neurosci 1988 Jan;11(1):1-2.

Cloning the gene for Werner syndrome: a disease with many symptoms of premature aging. Author(s): Lombard, D B; Guarente, L Language: eng Subject(s): Aging/physiology; Cloning, Molecular; DNA Helicases/chemistry/genetics; Human; Sequence Homology, Amino Acid; Werner Syndrome/genetics/physiopathology Source Trends Genet 1996 Aug;12(8):283-6.

Cloning the laboratory mouse. Abstract: A brief account is given of early attempts to clone mammals (mice) by transferring cells (nuclei) of preimplantation embryos into enucleated oocytes, zygotes or blastomeres of two-cell embryos. This is followed by a brief review of recent successes using adult somatic cells: mammary gland cells for sheep, muscle cells for cattle and cumulus cells for mice. We have developed a technique for cloning the laboratory mouse by transferring cumulus cell nuclei into enucleated oocytes. With this technique, we have produced a population of over 80 cloned animals, and have carried the process over four generations. Development and fertility of these appear normal. However, the yield is very low; only approximately 1% of injected oocytes are carried to term. The challenge is now to understand the reason for this high loss. Is it a problem of technique, genomic reprogramming, somatic mutation, imprinting or incompatible cell cycle phases? Author(s): Wakayama, T; Yanagimachi, R Language: eng Source Semin Cell Dev Biol 1999 Jun;10(3):253-8.

Cloning to reproduce desired genotypes. Abstract: Cloned sheep, cattle, goats, pigs and mice have now been produced using somatic cells for nuclear transplantation. Animal cloning is still very inefficient with on average less than 10% of the cloned embryos transferred resulting in a live offspring. However successful cloning of a variety of different species and by a number of different laboratory groups has generated tremendous interest in reproducing desired genotypes. Some of these specific genotypes represent animal cell lines that have been genetically modified. In other cases there is a significant demand for cloning animals characterized by their inherent genetic value, for example prize livestock, household pets and rare or endangered species. A number of different variables may influence the ability to reproduce a specific genotype by cloning. These include species, source of recipient ova, cell type of nuclei donor, treatment of donor cells prior to nuclear transfer, and the techniques employed for nuclear transfer. At present, there is no solid evidence that suggests cloning will be limited to only a few specific animals, and in fact, most data collected to date suggests cloning will be applicable to a wide variety of different animals. The ability to reproduce any desired genotype by cloning will ultimately depend on the amount of time and resources invested in research. Author(s): Westhusin, M E; Long, C R; Shin, T; Hill, J R; Looney, C R; Pryor, J H; Piedrahita, J A Language: eng Source Theriogenology 2001 Jan 1;55(1):35-49.

Cloning type-II restriction and modification genes. Abstract: We have cloned into Escherichia coli the genes for 38 type-II bacterial modification methyltransferases. The clones were isolated by selecting in vitro for protectively modified recombinants. Most of the clones modify their DNA fully but a substantial number modify only partially. In approximately one-half of the clones, the genes for the corresponding endonucleases are also present. Some of these clones restrict infecting phages and others do not. Clones carrying endonuclease genes but lacking methyltransferase genes have been found, in several instances, to be viable. Author(s): Lunnen, K D; Barsomian, J M; Camp, R R; Card, C O; Chen, S Z; Croft,

R; Looney, M C; Meda, M M; Moran, L S; Nwankwo, D O Language: eng Source Gene 1988 Dec 25;74(1):25-32.

Cloning vectors derived from animal viruses. Author(s): Rigby, P W Language: eng Subject(s): Animal; Cloning, Molecular; Gene Expression Regulation; Genes, Dominant; Genes, Viral; Genetic Vectors; Mice; Operon; Plasmids; Recombination, Genetic; Selection (Genetics); Virus Replication; Viruses/genetics Source J Gen Virol 1983 Feb;64 (Pt 2):255-66.

Cloning vectors derived from plasmids and phage of Bacillus. Author(s): Kreft, J; Hughes, C Language: eng Subject(s): Bacillus subtilis/genetics; Bacteriophages/genetics; Cloning, Molecular; Comparative Study; Escherichia coli/genetics; Genetic Vectors; Plasmids; Transformation, Genetic Source Curr Top Microbiol Immunol 1982;96:1-17.

Cloning with 2-micrometer DNA vectors and the expression of foreign genes in Saccharomyces cerevisiae. Author(s): Hollenberg, C P Language: eng Source Curr Top Microbiol Immunol 1982;96:119-44.

Cloning, 2001: hearing before a subcommittee of the Committee on Appropriations, United States Senate, One Hundred Seventh Congress, first session: special hearing, December 4, 2001, Washington, DC. Published/Created: Washington: U.S. G.P.O.: For sale by the Supt. of Docs., U.S. G.P.O., [Congressional Sales Office], 2002. Description: iii, 42 p; 24 cm. ISBN: 0160684188 Notes: Distributed to some depository libraries in microfiche. Shipping list no.: 2002-0203-P. Subjects: Human cloning--United States. Human cloning--Research--Law and legislation--United States.. Series: United States. Congress. Senate. S. hrg.; 107-444. LC Classification: KF26 .A653 2001j Govt. Doc. No.: Y 4.AP 6/2:S.HRG.107-444 GPO Item No.: 1033 1033 (online) 1033-A (MF) 1033-A (online)

Cloning, 2002: hearings before a subcommittee of the Committee on Appropriations, United States Senate, One Hundred Seventh Congress, second session: special hearings, January 24, 2002, Washington, DC, March 12, 2002, Washington, DC. Published/Created: Washington: U.S. G.P.O.: For sale by the Supt. of Docs., U.S. G.P.O., [Congressional Sales Office], 2002. Description: iii, 66 p; 24 cm. ISBN: 016069115X Notes: "Printed for the use of the Committee on Appropriations." Distributed to some depository libraries in microfiche. Shipping list no.: 2003-0030-P. Includes bibliographical references. Subjects: Human cloning--United States. Human cloning--Research--Law and legislation--United States.. Series: United States. Congress. Senate. S. hrg.; 107-541. LC Classification: KF26 .A653 2002a Govt. Doc. No.: Y 4.AP 6/2:S.HRG.107-541

Cloning, biobanks, multicultural meetings, medical turism....Medical ethics confronted with new challenges. Author(s): Hermeren, G Language: swe Source Lakartidningen 1997 Aug 6;94(32-33):2701-7.

Cloning, expression and regulation of angiotensin II receptors. Abstract: Complementary DNAs for angiotensin II type 1 receptor isoforms AT1A and AT1B were cloned by expression cloning from bovine adrenal and rat vascular smooth muscles. Human AT1 receptor was also cloned. Seven transmembrane structures emerged. The AT1 type receptor interacted with more than one type of G-proteins. The ligand binding site of AT1 involving Arg167, Lys199, and Asp263 has been identified by site directed mutagenesis. The regulation of the receptors occur at many stages. The isoform, AT2, was also expression cloned from rat pheochromocytoma cells. Although its ligand binding is not affected by stable GTP analogs, it is a seven transmembrane domain receptor. It mediates the modulations of phosphotyrosine phosphatase by angiotensin II and AT2 specific CGP42112A. The modulation was abolished by pertussis toxin. Thus, AT2 belongs to a new class of angiotensin receptors with unique signalling and regulatory mechanisms. AT1 mediates cellular growth. Interestingly, AT2 expression is inversely related to the mitogenic activity of cells. Author(s): Inagami, T; Yamano, Y; Bardhan, S; Chaki, S; Guo, D F; Ohyama, K; Kambayashi, Y; Takahashi, K; Ichiki, T; Tsuzuki, S Language: eng Source Adv Exp Med Biol 1995;377:311-7.

Cloning, expression and regulation of angiotensin II receptors. Abstract: Angiotensin II isoform 1 (AT1) receptor cDNAs were cloned by expression cloning from bovine adrenal and rat vascular smooth muscles. Human AT1 receptor was also cloned. Seven transmembrane structures emerged. A single type of receptor seems to interact with more than one type of G-protein. AT1 consists of subtypes AT1A and AT1B, and the regulation of the receptors occurs at many stages. The isoform AT2 was also expression cloned from rat pheochromocytoma cells. Although its ligand binding is not affected by GTP analogs, it is a seven transmembrane domain receptor. It mediates the inhibition of phosphotyrosine phosphatase by angiotensin II and AT2 specific CGP42112A; the inhibition was abolished by pertussis toxin. Thus, AT2 belongs to a new class of angiotensin receptors with unique signalling and regulatory mechanisms. Author(s): Inagami, T; Iwai, N; Sasaki, K; Yamano, Y; Bardhan, S; Chaki, S; Guo, D F; Furuta, H; Ohyama, K; Kambayashi, Y Language: eng Source Eur Heart J 1994 Dec;15 Suppl D:104-7.

Cloning, expression and regulation of angiotensin II receptors. Author(s): Inagami, T; Iwai, N; Sasaki, K; Yamamo, Y; Bardhan, S; Chaki, S; Guo, D F; Furuta, H Language: eng Subject(s): Amino Acid Sequence; Angiotensin II ; Cloning, Molecular; DNA; Gene Expression; physiology; Human; Receptors, Angiotensin. Source J Hypertens 1992 Aug;10(8):713-6.

Cloning, expression and structure of catechol-O-methyltransferase. Author(s): Lundstrom, K; Tenhunen, J; Tilgmann, C; Karhunen, T; Panula, P; Ulmanen, I Language: eng Subject(s): Amino Acid Sequence; Animal; Catechol O-Methyltransferase chemistry genetics metabolism; Cloning, Molecular; Crystallization; DNA, Complementary genetics; Human; Molecular Sequence Data Source Biochim Biophys Acta 1995 Aug 16;1251(1):1-10.

Cloning, expression, and regulation of tissue-specific genes in Drosophila Abstract: The family of esterase genes was studied in various Drosophila species. These genes are classified into tissue-specific and housekeeping ones. The expression of tissue-specific esterases in the male reproductive system of Drosophila species from the virilis and melanogaster groups was thoroughly examined. Modifier genes controlling activity level, time of synthesis, and distribution in cells of the tissue-specific esterase isozyme from the ejaculatory bulb were revealed. The structural gene coding for this enzyme was isolated, cloned, and sequenced. This gene was shown to be similar in different Drosophila species; the transcriptional level of tissue specificity of this gene was determined. The possibility of transforming the tissue-specific gene into a housekeeping one was demonstrated. In different Drosophila species, this gene can be expressed in different parts of the reproductive system. In transgenic males carrying the gene of another species the foreign gene is expressed as in the donor. Author(s): Korochkin, L I Language: rus Source Genetika 1995 Aug;31(8):1029-42.

Cloning, function and gene regulation of hematopoietin receptor superfamily; He, F C; Wu, Z Z Language: chi; Cloning, Molecular; Cytokines genetics ; Gene Expression Regulation; Receptors, Cell Surface genetics physiology; Receptors, Colony-Stimulating Factor; Receptors; Receptors, Interleukin-3. Source Sheng Li Ke Xue Jin Zhan 1992 Jul;23(3):234-9.

Cloning, functional expression and characterization of a human olfactory receptor. Abstract: The human olfactory system can recognize & discriminate a large number of different odorant molecules. The detection of chemically distinct odorants begins with binding of an odorant ligand to a specific receptor protein on the olfactory neuron cell surface. To address the problem of olfactory perception at a molecular level, we have cloned, functionally expressed and characterized the first human olfactory receptor (OR 17-40). Application of a mixture of hundred different odorants elicited a transient increase in intracellular calcium at HEK 293-cells which were transfected with a plasmid containing receptor encoding DNA & a membrane import sequence. By subdividing odorant mixture in smaller groups we identified a single component which represented the only effective substance: helional. Testing some structurally closely related molecules we found only one other compound which also could activate the receptor: heliotropyl acetone. All other compounds tested were ineffective. Findings represent beginning of molecular understanding of odorant recognition in humans. Author(s): Hatt, H; Gisselmann, G; Wetzel, C H Language: eng Source Cell Mol Biol (Noisy-le-grand) 1999 May;45(3):285-91.

Cloning, gene expression, and protein purification: experimental procedures and process rationale / Charles C. Hardin ... [et al.]. Published: New York: Oxford University Press, 2001. Related Names: Hardin, Charles C. Description: viii, 435 p.: ill.; 28 cm. ISBN: 0195132947 (alk. paper) Notes: Includes bibliographical references and index. Subjects: molecular cloning-gene expression-proteins-purification-laboratory manuals. LC Classification: QH442.2 .C567 2001 Dewey Class No.: 572.8 21

Cloning, organization and expression of the genes for eukaryotic ribosomal proteins; Begueret, J Language: fre Subject(s): Eukaryotic Cells ; Ribosomal Proteins Source Biochimie 1983 Jul;65(7):III-VIII.

Cloning, sequence analysis, and processing of the rat and human atrial natriuretic peptide precursors. Abstract: Complementary DNA sequences and structural genes encoding the atrial natriuretic peptide precursor (prepro-ANP) have been cloned. Analysis of DNA sequences, complementary to rat atrial prepro-ANP mRNA, has revealed that the various natriuretic peptides isolated from rat atrium reside at the carboxy terminus of a 152-amino-acid precursor protein. The human gene, comprised of three exons and two intervening sequences, encodes a protein of 151 amino acids highly homologous to the rat precursor. Although putative proteolytic processing sites can be identified throughout the prepro-ANP amino acid sequence, the natural form of the mature ANP has not been identified. Therefore, the sites and mechanisms of prepro-ANP processing to mature peptides forms are unknown. However, the successful cloning of the prepro-ANP gene and corresponding cDNAs provide the necessary molecular tools to address these fundamental questions relating to the regulation of ANP synthesis and processing in atrial and extraatrial tissues. Author(s): Lewicki, J A; Greenberg, B; Yamanaka, M; Vlasuk, G; Brewer, M; Gardner, D; Baxter, J; Johnson, L K; Fiddes, J C Language: eng Source Fed Proc 1986 Jun;45(7):2086-90.

Cloning, sequencing and characterization of the ccpA gene from Enterococcus faecalis. Abstract: Enzymes involved in the metabolism of complex carbon and energy sources are unnecessary under conditions of abundant, readily metabolisable nutrients such as glucose or fructose. The repression of these enzymes by glucose has been termed carbon catabolite repression. Mechanisms involved in the carbon catabolite repression in gram-positive bacteria are known to differ from those of gram-negative bacteria such as Escherichia coli. It appears to be mediated by transcriptional repression, requiring trans-acting CcpA, a member of the LacI-GalR family of bacterial regulatory proteins and a cis-acting consensus sequence, designated cre. Here, we report the cloning and characterisation of the chromosomal ccpA gene from Enterococcus faecalis JH2-2. This gene is predicted to encode a 333 amino acids protein with nearly 75% identity to CcpA of Lactobacillus casei. Author(s): Leboeuf, C; Auffray, Y; Hartke, A Language: eng Source Int J Food Microbiol 2000 Apr 10;55(1-3):109-13.

Cloning, structure, and expression of acid and solvent pathway genes of Clostridium acetobutylicum. Author(s): Papoutsakis, E T; Bennett, G N Language: eng Subject(s): Acids/metabolism; Base Sequence; Cloning, Molecular; Clostridium/enzymology/genetics/metabolism; Escherichia coli; Genes, Bacterial; Molecular Sequence Data; Multigene Family; RNA, Bacterial; Solvents/metabolism Source Biotechnology 1993;25:157-99.

Cloning, structure, chromosomal localization and promoter analysis of human 2-oxoglutarate dehydrogenase gene. Abstract: Human 2-oxoglutarate dehydrogenase (OGDH) is an E1-component of the OGDH multi-enzyme complex and catalyzes both the ThDP-dependent decarboxylation of 2-oxoglutarate and the subsequent reductive succinylation of the lipoyl moiety which is covalently bound to the E2 component, dihydrolipoamide succinyltransferase. The cDNA and genomic DNA encoding human OGDH has been cloned and sequenced. The cDNA contains a 3006-bp open reading frame encoding a 40-amino acid leader peptide and a 962-amino acid mature OGDH protein (Mr=108878). The gene contains 22 exons spanning approximately 85 kb. The putative ThDP-binding sequence motif is identified in both DNAs. The gene is localized to chromosome 7 at p13-p14 by fluorescence in situ hybridization. With the TATA- and CAAT-less 5'-flanking region (wild type, -3276/+212) of the OGDH gene-luciferase reporter vector construct and its nested deletion or linker-scanning mutant constructs the transient reporter expression assays in BHK-21 cells reveal the existence of two 10-bp cis acting elements (-53/-44 and -33/-24) and two trans-acting elements (-536/-496 and -93/-84). A nuclear factor that binds to the region from -63 to -24 including two cis-acting elements. Author(s): Koike, K Language: eng Source Biochim Biophys Acta 1998 Jun 29;1385(2):373-84.

Cloning, transgenesis, and genetic variance in animals. Author(s):Seidel, G E Jr Language: eng Subject(s): Agriculture; Animal; Animals, Genetically Modified; Chromosomes, artificial genetics; Cloning, Organism; Models, Animal; Research; Variation (Genetics) Source Cloning Stem Cells 2001;3(4):251-6.

Cloning. Author(s): Trent, R J Language: eng Subject(s): Animal; Animals, Domestic; Base Sequence; Cloning, Molecular; DNA/chemistry/genetics/physiology; Embryo Research; Ethics, Medical; Genetic Diseases, Inborn; Genetic Techniques; Germ Cells; Human; Organ Procurement; Patient Advocacy legislation & jurisprudence; reproductive techniques psychology standards; risk assessment; tissue donors. Source Baillieres Clin Obstet Gynaecol 1991 Sep;5(3):659-73.

Cloning. Published/Created: Larchmont, NY: Mary Ann Liebert, c1999-2001. Description: 3 v.: ill.; 28 cm. Vol. 1, no. 1 (1999)-v. 3, no. 2 (2001). ISSN: 1520-4553 Notes: Title from cover. Subjects: Cloning--Periodicals. Cloning, Organism--Periodicals. Biotechnology--Periodicals. Cloning, Molecular--Periodicals. Genetic Techniques--Periodicals. Genre/Form: Review Literature. LC Classification: QH442.2 .C559 Dewey Class No.: 660 13

Cloning: applications in humans 1. Technical aspects Abstract: The successful cloning experiments in mammals such as the sheep and mouse prompted speculations on clinical application in humans. Cloning is possible by nucleus transplantation and by embryo splitting. Nucleus transplantation does not result in a genetically completely identical individual because the mitochondrial DNA originates from the ovum donor. Embryo splitting may be regarded as the artificial production of a monozygotic multiplet. Possible applications of cloning in humans belong in the context of reproduction, transplantation of genetically identical tissue, and scientific research. Author(s): Geraedts, J P; de Wert, G M Language: dut Source Ned Tijdschr Tandheelkd 2001 Apr;108(4):145-50.

Cloning: applications in humans 2. Ethical considerations Abstract: Reproductive cloning in adults/children evokes unfavourable reactions. Direct objections are that cloning is unnatural, that it affects human dignity and violates the individual's right to genetic uniqueness. Consequential objections concern unjustified health risks for the progeny, unjustified psychosocial risks for the clone child and the risk of cloning for eugenetic purposes. There is consensus that reproductive cloning of existing persons is unjustify as yet because of the health risks for the offspring. Reproductive cloning of embryos is possible by means of nucleus transplantation and of embryo splitting. The ethical analysis of reproductive cloning of embryos depends on the purposes and applications. At least some of the moral objections against cloning of adults/children are not or not completely applicable to reproductive cloning of embryos. Conditions to be put to reproductive cloning of embryos are efficacy, safety and, at least for the time being, avoidance of asynchrony in transferring identical embryos. Ethical aspects of its application in the context of genetical reproductive techniques must be evaluated separately. Therapeutic cloning may be acceptable if alternatives are lacking. Author(s): de Wert, G M; Geraedts, J P Language: dut Source Ned Tijdschr Tandheelkd 2001 May;108(5):184-9.

Cloning: applications in humans. I. Technical aspects Abstract: The successful cloning experiments in mammals such as the sheep and mouse prompted speculations on clinical application in humans. Cloning is possible by nucleus transplantation and by embryo splitting. Nucleus transplantation does not result in a genetically completely identical individual because the mitochondrial DNA originates from the ovum donor. Embryo splitting may be regarded as the artificial production of a monozygotic multiplet. Possible applications of cloning in humans belong in the context of reproduction (treatment of couples with subfertility, with genetic problems or with a 'replica motive'), transplantation of genetically identical tissue, and scientific research. Author(s): Geraedts, J P; de Wert, G M Language: dut Source Ned Tijdschr Geneeskd 2000 May 13;144(20):921-6.

Cloning: applications in humans. II. Ethical considerations Abstract: Reproductive cloning in adults/children evokes unfavourable reactions. Direct objections are that cloning is unnatural, that it affects human dignity and violates the individual's right to genetic uniqueness. Consequential objections concern unjustified health risks for the progeny, unjustified psychosocial risks for the clone child and the risk of cloning for eugenetic purposes. There is consensus that reproductive cloning of existing persons is unjustifiable as yet because of the health risks for the offspring. Reproductive cloning of embryos is possible by means of nucleus transplantation and of embryo splitting. The ethical analysis of reproductive cloning of embryos depends on the purposes & applications. Some of the moral objections against cloning of adults /children are not or not completely applicable to reproductive cloning of embryos. Conditions to be put to reproductive cloning of embryos are efficacy, safety and, at least for the time being, avoidance of asynchrony in transferring identical embryos. Ethical aspects of application in context of genetical reproductive techniques must be evaluated separately. Therapeutic cloning may be acceptable if alternatives are lacking. Author(s): de Wert, G M; Geraedts, J P Language: dut Source Ned Tijdschr Geneeskd 2000 May 13;144(20):926-31.

Cloning: experience from the mouse and other animals. Abstract: Cloning mammals has been successful for many years by splitting an early embryo or transferring embryonic cell nuclei into enucleated oocytes. Cloning is now possible with adult somatic cells. At present, cloning efficiency--as determined by the proportion of live offspring developed from all oocytes that received donor cell nuclei--is low regardless of the cell type (including, embryonic stem (ES) cells) and animal species used. In all animals, except of Japanese black beef cattle, the vast majority (>97%) of cloned embryos perish before reaching full term. Even in the Japanese cattle, less than 20% of cloned embryos reach the adulthood. This low efficiency of cloning seems to be due largely to faulty epigenetic reprogramming of donor cell nuclei after transfer into recipient oocytes. Cloned embryos with major epigenetic errors die before or soon after implantation. Those with relatively 'minor' epigenetic errors may survive birth and reach adulthood. We found that almost all fetuses of inbred mice die at birth from respiratory problems, while those of hybrid mice do not, suggesting that genomic heterogeneity masks-to some extent-faulty epigenetic errors. Thus far, the majority of cloned mice that survived birth, had a normal life span and were fertile. However, these animals may not be totally free of health problems. Postpubertal obesity in certain strains of mice is one example. A trial and error approach may discover better cells for cloning, but it would be wiser to understand the molecular mechanisms of epigenetic nuclear programming and reprogramming to find the way to make cloning safer and more efficient. The relatively high cloning success rate in the Japanese black cattle may provide us a clue of solving the problem of high mortality of cloned offspring. Author(s): Yanagimachi, R Language: eng Source Mol Cell Endocrinol 2002 Feb 22;187(1-2):241-8.

Cloning: its relevance to monozygotic twins. Author(s):Weston, G; Cattrall, F; Trounson, A O; Healy, D L Language: eng Subject(s): Cloning, Organism; Ethics, Medical; Female; Human; Obstetrics; Pregnancy; Risk Assessment; Twins, Monozygotic Source Aust N Z J Obstet Gynaecol 2000 Aug;40(3):317-25.

Cloning: legal, medical, ethical, and social issues: hearing before the Subcommittee on Health and Environment of the Committee on Commerce, House of Representatives, One Hundred Fifth Congress, second session, February 12, 1998. Published/Created: Washington: U.S. G.P.O.: For sale by the U.S. G.P.O., Supt. of Docs., Congressional Sales Office, 1998. Description: iii, 141 p.: ill.; 24 cm. ISBN: 0160565103 Notes: Distributed to some depository libraries in microfiche. Shipping list no.: 98-0255-P. "Serial no. 105-70." Includes bibliographical references. Subjects: Human cloning--Human genetics--Medicine--Research--Moral and ethical aspects--United States. LC Classification: KF27 .E553 1998c Govt. Doc. No.: Y 4.C 73/8:105-70

Cloning: new breakthroughs leading to commercial opportunities. Abstract: Research on cloning animals, again, came to the forefront of public attention in 1997. Most scientists involved in biomedical and agricultural research have emphasized the benefits, of which there are many, of cloning to the public. Basic studies on nuclear transfer have and will continue to contribute to our understanding of how genomic activation and cell cycle synchrony affect nuclear reprogramming and cloning efficiencies, specifically. Also, more basic information on actual mechanisms and specific factors in the oocyte causing nuclear reprogramming is forthcoming. As new molecular approaches in functional genomics are combined with nuclear transfer experiments, new genes involved in nuclear reprogramming will be found. The commercial potentials of products stemming from discoveries in cloning are vast. Cloning will be a more efficient, faster and more useful way of making transgenic fetuses for cell therapies, adult animals for protein production and organs for xenotransplantation. Clearly there are new opportunities in animal cloning technology that will produce many benefits to society. Author(s): Stice, S L; Robl, J M; Ponce de Leon, F A; Jerry, J; Golueke, P G; Cibelli, J B; Kane, J J Language: eng Source Theriogenology 1998 Jan 1;49(1):129-38.

Cloning: potential benefits for human medicine. Author(s): Trounson, A O Language: eng Subject(s): Animal; Animals, Genetically Modified; Australia; Cloning, Organism legislation & jurisprudence; Ethics, Medical; Gene Therapy; Human; Risk Assessment; United States Source Med J Aust 1997 Dec 1-15;167(11-12):568-9.

Cloning: present and perspectives Abstract: Human embryonic cells obtained through somatic cloning would allow selfgrafting for therapeutical purposes. Data available from animal research indicate that this issue should be considered with great care. Author(s): Renard, J P; Legouy, E; Chastant, S; Heyman, Y; Vignon, X Language: fre Source Contracept Fertil Sex 1999 Jun;27(6):405-11.

Cloning: reproductive medicine or breeding program? Abstract: The presentation of clone-sheep Dolly in February 1997 which was the result of a long and costly research process by Ian Wilmut's team at Roslin Institute near Edinburgh brought world-wide headlines and a continuous debate. But neither cloning and cloning experiments nor the debates about it and the possible application on humans are as new as it is shown in the media. The following article gives some facts and arguments to the field of cloning. Author(s): Zulicke, F Language: ger Source Wien Med Wochenschr 1998;148(20):474-7.

Cloning: responsible science or technomadness? / edited by Michael Ruse & Aryne Sheppard. Published/Created: Amherst, N.Y.: Prometheus Books, 2001. Related Names: Ruse, Michael. Sheppard, Aryne. Description: 332 p.; 23 cm. ISBN: 1573928364 (pbk.: alk. paper) Notes: Includes bibliographical references and index. Subjects: Cloning--Social aspects. Series: Contemporary issues (Buffalo, N.Y.) LC Classification: QH442.2 .C5678 2001 Dewey Class No.: 174/.966 21

Cloning--a matter of life.. Author(s):De Jonge, Christopher Source J Androl 2003 Mar-Apr;24(2):164-6.

Cohen, Daniel, 1936- Cloning / Daniel Cohen. Edition Information: [Rev. and updated ed.] Published/Created: Brookfield, Conn.: Twenty-First Century Books, c2002. Description: 144 p.: ill.; 24 cm. ISBN: 0761328025 Summary: Examines the history, current developments, future, and ethical ramifications of cloning, recombinant DNA, and gene therapy. Notes: Includes bibliographical references (p. 141). Subjects: Cloning--Juvenile literature. Cloning. Genetic engineering. LC Classification: QH442.2 .C64 2002 Dewey Class No.: 571.8/9 21

Cohen, Daniel, 1936- Cloning / Daniel Cohen. Published/Created: Brookfield, Conn.: Millbrook Press, c1998. Description: 128 p.: ill.; 24 cm. ISBN: 0761303561 (lib. bdg.) Summary: Examines the history, current developments, future, and ethical ramifications of cloning, recombinant DNA, and gene therapy. Notes: Includes bibliographical references (p. 123-124) and index. Subjects: Cloning--Juvenile literature. Cloning. Genetic engineering. LC Classification: QH442.2 .C64 1998 Dewey Class No.: 571.8/9 21

Coincidence cloning. Taking the coincidences out of genome analysis. Abstract: The term "coincidence cloning" encompasses a wide range of methodologies, the aim of which is to isolate DNA sequences which occur in both of two input DNA sources. The nature of these input DNAs may be genomic or cDNA, cloned or uncloned, and as such the far reaching applicability of the techniques can be imagined. If the input DNAs are genomic then the product will be enriched for useful markers co-occurring between the two. If the input DNAs comprise one genomic resource and one cDNA resource the product will contain genes mapping to that particular genomic region. In this review a comparative description of the range of coincidence cloning methods is given, together with a discussion of their applications. Finally, consideration is given to the general limitations of these techniques. Author(s): Devon, R S; Brookes, A J Language: eng Source Mol Biotechnol 1996 Jun;5(3):243-52.

Coincident sequence cloning: a new approach to genome analysis. Abstract: Many analytical molecular genetic techniques developed over the past decade have evolved primarily to tackle the problem of targeting or isolating sequences of interest in complex mixtures. A new approach to this problem, Coincident Sequence Cloning (CSC), involves integrating a pair of DNA mixtures in such a way as to isolate any shared sequence components. Author(s): Brookes, A J; Porteous, D J Language: eng Source Trends Biotechnol 1992 Jan-Feb;10(1-2):40-4.

Committee meeting of Assembly Health Committee: testimony on cloning-related research: [May 4, 1998, Trenton, New Jersey] / meeting recorded and transcribed by the Office of Legislative Services, Public Information Office, Hearing Unit. Published/Created: Trenton, N.J.: The Unit, [1998] Related Names: New Jersey. Office of Legislative Services. Public Information Office. Hearing Unit. Description: 21, 6 p.; 28 cm. Notes: Cover title. Subjects: Cloning--Government policy--New Jersey. Cloning--Research--New Jersey. Human cloning--Government policy--New Jersey. LC Classification: QH442.2 .N49 1998

Complementation cloning of mammalian transcriptional regulators: the example of MHC class II gene regulators. Abstract: Cloning by complementation of mutant cell lines is a powerful way in which to identify and isolate important regulatory genes on the basis of functional assays. The recent cloning of two essential regulators of major histocompatibility complex (MHC) class II gene expression has not only advanced our understanding of the complex mechanisms controlling these genes, but also helps to illustrate the feasibility of this approach for the study of mammalian gene regulation. Author(s): Steimle, V; Mach, B Language: eng Source Curr Opin Genet Dev 1995 Oct;5(5):646-51.

Contribution of Japanese researchers to progress in studies on nephrology in the past 100 years: Molecular cloning of water channels in the kidney collecting duct and the etiology of nephrogenic diabetes insipidus. Author(s):Sasaki, Sei Language: jpn Subject(s): Amino Acid Sequence; Animal; Aquaporins/chemistry/genetics/physiology; Cloning, Molecular; Diabetes Insipidus, Nephrogenic/etiology; Genes, Dominant; Genes, Recessive; History of Medicine, 20th Cent.; Human; Japan; Molecular Sequence Data; Mutation; Nephrology/history Source Nippon Naika Gakkai Zasshi 2002 May 10;91(5):1437-9.

Cook, Robin, 1940- Shock / Robin Cook. Published/Created: New York: G.P. Putnam's Sons, c2001. Description: 370 p.; 24 cm. ISBN: 0399146008 (acid-free paper) Subjects: Women graduate students--Fiction. Fertility clinics--Fiction. Cloning--Fiction. Massachusetts--Fiction. LC Classification: PS3553.O5545 S56 2001 Dewey Class No.: 813/.54 21

Cooper, Margaret C. Code name, Clone / by Margaret C. Cooper; illustrations by Christopher Bigelow. Published/Created: New York, N.Y.: Walker, 1982. Related Names: Bigelow, Christopher, ill. Description: 121 p.: ill.; 22 cm. ISBN: 0802764746 (trade): 0802764754 (reinf.) Summary: In the future, cloned brothers Evonn and Stefan escape from behind the Iron Curtain to search for their father in the United States. Notes: Sequel to: Solution. Subjects: Science fiction. Cloning--Fiction. Series: Walker science fiction for young readers LC Classification: PZ7.C78736 Co 1982 Dewey Class No.: [Fic] 19

Cooper, Margaret C. Solution, escape / by Margaret C. Cooper; illustrated by Rod Burke. Published/Created: New York: Walker, 1981. Related Names: Burke, Rod. Description: 93 p.: ill.; 22 cm. ISBN: 0802764045: 0802764053 (lib. bdg.) : Summary: In the 21st century, 13-year-old Stefan is sent to a scientific research station where his movements are closely monitored. He discovers he is a clone, destined to be part of a fanatical scheme to control the government. Subjects: Science fiction. Cloning--Fiction. Series: Walker science fiction for young readers LC Classification: PZ7.C78736 So 1981 Dewey Class No.: Fic

Correlating the pharmacology and molecular biology of opioid receptors. Cloning and antisense mapping a kappa 3-related opiate receptor. Abstract: We cloned a kappa 3-related opioid receptor, and although it is still not clear whether this clone corresponds to the kappa 3 receptor itself or is a related gene product, the extensive antisense mapping and the antibody immunoreactivity strongly associate this clone with the kappa 3 receptor. Our approach also indicates the usefulness of antisense approaches in mapping and identifying orphan receptors. Perhaps it is most effective in identifying partial sequences prior to cloning them in their entirety. It also provides a mechanism of identifying proteins that are not expressed functionally. Author(s): Pasternak, G W; Pan, Y X; Cheng, J Language: eng Subject(s): Animal; Cloning, Molecular; Mice; Oligonucleotides, Antisense/pharmacology; Receptors, Opioid, kappa/drug effects/genetics/immunology; Receptors, Opioid, mu/drug effects Source Ann N Y Acad Sci 1995 May 10;757:332-8.

Could human cloning make us happy? Abstract: The idea of making mankind happy by cloning is entertained only by who reduce anything alive first to, a self-organizing system of cells", then to „ a set of elements" and finally to „the material for something". Such ontological blindness justifies all cases of manipulating what ever is alive including human life in its initial stages. From a biological point of view, cloning involves manipulating living genetic material. When judging this procedure, one must not solely consider the technical possibilities; we need to take into account the moral point of view as well. Tampering with genetic material - the human genome especially - is morally disputable as it distorts the values of natural biological order, which has not been created by man himself. The value of this natural order, and, in the case of man, human dignity - though invisible, unmeasurable, and unobservable at its beginning- constitutes the boundaries no scientist should transgress especially if he/she is a person with a moral conscience. Author(s): Siemianowski, A Language: pol Source Med Wieku Rozwoj 2001;5(1 Suppl 1):81-9.

Crichton, Michael, 1942- The lost world: a novel / by Michael Crichton. Edition Information: 1st trade ed. Published/Created: New York: Knopf: Distributed by Random House, 1995. Description: xi, 393 p.: ill.; 25 cm. ISBN: 0679419462 Subjects: Dinosaurs--Fiction. Cloning--Fiction. LC Classification: PS3553.R48 L67 1995b Dewey Class No.: 813/.54 20

Cupressaceae pollinosis: identification, purification and cloning of relevant allergens. Abstract: Allergy to Cupressaceae pollen is a worldwide pollinosis caused by several species. Pollen extracts prepared from allergenic species belonging to this family are characterised by low protein and high carbohydrate content. The allergenic components represented in the pollen extracts from different species of the Cupressaceae family show high levels of cross-reactivity when probed with human IgE from allergic subjects and share a number of common epitopes also identified by polyclonal rabbit antisera and monoclonal antibodies. A close relationship has also been described with the Taxodiaceae and Podocarpaceae families. Although both proteic and carbohydrate epitopes appear to be involved in IgE recognition and allergenic cross-reactivity, a large portion of the IgE reactivity of Cupressaceae-allergic patients seems to be associated with sugar moieties present on the relevant allergenic molecules. From this point of view, Cupressaceae/Taxodiaceae allergens constitute a particularly useful model to study IgE cross-reactivity, as they have been shown to display different levels of homology. Moreover, the availability of the purified allergens, together with their recombinant counterparts, may shed light on the actual role played by carbohydrate in allergic sensitisation, IgE recognition and allergenic cross-reactivity. Author(s): Di Felice, G; Barletta, B; Tinghino, R; Pini, C Language: eng Source Int Arch Allergy Immunol 2001 Aug;125(4):280-9.

Curran, Brendan. A terrible beauty is born: clones, genes and the future of mankind / Brendan Curran. Published/Created: New York, NY: Taylor & Francis, 2003. Description: p.; cm. ISBN: 0415287081 (hardcover: alk. paper) 041528709X (papercover: alk. paper) Notes. Includes bibliographical references and index. Subjects: Cloning--Moral and ethical aspects--Popular works. Bioethical Issues--Cloning, Organism--Genetic Engineering--Popular Works. Series: Science spectra; v. 3. LC Classification: QH442.2 .C87 2003 Dewey Class No.: 303.48/3 21

Current data on the cloning of virulence factors of pathogenic Streptococci Abstract: The data on the cloning of the main pathogenic determinants of group A streptococci including M-protein, erythrogenic toxin, streptokinase, streptolysin O are analyzed. The scientific importance and the possible ways to use the data obtained after cloning are discussed. The hypothesis on the operon system of a number of streptococcal virulence factors regulation is discussed. Construction of vector systems for streptococcal genes cloning is summarized. Author(s): Suvorov, A N; Golubkov, V I; Tomolian, A T Language: Source Mol Gen Mikrobiol Virusol 1988 Feb;(2):3-12.

Cystic fibrosis. 3. Cloning the cystic fibrosis gene: implications for diagnosis and treatment. Author(s): Porteous, D J; Dorin, J R Language: eng Source Thorax 1991 Jan;46(1):46-55.

DAP-kinase: from functional gene cloning to establishment of its role in apoptosis and cancer. Abstract: DAP-kinase is a pro-apoptotic Ca(2+) calmodulin-regulated serine/threonine kinase that participates in a wide array of apoptotic systems initiated by interferon-gamma, TNF-alpha, activated Fas, and detachment from extracellular matrix. It was isolated by an unbiased functional approach to gene cloning aimed at hitting central mediators of the apoptotic process. This 160 Kd protein kinase is localized to actin microfilaments and carries interesting modules such as ankyrin repeats and the death domain. The death promoting effects of DAP-kinase depend on its intact catalytic activity, the correct intracellular localization, and on the presence of the death domain. A few mechanisms restrain the killing effects of the protein in healthy cells. The enzyme's active site is negatively controlled by an adjacent CaM regulatory domain whose effect is relieved by binding to Ca(2+)-activated calmodulin. A second mode of autoinhibition engages the serine-rich C-terminal tail, spanning the last 17 amino acids of the protein. A link between DAP-kinase and cancer has been established. It was found that the mRNA and protein expression is frequently lost in various human cancer cell lines. Analysis of the methylation status of DAP-kinase's 5' UTR in DNA extracted from fresh tumor samples, showed high incidence of hypermethylation in several human carcinomas and B cell malignancies. The anti-tumorigenic effect of DAP-kinase was also studied experimentally in mouse model systems where the re-introduction of DAP-kinase into highly metastatic mouse lung carcinoma cells who had lost the protein, strongly reduced their metastatic capacity. Thus, it appears that loss of DAP-kinase confers a selective advantage to cancer cells and may play a causative role in tumor progression. A few novel kinases sharing high homology in their catalytic domains with DAP-kinase have been recently identified constituting altogether a novel family of death promoting serine/threonine kinases. Author(s): Cohen, O; Kimchi, A Language: eng Source Cell Death Differ 2001 Jan;8(1):6-15.

Degradation pathway(s) of chlorophyll: what has gene cloning revealed? Abstract: The mechanism responsible for the degreening of plants and the degradation of chlorophyll was unclear for many years. However, recent studies have identified the colorless intermediates and helped to construct a basic pathway for degradation. After the successive removal of phytol and Mg21 from the chlorophyll molecule by chlorophyllase and 'Mg dechelatase', pheophorbide a is cleaved and reduced to yield a colorless, open tetrapyrrole intermediate. After further modifications, this is finally transported to the vacuole. Cloning the genes for chlorophyllase isozymes and the reductase should help to elucidate the physiological roles of each enzyme at a molecular level. Author(s): Takamiya, K I; Tsuchiya, T; Ohta, H Language: eng Source Trends Plant Sci 2000 Oct;5(10):426-31.

Detection of genetic materials in mammalian somatic cell cloning Abstract: Following the production of cloned sheep, mice, goats, pigs, cats, rabbits and argali from somatic cells, the cellular and molecular biological tests of cloned animals and early embryos turn to be more important. In this review, chromosome analysis, nuclear DNA analysis, mitochondrial DNA analysis and telomeric DNA analysis that are applied in the detection of cloned mammal and early embryos are described. Author(s): Han, Zhi-Ming; Chen, Da-Yuan Language: chi Source Sheng Wu Hua Xue Yu Sheng Wu Wu Li Xue Bao (Shanghai) 2002 Nov;34(6):681-4.

Development and application of technology for large scale cloning of cattle. Abstract: Mammalian cloning technologies originally developed as methods of testing hypotheses about the mechanisms of cell differentiation. Embryo splitting procedures demonstrated that each of the cells in the early embryo are capable of developing into a complete new individual. Nuclear transplantation technologies have shown that loss of genetic sequences or even irreversible repression of gene function are also not mechanisms of cell differentiation. Therefore, both of these methods can be used for producing genetically identical animals. Nuclear transplantation has the advantage of being able to produce unlimited numbers of identical offspring. Highly efficient procedures have been developed for nuclear transplantation in mammals and several important characteristics of donor cells have been described. Unfortunately, the efficiency of producing cloned offspring is still low and many factors affecting the development of nuclear transfer embryos to term remain to be investigated. The tremendous potential of the technology for use in agriculture and medicine, however, will ensure that these problems are addressed and solved. Author(s): Robl, J M Language: eng Source Theriogenology 1999 Jan 15;51(2):499-508.

Development of animal cloning by nuclear transfer; Nagashima, H Language: jpn Subject(s): Animal; Cell Cycle; Cell Nucleus/transplantation; Chromosome Aberrations; Chromosome Disorders; Cloning, Organism Source Tanpakushitsu Kakusan Koso 1997 Nov;42(15):2522-31.

Development of gene cloning system in E. coli: cloning for a lethal gene by new expression system; TakahashiFujii, A; Ohshima, A; Sagawa, H Language: jpn Subject(s): Bacteriophage M13 genetics; Cloning, Molecular methods; DNA (Cytosine-5-)-Methyltransferase; Escherichia coli genetics; Genes, Lethal genetics; Genetic Vectors Source Tanpakushitsu Kakusan Koso 1999 Jul;44(9):1425-9.

Developments in cell biology and genetics: hearing before the Subcommittee on Health and the Environment of the Committee on Interstate and Foreign

Commerce, House of Representatives, Ninety-fifth Congress, second session ... May 31, 1978. Published/Created: Washington: U.S. Govt. Print. Off., 1978. Description: iii, 90 p.; 24 cm. Notes: "Serial no. 95-105." Includes bibliographies. Subjects: Molecular cloning--Cytology--Genetics--Research--United States. LC Classification: KF27 .I5543 19781 Dewey Class No.: 574.8

Developments in expression cloning. Abstract: Recent years have seen a dramatic expansion in the range of applications of expression cloning techniques. New vectors and detection methods promise to further broaden the applicability of function-based screening approaches to problems in gene discovery. A major theme in the past year has been the introduction of engineered reporter cells that heighten the sensitivity with which clones expressing cDNAs can be identified. Author(s): Seed, B Language: Source Curr Opin Biotechnol 1995 Oct;6(5):567-73.

Diazepam binding inhibitor peptide: cloning and gene expression. Abstract: Diazepam binding inhibitor (DBI) is a peptide, initially identified for its ability of displacing the binding of diazepam. The screening of lambda gt 10 cDNA libraries from rat brain with a 47merde-oxyoligonucleotide probe, complementary to a small portion of DBI coding region, allowed the isolation of cDNA clones encoding the entire aminoacid sequence of DBI. This sequence, when compared to that of mouse, human and bovine, revealed that DBI is a well conserved peptide, suggesting a similar function in different species. In order to characterize the function of DBI, studies on the regulation of DBI gene expression were undertaken. The expression of DBI mRNA occurs unevenly in the brain, as well as in peripheral tissues. Moreover, the biosynthesis of DBI is up-regulated in the cerebellum and cerebral cortex of rats made tolerant to diazepam, suggesting that changes in the biosynthesis of DBI might be one of the mechanisms eliciting tolerance to benzodiazepine. In peripheral tissues, the expression of DBI mRNA changes during development. In liver, the content of DBI mRNA was found maximal at postnatal day 1. In contrast, in kidney and heart a linear increase in levels of DBI mRNA was observed from postnatal day 1 to the adult stage, where it reached its maximum level. The tissue specific regulation of DBI mRNA expression, both pharmacologically or developmentally, leads to the hypothesis that DBI might have different functions in different tissues. This would be in line with recent findings that DBI might be also involved in the regulation of an important step of cell metabolism. Author(s): Mocchetti, I; Santi, M R Language: eng Source Neuropharmacology 1991 Dec;30(12B):1365-71.

Differential cloning approaches to the nervous system. Author(s): Milner, R J Language: eng Subject(s): Brain physiology; Cells, Cultured; Cloning, Molecular methods; Gene Library; Nerve Tissue Proteins genetics; nervous system embryology-physiology; Diseases genetics; Source Ann N Y Acad Sci 1990;579:273-80.

DiMagno, Lisa Margaret, 1966- Construction and characterization of cyanobacterial mutants with deletions in photosystem II components [microform] / Lisa Margaret DiMagno. Published/Created: 1992. Description: ix, 146 leaves: ill.; 28 cm. Notes: Thesis (Ph. D.)--University of Chicago, 1992. Microfilm. Chicago, Ill.: University of Chicago, Joseph Regenstein Library, Dept. of Photoduplication, 1993. 1 microfilm reel; 35 mm. Subjects: Cyanobacteria. Molecular cloning. LC Classification: Microfilm 94/2449 (Q)

Diphtheria toxin cloning and expression in Corynebacterium diphtheriae. Author(s): Greenfield, L Language: eng Subject(s): Amino Acid Sequence; Chromosome Mapping; Cloning, Molecular; Corynebacterium diphtheriae genetics; DNA, Bacterial; Diphtheria Toxin genetics; Genes, Bacterial. Source Targeted Diagn Ther 1992;7:273-305.

Direct cloning of lambda gt11 cDNA inserts into a plasmid vector. Author(s): Poulin, M L; Chiu, I M Language: eng Source Methods Mol Biol 1994;31:9-17.

Direct isolation of specific chromosomal regions and entire genes by TAR cloning. Author(s):Larionov, V Language: eng Subject(s): Chromosome Mapping; Chromosomes, Artificial, Yeast; Cloning, Molecular/methods; Genetic Vectors; Genome, Human; Human; Recombination, Genetic Source Genet Eng (N Y) 1999;21:37-55.

Dissection of signaling pathways and cloning of new signal transducers in tyrosine kinase-induced pathways by genetic selection. Abstract: We used genetic strategies which have been proven valuable to decipher signaling pathways in comparatively simple organisms such as Drosophila and Caenorhabditis elegans, to dissect signaling network activated by tyrosine kinases in mammals. The strategy was developed further towards a generally applicable expression cloning system to identify signal transducers in tyrosine kinase pathways. This system is based on the ability of downstream acting genes to rescue the transformation phenotype of partial loss-of-function mutants of BCR-ABL which still retain tyrosine kinase activity. Using this strategy we have previously shown that overexpression of c-Myc and Cyclin D1 can rescue a signaling defective SH2 mutant of BCR-ABL for transformation. In an unbiased approach to identify new compensating genes, a cDNA library was introduced by retroviral infection into fibroblasts which express the BCR-ABL SH2 mutant. CDNA clones, capable of rescuing the SH2 mutant for transformation should result in colony formation in soft agar. A PCR approach was used to recover these compensating genes from the genomic DNA of the transformed fibroblasts. Sequencing analysis of the initial cDNAs identified three known genes, the adapter molecule Shc, the kinases SPRK and p38 MAPK. These genes have been found to interact functionally with BCR-ABL for fibroblast and hematopoietic cell transformation. Currently, we are constructing and screening new libraries to identify novel genes which complement the BCR-ABL SH2 mutant. Our results demonstrate that this cloning approach is an effective means of identifying and characterizing signaling molecules that function in specific signaling pathways. This in turn may identify specific targets for mechanism-based therapeutic intervention to block altered signaling. Author(s): Mahlmann, S; McLaughlin, J; Afar, D E; Mohr, R; Kay, R J; Witte, O N Language: eng Source Leukemia 1998 Dec;12(12):1858-65.

DNA cloning vectors utilizing replication functions of the filamentous phages of Escherichia coli. Author(s): Geider, K Language: eng Subject(s): Cloning, Molecular; DNA Replication; DNA, Bacterial genetics; DNA, Recombinant rug effects; DNA, Single-Stranded/genetics; DNA, Viral genetics; Escherichia coli genetics; Genes, Viral; Genetic Vectors; Nucleic Acid Hybridization; Virus Replication Source J Gen Virol 1986 Nov;67 (Pt 11):2287-303.

DNA cloning: a practical approach / edited by D.M. Glover and B.D. Hames. Edition Information: 2nd ed. Published/Created: Oxford; New York: IRL Press, c1994- Related Names: Glover, David M. Hames, B. D. Description: p. cm. ISBN: 0199634777 (v. 1): 0199634769 (v. 1: pbk.): 0199634793 (v. 2): 0199634785 (v. 2: pbk.) : Contents: v. 1. Core techniques -- v. 2. Expression systems. Notes: Includes bibliographical references and indexes. Subjects: Molecular cloning--Recombinant DNA--Laboratory manuals. Series: Practical approach series; v. <148, 149 > LC Classification: QH442.2 .D59 1994 Dewey Class No.: 574.87/3282 20

DNA cloning: a practical approach / edited by D.M. Glover. Published/Created: Oxford; Washington, DC: IRL Press, c1985-c1987. Related Names: Glover, David M. Description: 3 v.: ill.; 23 cm. ISBN: 0947946187 (pbk.: v. 1) 0947946195 (pbk.: v. 2) Notes: Includes bibliographies and indexes. Subjects: Molecular cloning. Recombinant DNA. Series: Practical approach series LC Classification: QH442.2 .D59 1985 Dewey Class No.: 574.87/3282 19

DNA sequencing strategies: automated and advanced approaches / edited by Wilhelm Ansorge, Hartmut Voss, Jürgen Zimmermann. Published/Created: New York: Wiley; Heidelberg: Spektrum, c1997. Related Names: Ansorge, Wilhelm, 1944- Voss, Hartmut. Zimmermann, Jürgen. Description: xiv, 202 p.: ill.; 25 cm. ISBN: 0471136832 (alk. paper) Notes: Includes bibliographical references and index. Subjects: Nucleotide sequence--Automation--Molecular cloning--Automation--Laboratory manuals. LC Classification: QP625.N89 D53 1997 Dewey Class No.: 574.87/3282 20

Dolly, Polly and other 'ollys': likely impact of cloning technology on biomedical uses of livestock. Abstract: The idea of generating transgenic livestock which secrete into their milk large quantities of proteins for therapeutic use, was pioneered in the late 1980s with the disclosure of the production of a number of transgenic sheep. One particular animal, a sheep called Tracy, produced milk where over 50% of the protein consisted of human alpha 1 anti-trypsin. Sheep-derived protein has now entered clinical trials for cystic fibrosis (UK, USA) and congenital emphysema (UK). There are many other examples where this technology is making inroads into more traditional ways of making biopharmaceuticals. However, although robust, this technology has several limitations, including an inability to allow targeted insertion/modification of the animal genome, long timelines to production flocks/herds, and the rather unpredictable expression levels seen when different transgenic founders are compared. We believe that there is now a technical solution to all of these problems. Dolly is a high profile example of a new technology comprising the generation of identical animals from cultured somatic cells. This work has many implications. In the commercial context, the real benefits of this advance will be seen when genetically engineered somatic cells are shown to be suitable nuclear donors, and particularly when the manipulations are targeted to pre-determined sites in the host cell genome. The first objective has now been achieved with the birth of Polly, a cloned sheep which contains the human gene encoding Factor IX, a protein involved in preventing haemophilia. Author(s): Colman, A Language: eng Source Genet Anal 1999 Nov;15(3-5):167-73.

Donor cells for nuclear cloning: many are called, but few are chosen. Abstract: The few viable clones obtained at the end of a typical cloning experiment are genetic copies of the donor cell genome of a non-reproductive (somatic) or embryonic cell used for nuclear transfer. Nuclear totipotency has to be reestablished by erasing epigenetic constraints imposed on the donor genome during differentiation in a process which involves active chromatin remodeling. Various donor cell types and cell cycle combinations have proven to be capable of generating cloned offspring. However, an ideal nuclear donor may have not yet been found. This review summarizes current theoretical aspects of donor cell selection. It focuses on the impact of genetic and epigenetic differences between donor cell types on successful mammalian cloning. Author(s): Oback, Bjorn; Wells, David Language: eng Source Cloning Stem Cells 2002;4(2):147-68.

Drlica, Karl. Understanding DNA and gene cloning: a guide for the curious / Karl Drlica. Edition Information: 4th ed. Published/Created: Hoboken, NJ: Wiley, c2004. Description: xiv, 369 p.: ill.; 23 cm. ISBN: 0471434167 (pbk.: alk. paper) Notes: Includes bibliographical references (p. 325-335) and index. Subjects: Molecular cloning. Recombinant DNA. Genetic engineering. Genetics. LC Classification: QH442.2 .D75 2004 Dewey Class No.: 576.5 21

Drlica, Karl. Understanding DNA and gene cloning: a guide for the curious / Karl Drlica. Edition Information: 3rd ed. Published/Created: New York: Wiley, c1997. Description: xvi, 329 p.: ill.; 23 cm. ISBN: 047113774X (pbk.: alk. paper) Notes: Second ed. published in 1992. Includes bibliographical references (p. 287-300) and index. Subjects: Molecular cloning. Recombinant DNA. Genetic engineering. LC Classification: QH442.2 .D75 1997 Dewey Class No.: 574.87/3282 20

Drlica, Karl. Understanding DNA and gene cloning: a guide for the curious / Karl Drlica. Edition Information: 2nd ed. Published/Created: New York: Wiley, c1992. Description: xxii, 240 p.: ill.; 23 cm. ISBN: 0471622257 Notes: Includes bibliographical references (p. 204-211) and index. Subjects: Molecular cloning. Recombinant DNA. Genetic engineering. LC Classification: QH442.2 .D75 1992 Dewey Class No.: 574.87/328 20

Drlica, Karl. Understanding DNA and gene cloning: a guide for the curious / Karl Drlica; [with an introduction by Arthur Kornberg]. Published/Created: New York: Wiley, c1984. Description: xiii, 205 p.: ill.; 23 cm. ISBN: 0471879428 (pbk.) Notes: Includes index. Bibliography: p. 165-168. Subjects: Genetic engineering. DNA, Recombinant. Cloning, Molecular. Genetic Intervention. Molecular biology. LC Classification: QH442.2 .D75 1984 Dewey Class No.: 574.87/328 19

Dry cloning; Lohi, O Language: fin; Cloning, Molecular; Genetics, Medical/trends; Human; Human Genome Project; Molecular Biology/methods; Research/organization & administration Source Duodecim 1999;115(7):779-82.

DuPrau, Jeanne. Cloning / by Jeanne DuPrau. Published/Created: San Diego, CA: Lucent Books, c2000. Description: 112 p.: ill.; 24 cm. ISBN: 1560065834 (lib. alk. paper) Summary: Discusses the methods, regulation, and ethics of cloning in relation to agriculture, medicine, endangered species, and human beings. Notes: Includes bibliographical references (p. 98-105) and index. Subjects: Cloning--Juvenile literature. Series: Lucent Overview series LC Classification: QH442.2 .D87 2000 Dewey Class No.: 660.6/5 21

Ebon, Martin. The cloning of man: a brave new hope--or horror? / Edited by Martin Ebon. Published/Created: New York: New American Library, 1978. Description: 201 p.; 18 cm. Notes: Bibliography: p. 199-201. Subjects: Cloning--Social aspects. Genetic engineering--Social aspects. LC Classification: QH442.2 .E26 Dewey Class No.: 612.6

Effect of transplanted neurotransmitter receptors and ion channels in molecular cloning; Ni, Y Language: chi; Animal; Cloning, Molecular; DNA Probes; Genetic Engineering; Ion Channels/physiology; Oocytes; RNA, Messenger; Receptors, GABA-A/genetics; Receptors, Neurotransmitter/genetics; Translation, Genetic Source Sheng Li Ke Xue Jin Zhan 1990 Jan;21(1):61-4.

Efimov, Igor′ Markovich. The Judgment Day archives: a novel / Andrei Moscovit; translated by Robert Bowie. Published/Created: San Francisco: Mercury House, c1988. Description: vii, 402 p.; 23 cm. ISBN: 0916515451 : Notes: Translation of: Arkhivy Strashnogo suda. Subjects: Herbs--Therapeutic use--Soviet Union--Fiction. Cloning--Fiction. LC Classification: PG3479.7.F49 A8913 1988 Dewey Class No.: 891.73/44 19

Eiseman, Elisa. The National Bioethics Advisory Commission: contributing to public policy / Elisa Eiseman. Published/Created: Santa Monica, CA: RAND, 2003. Description: p.; cm. ISBN: 0833033646 (pbk.: alk. paper) Notes: "MR-1546." Includes bibliographical references. Subjects: National Bioethics Advisory Commission. Bioethics--Medical ethics--Government policy--United States. LC Classification: QH332 .E36 2003 Dewey Class No.: 174/.957/0973 21

Electrogenic Na+/HCO3- cotransporters: cloning and physiology. Abstract: Bicarbonate and CO2 comprise the major pH buffer of biological fluids. In the renal proximal tubule most of the filtered HCO3- is reabsorbed by an electrogenic Na/HCO3 cotransporter located at the basolateral membrane. This Na+ bicarbonate cotransporter (NBC) was recently cloned. This review highlights the recent developments leading to and since the cloning of NBC: NBC expression cloning, protein features, clone physiology, isoforms and genes, mRNA distribution, and protein distribution. With the NBC amino acid sequence 30-35% identical to the anion exchangers (AE1-3), a superfamily of HCO3- transporters is emerging. Physiologically, NBC is electrogenic, Na+ dependent, HCO3-dependent, Cl- independent, and inhibited by stilbenes (DIDS and SITS). NBC clones and proteins have been isolated from several tissues (other than kidney) thought to have physiologically distinct HCO3-transporters. For example, NBC occurs in pancreas, prostate, brain, heart, small and large intestine, stomach, and epididymis. Finally, there are at least two genes that encode NBC proteins. Possible future directions of research are discussed. Author(s): Romero, M F; Boron, W F Language: eng Source Annu Rev Physiol 1999;61:699-723.

Embryo cloning in domestic animals Abstract: The cloning of livestock is performed by the nuclear transfer of early embryonic stages into prepared oocytes in order to obtain a high number of genetic identical animals. As the most important technical steps there are maturation and enucleation of the oocytes, isolation of single blastomeres or karyoplasts of the donor embryo, transfer of the nucleus-containing membrane vesicle under the zona pellucida of the recipient cell, and fusion of the recipient cell and blastomere or karyoplast. Until now, the largest clone which is known exists of seven bulls. The efficiencies of the particular methodical steps have to be improved. More knowledge of the activation of oocytes, nucleus differentiation and availability of determined cell cycle-stages of mitosis is required. The combination of embryo cloning, cryopreservation of embryos and non-surgical embryo transfer is required for basic research and animal breeding. Author(s): Clement-Sengewald, A; Brem, G Language: ger Source Berl Munch Tierarztl Wochenschr 1992 Jan 1;105(1):15-21.

Embryo cloning. An accelerated development using animal models brings up the debate on human ethics; Andang, M; Ahrlund-Richter, L Language: swe; Animal; Blastocyst; Cloning, Molecular; Embryo Transfer; Ethics, Medical; Genetic Engineering; Human; Models, Biological Source Lakartidningen 1995 Mar 15;92(11):1078, 1083-6.

Embryonic and somatic cell cloning. Abstract: Revolutionary opportunities in biology, medicine and agriculture arise from the observation that offspring are obtained after nuclear transfer if somatic donor cells are induced to become quiescent. Exploitation of many of these opportunities will depend upon optimizing procedures for nuclear transfer. This may come about through an understanding of the means by which factors in the oocyte cytoplasm act upon the DNA of the transferred nucleus to regulate gene expression. Similarly, research will extend the procedure to other species. This technology may be used for embryo production, the introduction of genetic change and the derivation of cells needed to treat human diseases. Groups of genetically identical animals will be used in research to control genetic variation and to allow transfer of cells between individuals. In agriculture, production of a small number of clones will separate genetic and environmental effects, whereas production of larger numbers of offspring will disseminate genetic improvement from nucleus herds. Precise genetic modification will be achieved by site specific recombination in the donor cells before nuclear transfer. In all mammals it will become possible to define the role of any gene product and to analyse the mechanisms that regulate gene expression. Medical uses of these techniques will include the production of proteins needed to treat disease and the supply of organs such as hearts, livers and kidneys from pigs. As genome mapping projects identify loci associated with traits of commercial importance in agriculture then gene targeting will be used to study this effect. Finally, cells capable of differentiation into any of the tissues of a patient will provide treatment for diseases reflecting damage to a specific cell population that neither repairs nor replaces itself. Author(s): Wilmut, I; Young, L; Campbell, K H Language: eng Source Reprod Fertil Dev 1998;10(7-8):639-43.

Embryonic stem cells and therapeutic cloning Abstract: Increased interest in the therapeutic use of human stem cells has emerged following significant progress in ongoing research. The cloning of a sheep, the isolation of human embryonic stem cells, and the discovery that adult stem cells may be reprogrammed taken together give substance to hopes that novel principles of treatment may be developed for a variety of serious conditions. Embryonic stem cells are derived from pre-embryos at the blastocyst stage and may give rise to all bodily tissues and cells. Animal models have demonstrated that embryonic stem cells when transplanted into adult hosts may differentiate and develop into cells and tissues applicable for treatment of a variety of conditions, including Parkinson's disease, multiple sclerosis, spinal injuries, cardiac stroke and cancer. Transplanted embryonic stem cells are exposed to immune reactions similar to those acting on organ transplants, hence immunosuppression of the recipient is generally required. It is, however, possible to obtain embryonic stem cells that are genetically identical to the patient's own cells by means of therapeutic cloning techniques. The nucleus from a somatic cell is transferred into an egg after removal of the egg's own genetic material. Under specific condition the egg will use genetic information from the somatic cell in organising the formation of a blastocyst which in turn generates embryonic stem cells. These cells have a genetic composition identical to that of the patient and are suitable for stem cell therapy. Author(s): Sunde, A; Eftedal, I Language: nor Source Tidsskr Nor Laegeforen 2001 Aug 30;121(20):2407-12.

Engineering the human germline: an exploration of the science and ethics of altering the genes we pass to our children / edited by Gregory Stock and John Campbell. Published/Created: New York: Oxford University Press, 2000. Related Names: Stock, Gregory. Campbell, John H. (John Howland), 1938- Description: xvi, 169 p.; 24 cm. ISBN: 0195133021 (cloth: alk. paper) Notes: Includes bibliographical references and index. Subjects: Medical genetics--enetic engineering--Human cloning--Moral and ethical aspects. LC Classification: RB155 .E56 2000 Dewey Class No.: 174/.25 21

Epithelial sodium channel regulatory proteins identified by functional expression cloning. Abstract: We describe here our current strategy for identifying and cloning proteins involved in the regulation of the epithelial sodium channel (ENaC). We have set up a complementation functional assay in the Xenopus laevis oocyte expression system. Using this assay, we have been able to identify a channel-activating protease (CAP-1) that can increase ENaC activity threefold. We propose a novel extracellular signal transduction pathway controlling ionic channels of the ENaC gene family that include genes involved in mechanotransduction (degenerins), in peptide-gated channels involved in neurotransmission (FaNaCh), in proton-gated channels involved in pH sensing (ASIC) or pain sensation (DRASIC). Author(s): Vallet, V; Horisberger, J D; Rossier, B C Language: eng Source Kidney Int Suppl 1998 Sep;67:S109-14.

Essential molecular biology: a practical approach / edited by T.A. Brown. Edition Information: 2nd ed. Published/Created: Oxford; New York: Oxford University

Press, 2000-2001. Related Names: Brown, T. A. (Terence A.) Description: 2 v.: ill.; 24 cm. ISBN: 0199636435 (v. 1: hbk.) 0199636427 (v. 1: pbk.) 0199636451 (v. 2: hbk.) 0199636443 (v. 2: pbk.) Notes: Includes bibliographical references and indexes. Subjects: Molecular cloning. Recombinant DNA. Series: The Practical approach series LC Classification: QH442.2 .E77 2000 Dewey Class No.: 572.8/6 21

Essential molecular biology: a practical approach / edited by T.A. Brown. Published/Created: Oxford; New York: IRL Press at Oxford University Press, c1991. Related Names: Brown, T. A. (Terence A.) Description: 2 v.: ill.; 24 cm. ISBN: 019963114X (spiral hard: set) 0199631158 (pbk.: set) 0199631115 (pbk.: v. 1): 0199631107 (hard: v. 1): 0199631123 (spiral hard: v. 2) 0199631131 (pbk.: v. 2) Notes: Includes bibliographical references and indexes. Subjects: Molecular cloning. Recombinant DNA. Series: The Practical approach series LC Classification: QH442.2 .E77 1991 Dewey Class No.: 574.87/3282 20

Establishment, maintenance, and cloning of human dermal fibroblasts. Author(s): Jones, G E; Wise, C J Language: eng Subject(s): Cell Culture/methods; Clone Cells; Fibroblasts/cytology; Human; Skin/cytology Source Methods Mol Biol 1997;75:13-21.

ET-cloning: think recombination first. Author(s): Muyrers, J P; Zhang, Y; Stewart, A F Language: eng Subject(s): Animal; Bacterial Proteins metabolism; Chromosomes, Artificial, Bacterial genetics; Chromosomes, Bacterial genetics; Cloning, Molecular methods; DNA, Recombinant genetics; Escherichia coli genetics; Exodeoxyribonucleases metabolism; Genetic Engineering; Human; Mice; Recombination, Genetic Source Genet Eng (N Y) 2000;22:77-98.

Ethical and theological objections to animal cloning. Author(s): Linzey, Andrew Language: eng Subject(s): Animal Experimentation; Animal Rights; Animal Welfare; Cloning, Organism; Ethical Theory; Ethics; Religion; Stress, Psychological; Theology Source Bull Med Ethics 1997 Sep;No. 131:18-22.

Ethical aspects of regenerative medicine, with special reference to embryonic stem cells and therapeutic cloning Abstract: Regenerative medicine is expected to be new therapeutic means for treating incurable diseases but requires serious bioethical consideration. Embryonic stem (ES) cells, that are pleuripotent cells suitable to regenerative medicine, can be used in Japan for investigative use under a strict control by guide-lines. On the other hand, use of embryo produced by nuclear transfer has not been allowed in Japan and further serious consideration is required. Some other ethical aspects of regenerative medicine are also discussed. Author(s): Imura, Hiroo Source Nippon Rinsho 2003 Mar;61(3):379-84.

Ethical issues in human cloning: cross-disciplinary perspectives / edited by Michael C. Brannigan. Published/Created: New York, NY: Seven Bridges Press, c2001. Related Names: Brannigan, Michael C., 1948- Description: xii, 244 p.; 23 cm. ISBN: 1889119113 (pbk.) Notes: Includes bibliographical references. Subjects: Human cloning--Moral and ethical aspects. LC Classification: QH442.2 .E845 2001 Dewey Class No.: 174/.25 21

Ethical issues regarding human cloning: a nursing perspective. Abstract: Advances in cloning technology and successful cloning experiments in animals raised concerns about the possibility of human cloning in recent years. Despite many objections, this is not only a possibility but also a reality. Human cloning is a scientific revolution. However, it also introduces the potential for physical and psychosocial harm to human beings. From this point of view, it raises profound ethical, social and health related concerns. Human cloning would have an impact on the practice of nursing because it could result in the creation of new physiological and psychosocial conditions that would require nursing care. The nursing profession must therefore evaluate the ethics of human cloning, in particular the potential role of nurses. This article reviews the ethical considerations of reproductive human cloning, discusses the main reasons for concern, and reflects a nursing perspective regarding this issue. Author(s): Dinc, Leyla Source Nurs Ethics 2003 May;10(3):238-54.

Ethics and law in biological research / edited by Cosimo Marco Mazzoni. Published/Created: Boston: Kluwer Academic Publishers, 2002. Related Names: Mazzoni, Cosimo Marco. Description: x, 259 p.; 24 cm. ISBN: 9041117423 (alk. paper) Notes: Includes bibliographical references and index. Subjects: Cloning--Moral and ethical aspects. Bioethics. Medical ethics. Series: Nijhoff law specials; 52. LC Classification: QH438.7 .E835 2002 Dewey Class No.: 174/.957 21

Ethics and theology: a continuation of the national discussion on human cloning: hearing before the Subcommittee on Public Health and Safety of the Committee on Labor and Human Resources, United States Senate, One Hundred Fifth Congress, first session, on examining the ethical and theological implications of human cloning, June 17, 1997. Published/Created: Washington: U.S. G.P.O.: For sale by the U.S. G.P.O., Supt. of Docs., Congressional Sales Office, 1997. Description: iii, 69 p.; 24 cm. ISBN: 0160554756 Notes: Distributed to some depository libraries in microfiche. Shipping list no.: 98-0007-P. Includes bibliographical references. Subjects: Human cloning--Moral and ethical aspects--United States. Human cloning--Religious aspects. Series: United States. Congress. Senate. S. hrg.; 105-123. LC Classification: KF26 .L278 1997c Dewey Class No.: 174/.957 21 Govt. Doc. No.: Y 4.L 11/4:S.HRG.105-123 GPO Item No.: 1043-A 1043-B (MF)

Eugenics and human cloning Abstract: Because of legislative bans there are still no reports of human cloning. However eager public debate is currently running, concerning medical, legal, social and ethical aspects of human cloning. Arguments for and against human cloning are presented. An important argument against cloning is the danger of eugenic tendencies connected with cloning, which could lead to genetic discrimination. Author(s): Boloz, W Language: pol Source Med Wieku Rozwoj 2001;5(1 Suppl 1):53-62.

Experimental manipulation of gene expression / edited by Masayori Inouye. Published/Created: New York: Academic Press, 1983. Related Names: Inouye, Masayori. Description: xiii, 315 p.: ill.; 24 cm. ISBN: 0123723809 Notes: Includes index. Includes bibliographies. Subjects: Genetic engineering. Gene expression. Cloning. Cloning, Molecular. DNA, Recombinant. Genetic intervention. LC Classification: QH442 .E96 1983 Dewey Class No.: 575.1 19

Expression cloning of fungal enzyme genes; a novel approach for efficient isolation of enzyme genes of industrial relevance. Abstract: Expression cloning is a relatively new method for fast and efficient cloning of enzyme genes from fungi that are known to make complex enzyme mixtures. In contrast to traditional cloning methods that are usually dependent on knowledge of at least a partial amino acid sequence in order to synthesize appropriate DNA probes or primers, the expression cloning method solely relies on access to reliable and sensitive enzyme assays. A representative expression cDNA library is made in Saccharomyces cerevisiae form the donor strain and relevant cDNA clones are detected directly based on the encoded enzyme activity. Thus, time-consuming enzyme purification and characterization steps are avoided. The method has been applied on the characterization of extracellular enzyme genes from the filamentous fungus Aspergillus aculeatus and has resulted in the isolation of 20 different enzyme genes such as endoglucanases, xylanases, pectinases, proteases, hemicellulases and rhamnogalacturonan-degrading enzymes. All enzymes have been expressed in Aspergillus oryzae, purified and characterized. In the present review a description of the expression cloning technique will be given as well as examples of how the technique has been used in the exploration and characterization of a commercial enzyme product that is known to consist of a complex mixture of more than 25 different enzyme activities. Author(s): Dalboge, H Language: eng Source FEMS Microbiol Rev 1997 Aug;21(1):29-42.

Expression cloning of glycosyltransferases. Abstract: Cloning of cDNAs encoding glycosyltransferases enters into a new era because of the advent of expression cloning for glycosyltransferases. This mini-review summarizes a short historical view on the development of this method and surveys various improvements over the original method. This review also emphasizes the advantages of those recently developed improvements with a brief description on their background. Finally, the review provides some prospects based on these developments in the methodology. It is expected that increasing numbers of newly cloned glycosyltransferases will become critical tools for understanding the roles of carbohydrates. Author(s): Fukuda, M; Bierhuizen, M F; Nakayama, J Language: eng Source Glycobiology 1996 Oct;6(7):683-9.

Expression cloning of TGF-beta receptors. Abstract: Using a powerful expression cloning method in COS cells, we have cloned the TGF-beta types II and III receptors. The type III TGF-beta receptor is a membrane-bound proteoglycan with a core protein of about 110 kDa. Stable expression of the type III receptor in L6 myoblasts leads to an apparent increase in the ability of the type II receptor to bind iodinated TGF-beta 1. The cloned type II receptor has a predicted protein core of about 60 kDa with a cysteine-rich extracellular domain, a single transmembrane domain, and a functional serine/threonine kinase domain that is homologous to the activin receptor and to the C. elegans protein daf-1. These results implicate serine/threonine phosphorylation as an important mechanism of TGF-beta action. Author(s): Lin, H Y; Wang, X F Language: eng Source Mol Reprod Dev 1992 Jun;32(2):105-10.

Expression cloning of the epithelial sodium channel. Author(s): Canessa, C M; Horisberger, J D; Schild, L; Rossier, B C Language: eng Subject(s): Animal; Cloning, Molecular methods; Electrochemistry; Epithelium metabolism; Female; Gene Expression; In Vitro; Kinetics; Molecular Structure; Oocytes metabolism; Protein Conformation; Rats; Tissue Distribution; Xenopus Source Kidney Int 1995 Oct;48(4):950-5.

Expression cloning of the human V2 vasopressin receptor. Author(s): Barberis, C; Seibold, A; Ishido, M; Rosenthal, W; Birnbaumer, M Language: eng Subject(s): Amino Acid Sequence; Animal; Cloning, Molecular; Gene Expression/genetics; Human, Molecular Sequence Data; Rats; Receptors, Vasopressin/genetics; Sequence Homology, Amino Acid; Transformation, Genetic Source Regul Pept 1993 Apr 29;45(1-2):61-6.

Expression cloning of tumor suppressor genes: a guide for optimists. Author(s): Noda, M Language: eng Subject(s): Animal; Cells, Cultured; Cloning, Molecular/methods; Gene Expression; Genes, Tumor Suppressor; Human; In Vitro; Neoplasms/genetics Source Mol Carcinog 1990;3(5):251-3.

Expression cloning systems. Abstract: This review will cover the use of expression cloning in Xenopus oocytes, fission yeast, and mammalian cells. Of the systems covered herein, transient expression cloning systems in Xenopus oocytes and mammalian cells have proven to be the most effective and versatile, as demonstrated by the large number of cDNA clones isolated by these two methods in the past year. Of particular interest, are recent advances in the screening methodologies used in conjunction with transient expression in mammalian cells which have permitted the application of this system in the isolation of cDNAs encoding intracellular proteins. Author(s): Aruffo, A Language: eng Source Curr Opin Biotechnol 1991 Oct;2(5):735-41.

Expression cloning with pan kinesin antibodies. Author(s): Ginkel, L M; Wordeman, L Language: eng Source Methods Mol Biol 2001;164:21-41.

Expression of cloned genes in prokaryotic and eukaryotic cells / editors Takis S. Papas, Martin Rosenberg, Jack G. Chirikjian. Published/Created: New York: Elsevier, c1983. Related Names: Papas, Takis S. Rosenberg, Martin J. Chirikjian, Jack G. Description: xiv, 286 p.: ill.; 25 cm. ISBN: 0444008195 : Notes: Includes bibliographical references and index. Subjects: Gene expression. Molecular cloning. Genetic vectors. Gene expression regulation. Cloning, Molecular. Series: Gene amplification and analysis, 0275-2778; v. 3 LC Classification: QH450 .E955 1983 Dewey Class No.: 575.1 19

Facklam, Margery. From cell to clone: the story of genetic engineering / by Margery and Howard Facklam; illustrated with diagrs. by Paul Facklam and with photos. Edition Information: 1st ed. Published/Created: New York: Harcourt Brace Jovanovich, c1979. Related Names: Facklam, Howard, joint author. Facklam, Paul. Description: 128 p.: ill.; 24 cm. ISBN: 015230262X Summary: Discusses genetic engineering, particularly the history and techniques of cloning, and includes material on recombinant DNA research and test-tube babies. Notes: Includes index. Bibliography: p. 123-124. Subjects: Genetic engineering--Cloning--History. LC Classification: QH442 .F32 Dewey Class No.: 575.11

Familial Mediterranean fever: from the clinical syndrome to the cloning of the pyrin gene. Abstract: Familial Mediterranean fever (FMF) is a genetic disorder, restricted to peoples originating in the Middle East. The clinical syndrome is characterized by shortlived febrile episodes, accompanied by inflammation in one of the serous membranes, resulting in peritonitis pleuritis or synovitis. In many untreated FMF patients, systemic amyloidosis developed. Clinical presentation of amyloidosis in FMF is nephropathic, progressing from proteinuria, nephrosis to renal failure & end stage renal disease. Continuous colchicine treatment, introduced in 1972, prevents most febrile-inflammatory attacks of FMF, & inhibits the development of amyloidosis in this otherwise fatal disease. Recently, the gene that causes FMF was cloned. It is called the pyrin gene & encodes the pyrin protein. Five missense mutations were found so far in the gene. These give rise to 5 amino acid substitutions, all of them in the carboxyterminal part of the pyrin protein. The pyrin protein is expressed solely in neutrophiles white blood cells which are found in large numbers in the inflammatory sites during FMF attacks. It seems that the role of the wild type of the pyrin protein is to inhibit inflammation that can be provoked by a minor insult. The mutated pyrin protein in FMF is probably unable to inhibit these unnecessary inflammatory events. Author(s): Pras, M Language: eng Source Scand J Rheumatol 1998;27(2):92-7.

Farmer, Nancy, 1941- The house of the scorpion / Nancy Farmer. Edition Information: 1st ed. Published/Created: New York: Atheneum Books for Young Readers, c2002. Description: 380 p.; 24 cm. ISBN: 0689852223 Summary: In a future where humans despise clones, Matt enjoys special status as the young clone of El Patrón, the 142-year-old leader of a corrupt drug empire nestled between Mexico and the United States. Subjects: Cloning--Fiction. Science fiction. LC Classification: PZ7.F23814 Ho 2002 Dewey Class No.: [Fic] 21

Fibrosin, a novel fibrogenic protein: discovery, cloning and implications for fibrotic disorders. Abstract: Tissue fibrosis accounts for considerable chronic morbidity in certain connective tissue diseases. Knowledge of the pathobiology of this complication, particularly the identification of fibrogenic proteins, is now sufficiently advanced to consider novel therapeutic strategies to interfere with pathological fibrogenesis. Studies in a murine model of the helminthic infection, schistosomiasis mansoni, led to the discovery of a novel fibrogenic protein, fibrosin. Molecular analysis suggests that the domain of cloned fibrosin cDNA that encodes the fibrogenically active moiety of the protein is highly conserved in mice and humans, but lacks significant sequence homology with other known cDNAs. Native fibrosin, as well as a synthetic fibrosin peptide, stimulate a variety of relevant responses in fibroblasts. Several cell types can produce fibrosin, including CD4+ lymphocytes and fibroblasts themselves. Observations in experimental schistosomiasis and in scleroderma reveal two possible mechanisms by which fibrosin might be involved in fibrosis. These observations could have implications for diagnosis and novel therapeutic strategies for fibrosis that complicates connective tissue diseases. Author(s): Wyler, D J Language: eng Source Int Arch Allergy Immunol 1996 Dec;111(4):326-9.

Filamentous fusion phage cloning vectors for the study of epitopes and design of vaccines. Abstract: Foreign sequences can be inserted within the minor coat protein, pIII, of filamentous phage, creating a fusion protein that is incorporated into the virion; the virions, which we call "fusion phage," display the foreign amino acids encoded in the insert on their surface. Phage bearing specific antigenic determinants from a target gene can be purified in infectious form from a vast excess of phage bearing other determinants by affinity to antibody directed against the gene product. Fusion phage can be used as a source of antigen as a carrier-hapten conjugate for obtaining immunological reagents in rabbits, and for B epitope mapping. By generating a fusion phage library expressing virtually all possible short, amino acid sequences, it may be possible to study epitopes on immunologically important proteins without the use of synthetic peptides and without ever having cloned the genes. Author(s): Parmley, S F; Smith, G P Language: eng Source Adv Exp Med Biol 1989;251:215-8.

Filamentous phages as cloning vectors. Abstract: The virtue of Ff vectors goes beyond the fact that they deliver a single strand in a convenient package for sequencing and oligonucleotide-directed mutagenesis. Of all vectors in common use they are the easiest to propagate and process. Their genomes can be easily manipulated, and the knowledge acquired over a quarter century of basic research makes their behavior reasonably predictable. For this reason I have emphasized the general properties of Ff phage in this review and dealt at some length with applications that are still not fully developed, I hope this review will inspire readers to continue the tradition of imaginative exploitation of this unique class of viruses. Author(s): Smith, G P Language: eng Source Biotechnology 1988;10:61-83.

Firman, Keith. DNA cloning/sequencing workshop: a short course / Keith Firman. Published/Created: New York: E. Horwood, 1991. Description: 101 p.: ill.; 25 cm. ISBN: 0132180332 (student pbk. ed.) 0132180413 (lib. ed.) Series: Ellis Horwood series in biochemistry and biotechnology LC Classification: QH442.2 .F57 1991 Dewey Class No.: 660/.65 20

FISH as a tool for genome mapping and positional cloning. Author(s): Inazawa, J Language: eng Subject(s): Chromosome Mapping; Cloning, Molecular; Genetic Diseases, Inborn; In Situ Hybridization, Fluorescence; Neoplasms. Source Southeast Asian J Trop Med Public Health 1995;26 Suppl 1:86-8.

Flesh of my flesh: the ethics of cloning humans: a reader / edited by Gregory E. Pence. Published/Created: Lanham, Md.: Rowman & Littlefield, c1998. Related Names: Pence, Gregory E. Description: xv, 154 p.; 24 cm. ISBN: 0847689816 (cloth: alk. paper) 0847689824 (pbk.: alk. paper) Notes: Includes bibliographical references and index. Subjects: Cloning--Moral and ethical aspects. Medical genetics--Moral and ethical aspects. Human reproductive technology--Moral and ethical aspects. LC Classification: QH442.2 .F585 1998 Dewey Class No.: 174/.957 21

Forward genetic approach strikes gold: cloning of a mammalian clock gene. Author(s): Reppert, S M; Weaver, D R Language: eng Subject(s): Animal; Circadian Rhythm/genetics/physiology; Cloning, Molecular; Evolution; Human; Mammals; Mice; Models, Biological; Suprachiasmatic Nucleus/physiology; Trans-Activators/genetics; Vertebrates Source Cell 1997 May 16;89(4):487-90.

Fritz, Sandy. Genomics and cloning / by Sandy Fritz. Published/Created: Mankato, Minn.: Smart Apple Media, 2003. Description: p. cm. ISBN: 1583403655 (alk. paper) Summary: A discussion of the field of genetics, focusing on the study of the human genome, genetic engineering, and cloning. Subjects: Genomics--Genetic engineering--Cloning--Juvenile literature. Series: Hot science LC Classification: QH437.5 .F75 2003 Dewey Class No.: 572.8/6 21

From cloning to a commercial realization: human alpha interferon. Abstract: Recent applications of recombinant DNA techniques have enabled the cloning of several interesting human genes, leading to the production of rare biologicals in abundant quantities. We review here the discovery, early characterization, cloning, and expression of Interferon Alfa-2B (IFN alpha-2b or Intron A) as a therapeutic at Schering-Plough Research. IFN alpha-2a is marketed by Hoffman LaRoche under the trade name Roferon. The studies on the expression, purification, biology, and clinical aspects of this interferon offer a plethora of information on one of the earliest recombinant DNA based drugs to reach the market place. Author(s): Baron, E; Narula, S Language: eng Source Crit Rev Biotechnol 1990;10(3):179-90.

From mutation mapping to phenotype cloning. Author(s): Jonsson, J J; Weissman, S M Language: eng Subject(s): Animal; Chromosome Mapping; Cloning, Molecular; DNA/chemistry/genetics; Human; Mutation; Phenotype; Point Mutation; Polymerase Chain Reaction; Polymorphism, Restriction Fragment Length Source Proc Natl Acad Sci U S A 1995 Jan 3;92(1):83-5.

From positional cloning to gene identification and its function in genetic diseases Abstract: The huge development of genetic tools, together with the availability of powerful computers have enabled the mapping of several loci responsible for genetic diseases by the positional cloning technique, also called reverse genetics. The step consisting of identifying the gene itself remains a difficult step and the function of the gene identified may remain difficult to establish. In this chapter, we will examine the various steps allowing the identification of the gene, techniques allowing the creation of animal models of human diseases and the recent progresses of these techniques. Author(s): Soubrier, F Language: fre Source C R Seances Soc Biol Fil 1994;188(5-6):487-94.

Fukuyama, Francis. Our post-human future: consequences of the biotechnology revolution / Francis Fukuyama. Published/Created: London: Profile, 2002. Description: xiii, 256 p.; 24 cm. ISBN: 1861972970 Subjects: Human cloning--Moral and ethical aspects.

Fukuyama, Francis. Our posthuman future: consequences of the biotechnology revolution / Francis Fukuyama. Edition Information: 1st ed. Published/Created: New York: Farrar, Straus and Giroux, 2002. Description: 256 p.: ill.; 24 cm. ISBN: 0374236437 Notes: Includes bibliographical references (p. 243-256). Subjects: biotechnology--genetic engineering--human cloning--moral and ethical aspects. LC Classification: TP248.23 .F85 2002 Dewey Class No.: 174/.25 21

Functional cloning of a tumor suppressor gene, TSLC1, in human non-small cell lung cancer. Abstract: The identification of a tumor suppressor gene in non-small cell lung cancer (NSCLC) is one of the most important issues to elucidate the molecular mechanisms of this type of refractory cancer and to establish a novel strategy against it. Since NSCLC, like most other human cancers, develops as a sporadic disease, linkage analysis is not available for gene cloning. This review describes the functional cloning approaches to a tumor suppressor gene in sporadic cancers. Suppression of the malignant phenotype of cancer cells by fusion with a normal fibroblast was the first demonstration of the recessive phenotype of cancer cells in 1969. Evidence of tumor suppressor genes on the specific chromosomes was later provided by functional complementation of the cancer phenotype through microcell-mediated chromosome transfer. Further introduction of more restricted DNA fragments by YAC transfer provides a potent tool to localize the gene to a small segment, appropriate for the subsequent gene cloning. TSLC1, a novel tumor suppressor gene in NSCLC, was identified on chromosome 11q23.2 through a series of functional complementation of A549 cells in tumorigenicity. Two-hit inactivation of the TSLC1 by promoter methylation and gene deletion was observed in 40% of primary NSCLC tumors. The strong tumor suppressor activity of TSLC1, & its possible involvement in cell adhesion, suggest that the functional cloning approach could cast a new light on a group of genes that have not yet been characterized, but are important for general human carcino-genesis as well as tumor suppression. Author(s): Murakami, Yoshinori Language: eng Source Oncogene 2002 Oct 7;21(45):6936-48.

Functional cloning of genes involved in T-cell receptor-induced programmed cell death. Abstract: Programmed cell death (PCD) is a normal event under genetic control that regulates the life span of different cell types in multicellular organisms. Among other physiological processes, PCD plays a pivotal role in the regulation of the immune system. Using a functional selection strategy we have isolated and characterized genes involved in T-cell receptor-induced apoptosis. One, ALG-2, is a Ca(2+)-binding protein that participates in regulatory events that occur late in the apoptotic program, where several death signals converge. Another, ALG-3, is a mouse homologue of the chromosome 1 familial Alzheimer's disease gene PS2. ALG-3 codes for a truncated PS2 polypeptide that antagonizes the apoptotic role of PS2. A PS2 mutation associated with Alzheimer's disease generates a molecule with enhanced apoptotic activity indicating that it might accelerate the process of neurodegeneration that occurs in this disease. Author(s): D'Adamio, L; Lacana, E; Vito, P Language: eng Source Semin Immunol 1997 Feb;9(1):17-23.

Functional expression cloning of cDNA; Kubo, Y Language: jpn Subject(s): Animal; Cloning, Molecular/methods; DNA, Complementary; Ion Channels; Oocytes; Receptors, Cell Surface; Receptors, Cholinergic; Xenopus Source Nippon Seirigaku Zasshi 1997;59(7-8):279-300.

Galactosyltransferase I is a gene responsible for progeroid variant of Ehlers-Danlos syndrome: molecular cloning and identification of mutations. Abstract: A human cDNA encoding a novel galactosyltransferase was identified based on BLAST analysis of expressed sequence tags, and the cDNA clones were isolated, showing a type II membrane protein with 327 amino acids and 38% homology to the Caenorhabditis elegans sqv-3 gene involved in vulval invagination and oocyte development. This cDNA exhibited marked galactosyltransferase activity specific for p-nitrophenyl-beta-Dxylopyranoside, and also restored glycosaminoglycan (GAG) synthesis to galactosyltransferase I-deficient CHO mutant pgsB-761 cells. The enzyme product contained beta-1,4-linked galactosyl residues, indicating that the enzyme is galactosyltransferase I (UDP-D-galactose: D-xylose beta-1,4-D-galactosyltransferase; EC 2.4.1.133) involved in the synthesis of the GAG-protein linkage region of proteoglycans. Mutations of this gene were investigated in a case of Ehlers-Danlos syndrome (progeroid variant), since reduced activity of galactosyltransferase I had been reported in this disease by others. As expected, the patient gene contained two different mutations (A186D, L206P). The mutations showed, respectively, 10-50% and 0% of the enzyme activity compared with wild type, suggesting that galactosytransferase I (XGal-T1) is at least one of the genes responsible for Ehlers-Danlos syndrome (progeroid variant). Author(s): Furukawa, Koichi; Okajima, Tetsuya Language: eng Source Biochim Biophys Acta 2002 Dec 19;1573(3):377-81.

Gene [microform]. Published/Created: Amsterdam: Elsevier/North-Holland, [c1976]- Description: v.; 24 cm. Vol. 1, no. 1- Notes: Title from cover. "An international journal focusing on gene cloning and gene structure and function." Microfilm. Lausanne, Switzerland: Elsevier Sequoia: distributed by University Microfilms International. microfilm reels; 35 mm. SERBIB/SERLOC merged record Subjects: Genetic engineering--Molecular cloning--Recombinant DNA--Periodicals. LC Classification: Microfilm (o) 89/2585

Gene analysis techniques. Published/Created: New York, NY: Elsevier Science Pub. Co., c1984-c1989. Description: 6 v.: ill.; 28 cm. Vol. 1, no. 1 (Jan. 1984)-v. 6, no. 6 (Nov./Dec. 1989). ISSN: 0735-0651 ISSN: 0735-0561 Notes: Title from cover. SERBIB/SERLOC merged record Subjects: Molecular genetics--Technique--Periodicals. Molecular cloning--Technique--Nucleic acids--Analysis--Cloning, Molecular--methods--Genetic Techniques--Nucleic Acids--analysis--Periodicals. LC Classification: QH426 .G44 Dewey Class No.: 574.87/328/05 20

Gene cloning and analysis by RT-PCR / edited by Paul D. Siebert, James W. Larrick. Published/Created: [Natick, MA]: BioTechniques Books, c1998. Related Names: Siebert, Paul D. Larrick, James W. Description: xiv, 350 p.: ill.; 26 cm. ISBN: 1881299147 (pbk.) Notes: Includes bibliographical references and index. Subjects: Molecular cloning. Polymerase chain reaction. Reverse transcriptase. Series: BioTechniques molecular laboratory methods series, 1520-3182; 1 LC Classification: QH442.2 .G4623 1998 Dewey Class No.: 572.8/645 21

Gene cloning and analysis: a laboratory guide / compiled by the staff of the University of Leicester gene cloning course and edited by G.J. Boulnois. Published/Created: Oxford; Boston: Blackwell Scientific Publications, 1987. Related Names: Boulnois, G. J. (Graham J.) University of Leicester. Description: viii, 136 p.: ill.; 28 cm. ISBN: 0632019271 (pbk.) Notes: Updated version of: Basic cloning techniques / edited by R.H. Pritchard and I.B. Holland. 1985. Includes bibliographies. Subjects: Molecular cloning--Laboratory manuals. Cloning, Molecular--methods. LC Classification: QH442.2 .G46 1987 Dewey Class No.: 574.87/322 19

Gene cloning and analysis: current innovations / edited by Brian C. Schaefer. Published/Created: Norfolk, England: Horizon Scientific Press, c1997. Related Names: Schaefer, Brian C. Description: 214 p.: ill.; 24 cm. ISBN: 1898486069 1898486115 Notes: Includes bibliographical references and index. Subjects: Molecular cloning--Laboratory manuals. Series: Current innovations in molecular biology series; v. 4 LC Classification: QH442.2 .G462 1997 Dewey Class No.: 572.8 21

Gene cloning and expression technologies / edited by Michael P. Weiner, Quinn Lu. Published/Created: Westborough, MA: BioTechniques Press: Eaton Pub., c2002. Related Names: Weiner, Michael P., 1958- Lu, Quinn, 1962- Description: xvi, 603 p.: ill.; 26 cm. ISBN: 1881299201 Notes: Includes bibliographical references and index. Subjects: Molecular cloning--Gene expression--Methodology. Cloning, Organism--methods. Series: BioTechniques update series. LC Classification: QH442.2 .G4625 2002 Dewey Class No.: 572.8/633 21

Gene cloning in cell biology. Author(s): Paul, J Language: eng Subject(s): Animal; Coliphages; DNA Restriction Enzymes; DNA, Recombinant; Human; Hybridization, Genetic; Legislation; Nucleic Acid Hybridization; Plasmids; Transformation, Bacterial Source Cell Biol Int Rep 1978 Jul;2(4):311-26.

Gene cloning in Neurospora crassa. Author(s): Vapnek, D; Case, M Language: eng Subject(s): Cloning, Molecular; Genes; Genetic Vectors; Neurospora/genetics; Neurospora crassa/genetics; Replicon; Transformation, Genetic Source Curr Top Microbiol Immunol 1982;96:97-100.

Gene cloning in organisms other than E. coli / edited by P.H. Hofschneider and W. Goebel. Published/Created: Berlin; New York: Springer-Verlag, 1982. Related Names: Hofschneider, P. H. (Peter Hans), 1929- Goebel, W. (Werner), 1939- Description: 259 p.; 25 cm. ISBN: 0387111174 (U.S.) Notes: Includes bibliographies and index. Subjects: Molecular cloning. Series: Current topics in microbiology and immunology; 96 LC Classification: QR1 .E6 vol. 96 QH442 Dewey Class No.: 576 s 574.87/328 19

Gene cloning in plants: innovative approaches. Abstract: Although many technical advances have recently been made in the development of plant transformation systems and plant-gene identification and isolation approaches, some of these approaches are either technically or practically not feasible for many important gene systems; thus alternative approaches are required. Two innovative approaches for gene cloning in plants are discussed here, namely, shotgun rescue and chromosome-oriented approaches. Emphasis is placed on their methodology, scope and advantages. Finally, potential applications of these approaches are proposed with reference to the isolation of genes for which mutations are available, the development of artificial plant chromosomes and their possible uses and chromosome-mediated transformation. Author(s): Sawahel, W; Fukui, K Language: eng Source Biotechniques 1995 Jul;19(1):106-10, 112, 114-5.

Gene cloning in Streptomyces. Author(s): Chater, K F; Hopwood, D A; Kieser, T; Thompson, C J Language: eng Subject(s): Bacteriophages/genetics; Cloning, Molecular; DNA, Recombinant; Genes, Bacterial; Genetic Vectors; Plasmids; Streptomyces/genetics Source Curr Top Microbiol Immunol 1982;96:69-95.

Gene cloning to study viral antigens: prospects for novel influenza and rotavirus vaccines. Author(s): Both, G W; Jennings, P A Language: eng Source Prog Vet Microbiol Immunol 1987;3:179-213.

Gene cloning with small plasmids. Author(s): Collins, J Language: eng Subject(s): DNA Replication; DNA Restriction Enzymes/metabolism; DNA, Bacterial; DNA, Recombinant; Escherichia coli/genetics; Genes; Plasmids; RNA, Messenger/metabolism; Research; Safety; Transcription, Genetic; Transformation, Genetic; Translation, Genetic Source Curr Top Microbiol Immunol 1977;78:122-70.

Gene cloning: studies of nutritional regulation of gene expression. Author(s): Hesketh, J E; Partridge, K Language: eng Subject(s): Animal; Cloning, Molecular; DNA Restriction Enzymes; Gene Expression Regulation/physiology; Genetic Vectors; Human; Nutrition Source Proc Nutr Soc 1996 Mar;55(1B):575-81.

Gene delivery to mammalian cells / edited by William C. Heiser. Published/Created: Totowa, N. J.: Humana Press, 2004. Related Names: Heiser, William C. Description: v.; cm. ISBN: 1588290867 (v. 1: alk. paper) 1588290956 (v. 2: alk. paper) Contents: v. 1. Nonviral gene transfer techniques -- v. 2. Viral gene transfer techniques. Notes: Includes bibliographical references and indexes. Subjects: Transfection. Animal cell biotechnology. Cloning, Molecular. Gene Transfer Techniques. Cells, Cultured--cytology. Gene Targeting--methods. Mammals--genetics. Series: Methods in molecular biology (Clifton, N.J.); 245-246. LC Classification: QH448.4 .G42 2004 Dewey Class No.: 571.9/64819 21

Gene hunting for Fukuyama-type congenital muscular dystrophy by positional cloning; Toda, T Language: jpn Subject(s): Chromosome Mapping; Chromosomes, Artificial, Yeast; Cloning, Molecular methods; Cosmids; Genetic Markers; Genome, Human; Haplotypes; Homozygote; Linkage Disequilibrium; Muscular Dystrophies congenital; Mutagenesis, Insertional Source Tanpakushitsu Kakusan Koso 1997 Dec;42(17 Suppl):2779-85.

Gene manipulation and expression / edited by Robert E. Glass & Jaroslav Spizek. Published/Created: London; Dover, N.H.: Croom Helm, c1985. Related Names: Glass, Robert E., Ph. D. Spizek, Jaroslav. Description: 559 p., [2] p. of plates: ill.; 22 cm. ISBN: 0709938187 : Notes: Contains selected papers from a symposium held Sept. 1984 in Bechyne Castle, in South Bohemia, Czechoslovakia. Includes bibliographies and index. Subjects: Gene expression--Molecular cloning--Genetic translation--Recombinant DNA--Cloning, Molecular--DNA, Recombinant--Gene Expression Regulation--Translation, Genetic--Congresses. LC Classification: QH450 .G4628 1985 Dewey Class No.: 574.87/322 19

Gene transfer, recombination and gene cloning in Clostridium acetobutylicum. Abstract: Although Clostridium acetobutylicum has been used for over 70 years for the industrial production of solvents, it is only recently that studies on the genetics of this organism have been initiated. Recent advances in the development of genetic transfer systems as well as the cloning and expression of genes from this organism in Escherichia coli are reviewed. Author(s): Jones, D T; Woods, D R Language: eng Source Microbiol Sci 1986 Jan;3(1):19-22.

Gene. Published/Created: Amsterdam, Elsevier/North-Holland, 1976- Description: v. ill. 24 cm. v. 1- ISSN: 0378-1119 Notes: "An international journal focusing on gene cloning and gene structure and function." Latest issue consulted: Vol. 295, no. 1 (July 24, 2002). SERBIB/SERLOC merged record Vols. for 1996- includes separately paged section called: Gene-combis which is printed version of articles published electronically in Gene-combis on the World-Wide Web. Subjects: Genetic engineering--Periodicals. Molecular cloning--Periodicals. Recombinant DNA--Periodicals. Genetics, Biochemical--Periodicals. Recombination, Genetic--Periodicals. LC Classification: QH442 .G43 Dewey Class No.: 575.1

Generalized cloning vectors for Bacillus subtilis. Author(s): Errington, J Language: eng Subject(s): Bacillus subtilis/genetics; Bacteriophages/genetics; Cloning, Molecular/methods; Genes, Bacterial; Genetic Vectors; Plasmids Source Biotechnology 1988;10:345-62.

Genetic analysis, biomolecular engineering. Published/Created: Amsterdam; New York: Elsevier, 1995-1999. Description: 4 v.: ill. (some col.); 28 cm. Vol. 12, no. 1 (Mar. 1995)-v. 15, no. 6 (15 Dec. 1999). ISSN: 1050-3862 Notes: SERBIB/SERLOC merged record Subjects: Genetic engineering--Molecular genetics--Molecular cloning--Nucleic acids--Analysis--Biotechnology--Genetic Engineering--Genetic Techniques--Periodicals. LC Classification: QH426 .G44 Dewey Class No.: 574.87/328/05 20

Genetic analysis, techniques and applications. Published/Created: New York, NY: Elsevier Science Pub. Co., c1990-c1994. Description: 5 v.: ill.; 28 cm. Vol. 7, no. 1 (Feb. 1990)-v. 11, no. 5-6. ISSN: 1050-3862 ISSN: 0735-0651 Notes: Title from cover. SERBIB/SERLOC merged record Subjects: Molecular genetics-- Molecular cloning-- Nucleic acids--Analysis-- Biotechnology--Periodicals. Cloning, Molecular--methods--Periodicals. Genetic Techniques--Nucleic Acids--analysis--Periodicals. LC Classification: QH426 .G44 Dewey Class No.: 574.87/328/05 20

Genetic engineering of human interferons from lymphoblastoid cells: I. Cloning of interferon species expressed in Sendai-induced Namalva cells. Author(s): Cohen, S; Velan, B; Shafferman, A Language: eng Source Adv Biotechnol Processes 1985;4:1-23.

Genetic exchange: a celebration and a new generation / editors, Uldis N. Streips ... [et al.]. Published/Created: New York: M. Dekker, c1982. Related Names: Streips, Uldis N., 1942- Description: xvi, 369 p.: ill.; 24 cm. ISBN: 0824714180 Notes: Includes bibliographical references and indexes. Subjects: Bacterial transformation. Transformation, Genetic--Congresses. Cloning, Molecular--Congresses. Series: Genetic and cellular technology; v. 1 LC Classification: QH448.4 .G45 1982 Dewey Class No.: 589.9/087328 19

Genetics of childhood disorders: VII. Cloning genes of interest. Author(s):Sparks, R; Lombroso, P J; Gruen, J R Language: eng Subject(s): Base Sequence; Child; Child Development; Cloning, Molecular/methods; Gene Library; Genomic Library; Human; Mental Disorders/genetics; Molecular Sequence Data Source J Am Acad Child Adolesc Psychiatry 1999 Oct;38(10):1316-9.

Gish, Duane T. Manipulating life, where does it stop?: Genetic engineering / by Duane T. Gish and Clifford Wilson. Published/Created: San Diego, Calif: Master Books, c1981. Related Names: Wilson, Clifford A. Description: xv, 220 p.; 22 cm. ISBN: 0890510717 Notes: Includes bibliographical references. Subjects: Abortion--Genetic engineering--Religious aspects. Cloning--Artificial insemination, Human--Moral and ethical aspects. LC Classification: QH442 .G57 Dewey Class No.: 574/.2 19

Glover, David M. Gene cloning: the mechanics of DNA manipulation / David M. Glover. Published/Created: London; New York: Chapman and Hall, 1984. Related Names: Glover, David M. Genetic engineering, cloning DNA. Description: viii, 222 p.: ill.; 25 cm. ISBN: 0412266008 0412254301 (pbk.) Notes: Rev. ed. of: Genetic engineering, cloning DNA. 2nd ed. 1980. Includes bibliographies and index. Subjects: Molecular cloning. Recombinant DNA. Series: Outline studies in biology (Chapman and Hall) LC Classification: QH442.2 .G56 1984 Dewey Class No.: 574.87/3282 19

Glycoproteins: carbohydrates to cloning. Abstract: This article is dedicated to Professor Saul Roseman and briefly outlines some of the early studies on sialyltransferases, on glycoproteins such as pig submaxillary mucins and, more recently, on a series of unusual proteins and glycoproteins high in proline, the so-called proline-rich proteins. Hopefully, it represents, in an inadequate manner, my appreciation for 'The man and his works'. During the Roseman Symposium at the 11th International Symposium on Glycoconjugates in Toronto, several of his former students and postdocs tried to describe what it was like in the Roseman laboratory. Clearly, the time I was in Saul's lab was like no other time in my career. Thanks for everything, Saul. Author(s): Carlson, D M Language: eng Source Glycobiology 1991 Nov;1(5):463-7.

God and the embryo: religious voices on stem cells and cloning / Brent Waters and Ronald Cole-Turner, editors. Published: Washington, D.C.: Georgetown University Press, c2003. Related Names: Waters,

Brent. Cole-Turner, Ronald, 1948- Description: p. cm. ISBN: 087840998X (pbk.: alk. paper) Notes: Includes bibliographical references and index. Subjects: Human cloning--Stem cells--Research--Religious aspects. LC Classification: QH442.2 .G63 2003 Dewey Class No.: 241/.64957 21

Gonadotropin-releasing hormone receptor: cloning, expression and transcriptional regulation. Abstract: In summary, isolation of GnRH receptor cDNA, its gene, and identification of regulatory elements in the flanking region of the gene have added to our knowledge regarding the tissue-specific expression of the GnRH receptor gene, and the mechanisms that mediate and influence its transcriptional regulation. However, the interactions of the different regulatory factors (nuclear factors) and the effects of these interactions on the regulation of the GnRH receptor gene remain unclear. Due to existence of multiple promoters and transcriptional start sites in human GnRH receptor gene and the lack of a human gonadotrope cell line, the precise promoter and transcriptional start sites in human pituitary, extra-pituitary tissues and tumors have not yet been identified. Author(s): Kakar, Sham S; Malik, M Tariq; Winters, Stephen J Language: eng Source Prog Brain Res 2002;141:129-47.

Goodnough, David. The debate over human cloning / David Goodnough. Published: Berkeley Heights, NJ: Enslow Pub., c2003. Description: 64 p.: col. ill.; 24 cm. ISBN: 0766018180 (hardcover) Summary: Provides an overview of the technology and history of cloning and presents arguments for and against human cloning. Contents: The cloning debate -- Cloning -- Why clone? -- Why not to clone -- Brave new world. Notes: Includes bibliographical references (p. 63) and index. Subjects: Human cloning--Juvenile literature. Genetic engineering. Series: Hot pro/con issues. LC Classification: QH442.2 .G66 2003 Dewey Class No.: 174/.25 21

Gormley, Beatrice. Best friend insurance / by Beatrice Gormley; illustrated by Emily Arnold McCully. Edition Information: 1st ed. Published/Created: New York: E.P. Dutton, c1983. Related Names: McCully, Emily Arnold, ill. Description: 147 p.: ill.; 21 cm. ISBN: 0525440666 (lib. bdg.) : Summary: Just when Maureen feels she has been demoted to second-best status by her former best friend Tracey, a rather unusual insurance agent turns up, guaranteeing replacement of lost friends within twenty-four hours. Subjects: Friendship--Fiction. Cloning--Fiction. LC Classification: PZ7.G6696 Be 1983 Dewey Class No.: [Fic] 19

G-protein linked receptors: a family probed by molecular cloning and mutagenesis procedures. Author(s): Houslay, M D Language: eng Subject(s): Binding Sites; Cloning, Molecular; GTP-Binding Proteins/chemistry/genetics; Gene Expression/physiology; Human; Mutagenesis; Receptors, Cell Surface/chemistry/genetics Source Clin Endocrinol (Oxf) 1992 Jun;36(6):525-34.

Graham, Ian, 1953- Genetics: the study of heredity / by Ian Graham. Edition Information: U.S. ed. Published/Created: Milwaukee: G. Stevens Pub., 2002. Description: 33, [3] p.: col. ill.; 25 cm. ISBN: 0836832310 (lib. bdg.) Summary: Examines the field of genetics, discussing DNA research, heredity, cloning, genetic engineering, and the possibility of gene therapy. Notes: Includes bibliographical references (p. [35]) and index. Series: Investigating science LC Classification: QH437.5 .G73 2002 Dewey Class No.: 576.5 21

Hackett, Perry B. An introduction to recombinant DNA techniques: basic experiments in gene manipulation / Perry B. Hackett, James A. Fuchs, Joachim W. Messing. Published/Created: Menlo Park, Calif.: Benjamin/Cummings, c1984. Related Names: Fuchs, James A. Messing, Joachim W. Description: xiii, 194 p.: ill.; 28 cm. ISBN: 0805336729 (pbk.) : Notes: Includes index. Bibliography: p. 189-191.

Subjects: Recombinant DNA--Experiments--Laboratory manuals. Genetic engineering--Experiments--Laboratory manuals. DNA, Recombinant--Genetic intervention--Cloning, Molecular--Methods--Experiments. Series: The Benjamin/Cummings series in the life sciences LC Classification: QH442 .H33 1984 Dewey Class No.: 574.87/3282 19

Haemochromatosis: understanding the mechanism of disease and implications for diagnosis and patient management following the recent cloning of novel genes involved in iron metabolism. Abstract: Haemochromatosis, a common recessive genetic disorder in people of Northern European descent, is an iron storage disorder characterized by excessive hepatic iron accumulation resulting from disruption of the regulation of intestinal iron absorption. The identification of novel genes involved in the control of iron absorption from the diet has allowed improved understanding of iron metabolism in health and disease. In particular, the identification of the haemochromatosis gene (HFE) and more recently the transferrin receptor 2 gene (TfR2) together with the specific mutations in these genes which result in hepatic iron overload, has enhanced our understanding of the pathophysiology of haemochromatosis. However, because of the wide variation in phenotypic expression of the disease, there now exists a considerable challenge to diagnosis and patient management. Author(s): Fletcher, L M; Halliday, J W Language: eng Source J Intern Med 2002 Mar;251(3):181-92.

Halam, Ann. Taylor Five / Ann Halam. Published/Created: New York: Wendy Lamb Books, c2004. Description: p. cm. ISBN: 0385730942 (trade) 0385901143 (library binding) Summary: Fourteen-year-old Taylor is still dealing with the fact that she is a clone produced by the same company that funds the Orangutan Reserve which is her home on the island of Borneo, when the Reserve is attacked and she flees with her younger brother and Uncle, the Reserve's mascot. Subjects: Cloning--Brothers and sisters--Orangutan--Fiction. Borneo--Fiction LC Classification: PZ7.H1283 Tay 2004 Dewey Class No.: [Fic] 21

Hapten-specific B lymphocytes: enrichment, cloning, receptor analysis, and tolerance induction. Author(s): Nossal, G J; Pike, B L; Stocker, J W; Layton, J E; Goding, J W Language: eng Source Cold Spring Harb Symp Quant Biol 1977;41 Pt 1:237-43.

Hematopoietic hormones: from cloning to clinic. Abstract: In order to maintain adequate circulating numbers of blood cells, the bone marrow must produce billions of cells each day and must be able to rapidly increase production by 10-20-fold in response to infection and hemorrhage. The existence of circulating factors that regulate this process has been suspected for over 100 years. Recently, the genes encoding these growth factors were cloned and their functions are now identified. Interleukin-3 (IL-3) acts on the most primitive hematopoietic stem cell, driving this self-renewing cell to produce progeny of all hematopoietic lineages. Granulocyte-macrophage colony-stimulating factor (GM CSF) stimulates the granulocyte-macrophage progenitor cell, as well as cells committed to the erythroid lineage, to differentiate. G-CSF and M-CSF stimulate the most differentiated myeloid progenitors to produce granulocytes and monocytes/macrophages, respectively. Erythropoietin stimulates the differentiation of late erythroid progenitors. In the lymphoid progenitor lineage, IL-2 stimulates T cell differentiation; IL-4 and IL-6 stimulate differentiation of B cells. The colony-stimulating factors also enhance function and cause activation of the mature cells whose production they induce. In clinical trials, these hormones have successfully ameliorated anemia in renal failure, chronic disease, and in prematurity. They have improved pancytopenias in aplastic anemia, myelodysplastic syndromes, and congenital cytopenias, and they have hastened recovery from chemotherapy and bone marrow transplantation. Author(s): Perrine, S P; Vichinsky, E; Faller, D V Language: eng Source Am J Pediatr Hematol Oncol 1989 Fall;11(3):268-75.

Hemopoietic colonies: in vitro cloning of normal and leukemic cells. Author(s): Metcalf, D Language: eng Source Recent Results Cancer Res 1977;(61):Title page, 1-227.

Heterogeneity and regulation of central dopamine receptor subtypes studied by cDNA cloning methodology. Author(s): Sokoloff, P; Martres, M P; Giros, B; Bouthenet, M L; Schwartz, J C Language: eng Source Neurochem Int 1992 Mar;20 Suppl:27S-32S.

High-speed positional cloning based on restriction landmark genome scanning. Abstract: Restriction landmark genome scanning (RLGS) was developed as a method of genome analysis that is based on the concept that restriction enzyme sites can be used as landmarks. In this article, we demonstrate how this method can be used for the systematic, successful positional cloning of mouse mutant reeler gene. The major advantage of the RLGS method is that it allows the scanning of several thousand spots/loci throughout the genome with one RLGS profile. High-speed positional cloning based on the RLGS method includes (1) high-speed construction of a linkage map (RLGS spot mapping), (2) high-speed detection of RLGS spot markers tightly linked to the mutant phenotype (RLGS spot bombing method), and (3) construction of YAC contigs covering the region where tightly linked spot markers are located (RLGS-based YAC contig mapper). We introduced a series of these procedures by using them to positionally clone the reeler gene. High-speed construction of the whole genetic map and spots/loci (less than 1 cM) within the closest flanking markers is demonstrated. The RLGS-based YAC contig mapper also efficiently yielded the YAC physical contig map of the target region. Finally, we cloned the reeler gene, which is the causal gene for the perturbation of the three-dimensional brain architecture due to the abnormal migration of neuroblasts in reeler mouse. Since the RLGS method itself can be used for any organism, we conclude that the total RLGS-based positional cloning system can be used to identify any mutant gene of any organism. Author(s): Okazaki, Y; Hayashizaki, Y Language: eng Source Methods 1997 Dec;13(4):359-77.

High-technology route to virus vaccines: proceedings of the First Annual Southwest Foundation for Biomedical Research International Symposium, Houston, Texas, 8-10 November 1984 / editors, Gordon R. Dreesman, Judith Gunn Bronson, Ronald C. Kennedy. Published/Created: Washington, D.C.: American Society for Microbiology, 1985. Related Names: Dreesman, Gordon R. Bronson, Judith Gunn. Kennedy, Ronald C. Description: vii, 180 p.: ill.; 23 cm. ISBN: 0914826816 (pbk.) : Notes: Includes bibliographies and indexes. Subjects: viral vaccines--immunological adjuvants--peptides--physiological effect--molecular cloning--immunoglobulin idiotypes--viral vaccines--congresses. LC Classification: QR189.5.V5 S68 1984 Dewey Class No.: 615/.372 19

HLA-DR antigens: structure, separation of subpopulations, gene cloning and function. Author(s): Shackelford, D A; Kaufman, J F; Korman, A J; Strominger, J L Language: eng Source Immunol Rev 1982;66:133-87.

Homzie, Hillary. In the doghouse / by H.B. Homzie. Edition Information: 1st Aladdin Paperbacks ed. Published/Created: New York: Aladdin Paperbacks, 1999. Description: p. cm. ISBN: 0689823436 (pbk.) Summary: Twins Barton and Nancy try to use their alien clones to buy tickets for a rock concert, but their plan leads to chaos. Subjects: Extraterrestrial beings--Cloning--Twins--Brothers and sisters--Fiction. Series: Homzie, Hillary. Alien clones from outer space; #2. LC Classification: PZ7.H7458 In 1999 Dewey Class No.: [Fic] 21

Homzie, Hillary. Two heads are better than one / by H.B. Homzie. Edition Information: 1st Aladdin Paperbacks ed. Published/Created: New York: Aladdin Paperbacks, 1999. Description: p. cm. ISBN: 0689823428 (pbk.) Summary: Seventh grader Barton's life is terribly complicated when alien clones of him and his sister Nancy arrive on Earth and start acting in their place. Subjects: Extraterrestrial beings--Fiction. Cloning--Fiction. Science fiction. Series: Homzie, Hillary. Alien clones from outer space; #1. LC Classification: PZ7.H7458 Tw 1999 Dewey Class No.: [Fic] 21

Horowitz, Anthony, 1955- Point Blanc: a stormbreaker adventure / Anthony Horowitz. Edition Information: 1st American ed. Published/Created: New York: Philomel Books, 2002. Description: p. cm. ISBN: 039923621X Summary: Fourteen-year-old Alex continues his work as a spy for the British MI6, investigating an exclusive school for boys in the French Alps. Notes: Sequel to: Stormbreaker. Subjects: Spies--Fiction. Cloning--Fiction. Schools--Fiction. Orphans--Fiction. LC Classification: PZ7.H7875 Po 2002 Dewey Class No.: [Fic] 21

Host: vector systems for gene cloning in Clostridium. Abstract: The genus Clostridium contains some of the most toxic bacteria known to man (e.g., Clostridium tetani, Clostridium botulinum) as well as solvent-producing organisms of biotechnological interest (e.g., Clostridium acetobutylicum). Developments within the areas of plasmid vector construction and plasmid delivery methodology now provide the opportunity for applying recombinant DNA technology to this important group of bacteria. Author(s): Minton, N P; Oultram, J D Language: eng Source Microbiol Sci 1988 Oct;5(10):310-5.

Host: vector systems for gene cloning in Pseudomonas. Author(s): Bagdasarian, M; Timmis, K N Language: eng Subject(s): Cloning, Molecular; Genes, Bacterial; Genetic Engineering; Genetic Vectors; Plasmids; Pseudomonas/genetics Source Curr Top Microbiol Immunol 1982;96:47-67.

Houdebine, Louis-Marie. Animal transgenesis and cloning / Louis-Marie Houdebine; translated by Louis-Marie Houdebine ... [et al.]. Published/Created: Chichester, UK; Hoboken, NJ: John Wiley & Sons, c2003. Description: xiv, 220 p.: ill.; 23 cm. ISBN: 0470848286 (pbk.: alk. paper) 0470848278 (hbk.: alk. paper) Contents: From the gene to the transgenic animal -- Techniques for cloning and transgenesis -- Applications of cloning and transgenesis -- Limits and risks of cloning, gene therapy and transgenesis. Notes: Includes bibliographical references (p. [199]-216) and index. Subjects: Transgenic animals. Cloning. Animals, Genetically Modified. Cloning, Organism. Gene Therapy. Genetic Engineering--methods. LC Classification: QH442.6 .H6813 2003 Dewey Class No.: 636.08/21 21

Howard, Ted. Who should play God? [sound recording]. Published/Created: Washington, D.C.: National Public Radio, p1978. Related Names: Waters, Mike. National Public Radio (U.S.) Description: 1 sound cassette (59 min.): analog. Summary: Ted Howard, coauthor of Who should play God?, talks about genetic engineering. Focusing on the future social application of this research, he explores the involvement of government, business, and science in promoting genetic engineering. Mr. Howard explores some current uses of this research and talks about cloning, artificial insemination, and genetic screening tests for employment. Notes: Narration title: Who shall play God? Originally broadcast June 6, 1978, on the National Public Radio program entitled Options. Recorded at the University of Wyoming in Laramie, Wyo. Subjects: Genetic engineering. LC Classification: QH442 RYA 7923

Howe, C. J. Gene cloning and manipulation / Christopher Howe. Published/Created: Cambridge [England]; New York: Cambridge University Press, 1995. Description: xiii, 210 p.: ill.; 27 cm. ISBN:

0521403413 (hbk.) 0521407001 (pbk.) Notes: Includes bibliographical references (p. 195-200) and index. Subjects: Molecular cloning. LC Classification: QH442.2 .H69 1995 Dewey Class No.: 575.1/0724 20

Hughes, R. N. A functional biology of clonal animals / Roger N. Hughes. Published/Created: London; New York: Chapman and Hall, 1989. Description: xii, 331 p.: ill.; 24 cm. ISBN: 0412331306 Notes: Includes index. Bibliography: p. [276]-306. Subjects: Cloning. Series: Functional biology series LC Classification: QH475 .H84 1989 Dewey Class No.: 574.1/62 19

Human 25-hydroxyvitamin D-1alpha-hydroxylase: cloning, mutations, and gene expression. Abstract: The rate-limiting, hormonally regulated step in the bioactivation of vitamin D is the 1alpha-hydroxylation of 25-hydroxyvitamin D, which occurs in the kidney and other tissues and is catalyzed by the mitochondrial cytochrome P450 enzyme, P450c1alpha. After many years of effort, the cDNA and gene encoding this enzyme were cloned from mouse, rat, and human tissue in late 1997. The human gene encoding the 1alpha-hydroxylase is 5 kb in length, located on chromosome 12, and comprises nine exons and eight introns; its intron/exon organization is very similar to that of the other four mitochondrial P450 enzymes cloned to date. Mutations in P450c1alpha cause 1alpha-hydroxylase deficiency, also known as vitamin D-dependent rickets type 1, a rare autosomal recessive disease characterized by rickets and impaired growth due to failure of renal synthesis of 1,25(OH)2D. To date, 31 patients have been studied and 20 distinct mutations in the gene identified, including 13 mis-sense mutations, none of which encode a protein with significant enzyme activity. Recent studies in animals demonstrate that regulation of P450c1alpha gene expression by parathyroid hormone (PTH), low calcium diet, low phosphorus diet, and 1,25(OH)2D occurs at the level of its mRNA. Transcriptional activity of the mouse and human P450c1alpha gene promoters can be stimulated by PTH, cAMP, and forskolin and suppressed by 1,25(OH)2D. Author(s): Portale, A A; Miller, W L Language: eng Source Pediatr Nephrol 2000 Jul;14(7):620-5.

Human and mouse monoclonal antibodies by repertoire cloning. Abstract: Antibody repertoires, the wide range of antibody molecules produced by animals, can now be established in bacteria by cloning and expression of antibody genes. Beginning with immunized animals, antigen can be used to select, from the repertoire, clones which secrete specific monoclonal antibody. In the future, immunization may become unnecessary. The method may provide a general route, which has so far eluded biotechnologists, to human monoclonal antibodies. Author(s): Burton, D R Language: eng Subject(s): Animal; Antibodies, Monoclonal/genetics; Antibody Diversity; Cloning, Molecular; Forecasting; Human; Mice Source Trends Biotechnol 1991 May;9(5):169-75.

Human and rabbit paraoxonases: purification, cloning, sequencing, mapping and role of polymorphism in organophosphate detoxification. Abstract: Human and rabbit paraoxonases/arylesterases were purified to homogeneity by chromatographic and gel electrophoretic/isofocusing procedures coupled with activity stains. N-terminal and peptide sequence analysis suggested retention of the secretion signal sequence and allowed design of oligonucleotide probes. The probes were used to isolate a 1294-bp rabbit paraoxonase cDNA clone, which, in turn, was used to isolate three human cDNA clones. Comparison of rabbit and human protein and cDNA sequences indicated a high degree of sequence conservation (approximately 85% identity) and verified that paraoxonase retains its signal sequence (except for the N-terminal Met). The rabbit cDNA encodes a protein of 359 amino acids and the human a protein of 355 amino acids. In situ hybridization demonstrated, as expected, that the paraoxonase gene maps to the long arm of human chromosome 7. Arginine at position 192 specifies high activity paraoxonase and glutamine low activity human paraoxonase. Variation in protein levels explains the variation of enzyme activity observed within a genetic class. Toxicity studies showed that raising rat plasma paraoxonase levels by i.v. administration of partially purified rabbit paraoxonase protected animals against cholinesterase inhibition by paraoxon and chlorpyrifos oxon. Protection correlated with the relative rates of hydrolysis of these two compounds. Author(s): Furlong, C E; Costa, L G; Hassett, C; Richter, R J; Sundstrom, J A; Adler, D A; Disteche, C M; Omiecinski, C J; Chapline, C; Crabb, J W Language: eng Source Chem Biol Interact 1993 Jun;87(1-3):35-48.

Human CD40 ligand: molecular cloning, cellular distribution and regulation of IgE synthesis. Author(s): Gauchat, J F; Mazzei, G; Life, P; Henchoz, S; Peitsch, M C; Aubry, J P; Jomotte, T; Bonnefoy, J Y Language: eng Source Res Immunol 1994 Mar-Apr;145(3):240-4; discussion 244-9.

Human cloning 2001. Abstract: This review summaries human cloning from a clinical perspective. Natural human clones, that is, monozygotic twins, are increasing in the general community. Iatrogenic human clones have been produced for decades in infertile couples given fertility treatment such as ovulation induction. A clear distinction must be made between therapeutic cloning using embryonic stem cells and reproductive cloning attempts. Unlike the early clinical years of in vitro fertilization, with cloning there is no animal model that is safe and dependable. Until there is such a model, 'Dolly'-style human cloning is medically unacceptable. Author(s): Healy, David L; Weston, Gareth; Pera, Martin F; Rombauts, Luk; Trounson, Alan O Language: eng. Source Hum Fertil (Camb) 2002 May;5(2):75-7.

Human cloning and human dignity: an ethical inquiry. Published/Created: Washington, D.C.: The President's Council on Bioethics, 2002. Related Names: President's Council on Bioethics (U.S.). Description: xxxix, 299 p.; 28 cm. Contents: Letter of transmittal to the President -- Members of the President's Council on Bioethics -- Council staff and consultants -- Preface -- Executive summary -- The meaning of human cloning: an overview -- Historical aspects of cloning -- On terminology -- Scientific background -- The ethics of cloning-to-produce-children -- The ethics of cloning-for-biomedical-research -- Public policy options -- Policy recommendations -- Bibliography -- Glossary of terms -- Appendix: personal statements. Notes: "July 2002." Includes bibliographical references (p. 235-242). Subjects: Human cloning--Moral and ethical aspects--United States. Human cloning--Government policy--United States. LC Classification: QH442.2 .H845 2002

Human cloning from the perspective of The Council of Europe bioethical standards Abstract: Allegations negating the role of law in the resolving the controversial problem of human cloning are unjustified. Human rights, implying the magnitude and value of human person and deeply rooted inherent dignity of the human being, constitute the very foundation of every legal order. Every lawyer, as well as every specialist in medicine or biology must be aware of his own human dignity and the ethical consequences. Among the standards of the Council of Europe in the context of human cloning there must be mentioned the Additional Protocol of January 12, 1998 on the Prohibition of Cloning Human Beings. It is the integral element of the normative system of the mother convention, the European Bioethical Convention of April 4, 1997. It has the distinguished place - within this system as far as its substantial provisions exclude the possibilities of limitations and derogation. The system of the Convention and the Protocol must be viewed in the light of a broader normative environment, including the integral system of the European Convention on Human Rights and the set of recommended bioethical standards. The absolute prohibition embodied in the Protocol is limited to all the methods leading to the creation of genetically identical human beings. The protocol does not directly regulate cloning of human tissues and cells, including embryonic stem cells. However, some conclusions may be taken from the Convention and from the recommended standards. It is the assumption of the Convention and the Protocol that the guarantees embodied there must be apprehended as practical and effective ones, justifiable and not excluding the use of proper sanctions by the State. It is an unacceptable view that scientists are excluded from the sphere of the functioning of the above-mentioned prohibition. The real sense of this prohibition is to stop the unlimited liberty and arbitrary practice of such scientists. Author(s): Jasudowicz, T Language: pol Source Med Wieku Rozwoj 2001;5(1 Suppl 1):213-25.

Human cloning in the activities of the European Union Abstract: The European Union has been concerned with human cloning since the late 80. It resulted from inclusion of biotechnology into the sphere of European integration. The attitude of the European Union in the domain of human cloning was shaped, in principle in the second part of the 90. As the Community law stands at present, the European Union is not able to regulate all aspects of the cloning of human beings. It has no general power to decide in that sphere, especially, as far as bioethic aspects are concerned. The cloning of human beings in the European Union is understood as a process aiming at producing new human being, genetically identical with another live or dead human being. Thus the notion of human cloning is reduced to reproductive cloning. Three instruments are at the disposal of the European Union in the domain of human cloning. The first is prohibition of reproductive cloning as a general principle of Community law. However, that principle is not the result of judicial activity of the European Court of Justice (as general principles normally are), but the logical consequence of views formally expressed by the European Parliament, the Council of the Europe as well as the Commission. The principle was finally included in the Charter of fundamental rights of the European Union. The second instrument is an imperative prohibition of patent granting to biotechnological inventions on human reproductive cloning. Last, but not least, the Union applies a prohibition of financing scientific research connected with human cloning from the budget of the European Communities within the V Framework Programme in the field of research and technological development. Author(s): Mik, C Language: pol Source Med Wieku Rozwoj 2001;5(1 Suppl 1):195-212.

Human cloning or cannibalism Abstract: In this article I develop the idea presented in my previous work that human cloning would be of little practical use since almost any aim that one would like to attain by multiple cloning of a concrete man or a group of people, are unattainable or it might be achieved by easier, cheaper and more efficient traditional methods. For this reason cloning of a man is unlikely to occur on a larger scale and only few people will decide to clone themselves. In this sense no social effects of human cloning will be disastrous for the human population. Yet investigations in human genetics are very important since they may provide medical applications far more important than human cloning. It is argued that the main trend of modern medicine: organ transplantation from an alien donor, will become socially dangerous in near future since the number of donors will be drastically smaller than the number of potential patients waiting for transplantations. This in turn may cause social conflicts and a form of medical cannibalism may arise. These problems and conflicts will be avoided if organ transplantation from an alien donor is replaced by organ cloning, i.e. by transplanting an organ developed from the patient. Author(s): Sokolowski, L M Language: pol Source Med Wieku Rozwoj 2001;5(1 Suppl 1):91-101.

Human cloning, as ideological construction of technology Abstract: The article treats the issue of cloning as a part of the dominant ideology of the so-called technical progress. The author shows that the notion of "technical progress" is an ideological construction full of hidden valuation. The whole class of technicians is simply materially interested in disseminating this ideology. According to the author the so-called technical progress doesn't improve the quality of life as far as psychological satisfaction is concerned. The cult of technology is rooted in the crisis of humanistic values and civilization. It implies avoiding serious existential questions and civilization challenges. Technology itself appears in this statement only as a result of concrete conditions in concrete historical moment, but not as an independent process. Cloning can be only the contemporary answer to the collapse of the art of breeding and educating consecutive generations. It can be only the next step of the fall of western civilization or the so-called revolt of the masses. Author(s): Swolkien, O Language: pol Source Med Wieku Rozwoj 2001;5(1 Suppl 1):135-40.

Human cloning: hearing before the Subcommittee on Crime of the Committee on the Judiciary, House of Representatives, One Hundred Seventh Congress, first session on H.R. 1644 and H.R. 2172, June 7 and June 19, 2001. Published/Created: Washington: U.S. G.P.O.: For sale by the Supt. of Docs., U.S. G.P.O., [Congressional Sales Office], 2001. Description: iv, 95 p.; 24 cm. ISBN: 0160666376 Notes: Distributed to some depository libraries in microfiche. Shipping list no.: 2002-0069-P. "Serial no. 40." Subjects: Human cloning--Human reproductive technology--Stem cells--Research--Moral and ethical aspects--United States. LC Classification: KF27 .J858 2001 Govt. Doc. No.: Y 4.J 89/1:107/40

Human cloning: judicial and legislative framework Abstract: The application, either theoretical or fantastical, of the somatic cloning which resulted in the birth of Dolly the sheep to human reproduction has led to international uproar. A paragraph specifically banning reproductive cloning (defined as "any intervention seeking to produce genetically identical human individuals in the sense of individuals sharing the same nuclear gene set") was added to the Council of Europe Convention on Human Rights and Biomedicine in January 1998. In several responses both at national and international level, reproductive cloning is banned either directly or indirectly. The challenge for the future is the consideration of integrating therapeutic cloning in legislations both nationally and in international declarations. Author(s): Shenfield, F Language: fre Source Gynecol Obstet Fertil 2000 Sep;28(9):643-8.

Human cloning: must we sacrifice medical research in the name of a total ban?: hearing before the Committee on the Judiciary, United States Senate, One Hundred Seventh Congress, second session, February 5, 2002. Published: Washington: U.S. G.P.O.: For sale by the Supt. of Docs., U.S. G.P.O. [Congressional Sales Office], 2002. Description: iii, 97 p.: ill.; 24 cm. ISBN: 0160695201 Notes: "Printed for the use of the Committee on the Judiciary." Distributed to some depository libraries in microfiche. Shipping list no.: 2003-0107-P. "Serial no. J-107-55." Subjects: Human cloning--Research--United States--Moral and ethical aspects. Series: United States. Congress. Senate. S. hrg.; 107-812. LC Classification: KF26 .J8 2002a Govt. Doc. No.: Y 4.J 89/2:S.HRG.107-812

Human cloning: religious responses / Ronald Cole-Turner, editor. Edition Information: 1st ed. Published/Created: Louisville, Ky.: Westminster John Knox Press, c1997. Related Names: Cole-Turner, Ronald, 1948- Description: xiii, 151 p.; 22 cm. ISBN: 0664257712 Notes: Includes bibliographical references. Subjects: Human cloning--Religious aspects--Christianity--Moral and ethical aspects. LC Classification: QH442.2 .H86 1997 Dewey Class No.: 241/.64957 21

Human cloning: science, ethics, and public policy / edited by Barbara MacKinnon. Published/Created: Urbana: University of Illinois Press, c2000. Related Names: MacKinnon, Barbara, 1937- Description: 171 p.: ill.; 22 cm. ISBN: 0252024915 (cloth: acid-free paper) Notes: Includes bibliographical references (p. [161]-163) and index. Subjects: Human cloning--Moral and ethical aspects. LC Classification: QH442.2 .H875 2000 Dewey Class No.: 174/.25 21

Human dehydroepiandrosterone sulfotransferase: molecular cloning of cDNA and genomic DNA. Abstract: Human tissues contain at least three well-characterized cytoplasmic sulfotransferase (ST) enzymes, dehydroepiandrosterone (DHEA) ST and two of phenol ST (PST). DHEA ST catalyzes the sulfation of DHEA and other steroids. We cloned and expressed two cDNAs for human liver DHEA ST. The cloning strategy involved the design of PCR primers directed against two conserved domains in ST proteins. These primers were used to generate a specific PCR product that was then used successfully to clone cDNAs for DHEA ST from a human liver cDNA library. Two cDNAs were isolated that were approximately 1.1 and 1.8 kb in length. These two clones had identical open reading frames. Both cDNAs produced enzymatically active DHEA ST protein in a mammalian expression system. Northern blot analysis confirmed the presence of 1.1 and 1.8 kb transcripts in human liver. cDNAs for a number of eukaryotic enzymes have now been cloned, and they share significant sequence homology. These ST cDNAs appear to fall into distinct groups on the basis of amino acid sequences of the proteins that they encode, thus demonstrating that the enzymes comprise a gene superfamily. We have also isolated, a genomic clone for human DHEA ST that contains approximately 3 kb of 5'-flanking sequence, exon 1 and 1.7 kb of intron 1. Characterization of the structure and regulatory elements of this gene should help to elucidate mechanisms involved in the regulation of DHEA ST in humans. Author(s): Otterness, D M; Weinshilboum, R Language: eng Source Chem Biol Interact 1994 Jun;92(1-3):145-59.

Human somatic cell nuclear transfer (cloning). The Ethics Committee of the American Society for Reproductive Medicine. Language: eng Subject(s): Cell Nucleus/transplantation; Child; Cloning, Organism/methods; Ethics, Medical; Family; Human; Social Change Source Fertil Steril 2000 Nov;74(5):873-6.

Human tumor cloning assays: applications in clinical oncology and new antineoplastic agent development. Abstract: The human tumor cloning assay (HTCA) has been available for preclinical and clinical applications for the last 11 years. This article examines the usefulness of that assay both in the practice of clinical oncology and in the development of new antineoplastic agents. In the area of prediction of response of an individual patient's tumor to a particular antineoplastic agent, in a total of 2274 correlations in a variety of clinical trials, the assay has shown a remarkably good ability to predict whether a patient's tumor would respond to a particular agent (percent true positives 69%; percent true negatives 91%). However, despite this ability to predict these responses, the assay has not yet been accepted for general clinical use, because there have not been rigorous trials to prove the assay will improve patient survival. These rigorous clinical trials are now underway. In the area of drug development, the HTCA has been used to screen for new antineoplastic agents as well as to pinpoint tumor types against which the new agent will be active in phase II clinical trials. As will be seen in this review, the HTCA has been most successful in this area. Author(s): Von Hoff, D D Language: eng Source Cancer Metastasis Rev 1988 Dec;7(4):357-71.

Human tumor cloning: drug sensitivity testing clinical correlations. Author(s): Alberts, D S; Salmon, S E; Moon, T E Language: eng Subject(s): Animal; Antineoplastic

Agents/pharmacology; Cell Survival/drug effects; Cells, Cultured; Clinical Trials; Clone Cells; Dose-Response Relationship, Drug; Drug Evaluation, Preclinical; Drug Therapy, Combination; Human; Leukemia, Experimental drug therapy; Neoplasms drug therapy pathology; Prospective Studies Source Prog Clin Cancer 1982;8:147-79.

Human tumor cloning: proceedings of the Fourth Conference on Human Tumor Cloning: Tucson, Arizona, January 8-10, 1984 / edited by Sydney E. Salmon, Jeffrey M. Trent. Published/Created: Orlando: Grune & Stratton, c1984. Related Names: Salmon, S. E. (Sydney E.), 1936- Trent, Jeffrey M. Description: xviii, 700 p.: ill.; 24 cm. ISBN: 0808916718 Notes: Includes bibliographical references and index. Subjects: Cancer cells--Cloning--Congresses. LC Classification: RC267 .C596 1984 Dewey Class No.: 616.99/4027 19

Humanity and cloning, a therapeutic tool or a dream of immortality?; Thibault, C Language: fre Subject(s): Animal; Cloning, Organism; Ethics, Medical; Human Source Contracept Fertil Sex 1999 Jan;27(1):12-6.

Hunter syndrome: cloning of the gene, mutations and carrier detection. Author(s): Adinolfi, M Language: eng Subject(s): DNA, Recombinant genetics; Female; Gene Therapy trends; Heterozygote Detection; Human; In Situ Hybridization; Lysosomal Storage Diseases; Male; Mucopolysaccharidosis II. Source Dev Med Child Neurol 1993 Jan;35(1):79-85.

Hyde, Margaret O. (Margaret Oldroyd), 1917- Cloning and the new genetics / Margaret O. Hyde & Lawrence E. Hyde. Published/Created: Hillside, N.J., U.S.A.: Enslow Publishers, c1984. Related Names: Hyde, Lawrence E. Description: 128 p.: ill.; 25 cm. ISBN: 0894900846 : Notes: Includes index. Bibliography: p. 121-122. Subjects: Cloning. Genetics. LC Classification: QH442.2 .H93 1984 Dewey Class No.: 575.1 19

Identification and cloning of differentially expressed genes. Abstract: Only a few of the methods currently used for identification of differentially expressed genes take advantage of the fact that (near) complete sets of cDNA clones and sequences representing all human and mouse genes will be available for high throughput survey of gene expression. Accordingly, strategies based on hybridization of complex (cDNA or RNA) probes to cDNA microarrays, either on glass slides or on chips, are likely to become increasingly more advantageous. Recognizing, however, that the power of these methods depends upon the availability of such resources, strategies are being pursued to facilitate completion of the ongoing efforts to identify all human and mouse genes. Author(s): Soares, M B Language: eng Subject(s): Animal; Cloning, Organism/methods; DNA, Complementary; Gene Expression; Gene Library; Human; Mice; Polymerase Chain Reaction/methods; RNA Probes Source Curr Opin Biotechnol 1997 Oct;8(5):542-6.

Identification and cloning of RA-regulated genes by mRNA-differential display. Author(s):White, J A; Petkovich, M Language: eng Subject(s): Animal; Cloning, Molecular; Gene Expression Regulation; Proteins/genetics; RNA, Messenger; Tretinoin/metabolism Source Methods Mol Biol 1998;89:389-404.

Identification and expression of G-protein coupled receptors / edited by Kevin R. Lynch. Published/Created: New York: Wiley-Liss, c1998. Related Names: Lynch, Kevin R. Description: xii, 222 p.: ill.; 26 cm. ISBN: 0471183105 (alk. paper) 047119493X (pbk.: alk. paper) Notes: "A John Wiley & Sons, Inc., publication." Includes bibliographical references and index. Subjects: G proteins--Recpetors--Research--Methodology. Molecular cloning. Series: Receptor biochemistry and methodology (Unnumbered) LC Classification: QP552.G16 I335 1998 Dewey Class No.: 572/.6 21

Identification of causative genes for neurological diseases by positional cloning approach; Tsuji, S Language: jpn Subject(s): Cloning, Molecular/methods; Dentate Gyrus/pathology; Genome, Human; Globus Pallidus/pathology; Human; Neurodegenerative Diseases/genetics/pathology; Trinucleotide Repeats Source Tanpakushitsu Kakusan Koso 1997 Dec;42(17 Suppl):2773-8.

Identification, cloning, and purification of protein antigens of Treponema pallidum. Abstract: Difficulties in culturing the bacterium Treponema pallidum have greatly hindered syphilis research. In recent years, several laboratories have begun applying recombinant DNA technology to the study of this organism. Recent work is summarized concerning the expression of T. pallidum DNA in Escherichia coli. A number of E. coli clones expressing treponemal protein antigens have been identified. In one instance, a recombinant protein was purified to homogeneity and shown to be identical to a highly immunogenic, native T. pallidum membrane protein of molecular weight 39,000, which was designated the basic membrane protein (BMP) of this organism. In addition, recent experiments are described that were designed to identify cell-surface proteins that would serve as the primary focus of our cloning efforts. Results obtained with use of several different approaches strongly suggest that the outer membrane of T. pallidum is an antigenically inert structure largely devoid of protein. However, a class of low-molecular-weight protein antigens have been identified that are actively secreted into the extracellular medium. Attempts currently are being made to clone these secreted proteins and investigate their roles in the pathogenesis and immunobiology of syphilis. Author(s): Stamm, L V; Dallas, W S; Ray, P H; Bassford, P J Jr Language: eng Source Rev Infect Dis 1988 Jul-Aug;10 Suppl 2:S403-7.

Identifying and cloning Drosophila genes by single P element insertional mutagenesis. Author(s): Cooley, L; Berg, C; Kelley, R; McKearin, D; Spradling, A Language: eng Subject(s): Animal; Chromosome Mapping; Cloning, Molecular; Crosses, Genetic; DNA Transposable Elements; Drosophila/genetics; Female; Genes; Male; Mutation Source Prog Nucleic Acid Res Mol Biol 1989;36:99-109.

Identifying human disease genes by positional cloning. Author(s): Collins, F S Language: eng Subject(s): Amino Acid Sequence; Base Sequence; Chromosome Mapping; Cloning, Molecular; Cystic Fibrosis; DNA; Genes, Neurofibromatosis 1; Genetic Diseases, Inborn; Human; Molecular Sequence Data; Neurofibromatosis 1. Source Harvey Lect 1990-91;86:149-64.

Ignorance is not bliss: why a ban on human cloning is unacceptable. Author(s):Smith, S H Language: eng Subject(s): Cloning, Organism/legislation & jurisprudence; Ethics, Medical; Government Regulation; Human; Organ Procurement; Public Opinion; Public Policy; Reproductive Techniques; United States Source Health Matrix Clevel 1999 Summer;9(2):311-34.

Impact of cloning in Bacillus subtilis on fundamental and industrial microbiology. The Eighth Griffith Memorial Lecture. Author(s): Young, F E Language: eng Subject(s): Bacillus subtilis; Cloning, Molecular; Containment of Biohazards; DNA, Bacterial; DNA, Recombinant; Escherichia coli; Genes, Structural; Industry; Plasmids; Transformation, Bacterial Source J Gen Microbiol 1980 Jul;119(1):1-15.

Impact of transgenes and cloning on xenografts Abstract: Transgenesis can theoretically add a foreign gene or specifically replace an endogenous gene by another gene. Gene addition in mammals is generally achieved by DNA microinjection into one-cell embryos. Gene replacement implies homologous recombination in cultured cells which must be selected and remain capable of generating a living organism. The use of totipotent cells can, though currently in the mouse only, lead to the generation of chimeric animals transmitting the genetic modification to offsprings. The embryo cloning technique has recently allowed the use of somatic fetal cells, in which gene replacement occurred to generate living sheeps. This technique is being extended to other domestic ruminants, to pigs and rabbits. The mouse, rat and rabbit are being used as models to define the genes which should be added or inactivated to reduce rejections of xenografts. Transgenic pigs harbouring the human CD59 or the DAF genes have been obtained by several groups. Heart, kidney and isolated cells from these animals are more resistant to hyperacute rejection when grafted to experimental primates. Additional genes are to be added in the future to inhibit the other rejection mechanisms as soon as their action has been demonstrated in laboratory animals. Experiments in progress aim at inactivating. the gene for Gal: 1-3-galactosyl transferase by homologous recombination in the pig genome. Transgenesis might also be used to prevent expression of endogenous pig retroviral vectors and to prepare recombinant proteins having antirejection activity from the milk of animals. Author(s): Houdebine, L M Language: fre Source Pathol Biol (Paris) 2000 May;48(4):383-6.

Improved suppressor tRNA cloning vectors and plasmid-phage recombination. Author(s): Huang, H V; Little, P F; Seed, B Language: eng Subject(s): Animal; Bacteriophages genetics; Cell Line; Cloning, Molecular; Genetic Vectors; Gram-Negative Bacteria; Plasmids; RNA, Transfer; Recombination, Genetic; Suppression, Genetic Source Biotechnology 1988;10:269-83.

Inducible nitric oxide synthase: from cloning to therapeutic applications. Abstract: Elucidation of the many roles of nitric oxide (NO) in homeostasis and disease states has uncovered many areas where manipulation of NO production would be of therapeutic benefit. Recent advances in gene transfer technology and the cloning of the inducible nitric oxide synthase (iNOS) gene have led to the development of strategies for gene therapy to increase NO production for the treatment of disorders ranging from vascular restenosis to impaired wound healing. This review summarizes the current status of iNOS gene therapy research. Author(s): Schwentker, Ann; Billiar, Timothy R Language: Source World J Surg 2002 Jul;26(7):772-8.

Initiation of autoimmune type 1 diabetes and molecular cloning of a gene encoding for islet cell-specific 37kd autoantigen. Abstract: Insulin-dependent diabetes mellitus (IDDM) is believed to be an autoimmune disease, characterized by lymphocytic infiltration of the islets and the presence of islet cell autoantibodies. Autoimmunity may result from an intrinsically abnormal immune system, primary alterations of the target beta cell or both. However, the initial event that causes the beta cell-specific autoimmunity remains unknown. Our recent experimental results showed that islet grafts from neonatal BB rats remained intact without insulitis when transplanted into the renal subcapsular space of acutely diabetic BB rats. In contrast, islet grafts from adult BB rats (which had been treated with silica for the prevention of insulitis) revealed severe insulitis and were rapidly destroyed. These results suggest that the delayed expression of a beta cell-specific autoantigen may result in the initiation of beta cell-specific autoimmunity. The islet cell-specific 37 kd autoantigen is not expressed early in the life of BB rats, but is expressed at around 30 days of age. This islet-specific autoantigen might be recognized and attacked by the immune effectors. In contrast, nondiabetic Wistar Furth rats express the autoantigen from birth. Author(s): Jun, H S; Yoon, J W Language: eng Source Adv Exp Med Biol 1994;347:207-20.

Interleukin 4--special reference to the cloning of genes; Naito, T; Hama, K; Honjo, T Language: jpn Source Nippon Rinsho 1986 Aug;44(8):1892-901.

Interleukin-1 receptors: cloning studies and role in central nervous system disorders. Author(s):Loddick, S A; Liu, C; Takao, T; Hashimoto, K; De Souza, E B Language: eng Subject(s): Animal; Brain/immunology/metabolism; Caspase 1; Central Nervous System Diseases/immunology/physiopathology; Cloning, Molecular; Cysteine Endopeptidases/metabolism; Human; Nerve Degeneration; Receptors, Interleukin-1/biosynthesis/physiology; Recombinant Proteins/biosynthesis; Signal Transduction Source Brain Res Brain Res Rev 1998 May;26(2-3):306-19.

International journal of cell cloning. Published/Created: Dayton, OH: AlphaMed Press, c1983-c1992. Description: 10 v.: ill., ports.; 26 cm. Vol. 1, no. 1 (Apr. 1983)-v. 10, no. 6 (Nov. 1992). ISSN: 0737-1454 Notes: Title from cover. SERBIB/SERLOC merged record Subjects: Cloning--Clone cells--Cell Differentiation--Cell Division--Hematopoietic Stem Cells--Periodicals. LC Classification: QH442.2 .I57 Dewey Class No.: 574.87/6/05 19

Introduction of cloning plasmids into Agrobacterium tumefaciens. Author(s): Shaw, C H Language: eng Subject(s): Agrobacterium tumefaciens/genetics; Cloning, Molecular/methods; Escherichia coli/genetics; Gene Transfer Techniques; Plasmids/genetics Source Methods Mol Biol 1995;49:33-7.

Introduction to positional cloning. Abstract: The most common procedure for isolating genes that predispose to a certain disease is positional cloning. This process is initiated by mapping the responsible gene to its location on a chromosome linked to the disease. Successive cloning of the region results in the isolation of candidate genes. Finally, identification of mutations by mutational analysis points to the correct gene. Author(s): Papadopoulos, N Language: eng Score Source Clin Exp Allergy 1995 Nov;25 Suppl 2:116-8.

Inward rectifier potassium channels. Cloning, expression and structure-function studies. Abstract: A PCR-based cloning strategy was used to identify novel subunits of the two-transmembrane domain inward rectifier potassium channel family from rat brain, heart, and skeletal muscle. When expressed in Xenopus oocytes, two of these clones (Kir4.1 and Kir2.3) gave rise to inwardly rectifying potassium currents. Two-electrode voltage clamp commands to potentials negative to EK evoked inward potassium-selective currents which rapidly reached a peak amplitude and then relaxed to a steady-state level. Differences in the extent of current relaxation, the degree of rectification, and the voltage-dependent block by external cesium were detected. Two other members of this family (Kir5.1 and Kir3.4) did not produce macroscopic currents, when expressed by themselves, yet both subunits modified the currents when coexpressed with other specific members of the Kir family. Expression of chimeric subunits between Kir4.1 and either Kir5.1 or Kir3.4 suggested that the transmembrane domains determine the specificity of subunit hetero-polymerization, while the C-terminal domains contribute to alterations in activation kinetics and rectification. Expression of covalently linked subunits demonstrated that the relative subunit positions, as well as stoichiometry, affect heteromeric channel activity. Author(s): Lagrutta, A A; Bond, C T; Xia, X M; Pessia, M; Tucker, S; Adelman, J P Language: eng Source Jpn Heart J 1996 Sep;37(5):651-60.

Is cloning the absolute evil?; Hottois, G Language: eng Subject(s): Bioethics; Cloning, Organism; Genetic Determinism; Genetics, Behavioral; Personal Autonomy; Risk Assessment; Social Control, Formal Source Hum Reprod Update 1998 Nov-Dec;4(6):787-90.

Isolation of genes involved in iron acquisition by cloning and complementation of Escherichia coli mutants. Author(s): West, S E Language: eng Source Methods Enzymol 1994;235:363-72.

Isolation of pharmacologically distinct GABA-benzodiazepine receptors by protein chemistry and molecular cloning. Author(s): Olsen, R W; Bureau, M; Khrestchatisky, M; MacLennan, A J; Chiang, M Y; Tobin, A J; Xu, W; Jackson, M; Sternini, C; Brecha, N Language: eng Subject(s): Affinity Labels; Animal; Cloning, Molecular; Receptors, GABA-A/chemistry/drug effects/isolation & purification Source Adv Biochem Psychopharmacol 1990;46:35-49.

Isolation of yeast mutants defective in phospholipid metabolism and cloning of genes for phospholipid synthesis; Nikawa, J; Yamashita, S Language: jpn Subject(s): Cloning, Molecular; Genes, Fungal; Mutation; Phospholipids biosynthesis; Saccharomyces cerevisiae genetics Source Seikagaku 1984 Sep;56(9):1125-32.

Isolation, localization, and cloning of a kainic acid binding protein from frog brain. Abstract: Excitatory amino acids (EAA) are major neurotransmitters in the vertebrate central nervous system. EAA receptors have been divided into three major subtypes on the basis of electrophysiological and ligand binding studies: N-methyl-D-aspartate, kainate, and quisqualate receptors. To understand their molecular properties, we undertook a project aimed at isolation and cloning of these receptor subtypes. We purified a kainate binding protein (KBP) from frog brain, in which kainate binding sites are about fortyfold more abundant than in rat brain, using domoic acid affinity chromatography, and made monoclonal and polyclonal antibodies to the purified protein. These antibodies immunoprecipitate the frog KBP but not KBPs from other species. Immunocytochemical analyses show that KBP has a synaptic and extrasynaptic localization in frog optic tectum, with most labeling being extrasynaptic. The cDNA encoding frog brain KBP was isolated by screening a frog brain cDNA library with oligonucleotide probes that were based on the amino acid sequence of the purified protein. The deduced amino acid sequence of the KBP has a hydrophobic profile similar to those of other ligand-gated ion channel subunits, such as the nicotinic acetylcholine receptor, the GABAA receptor, and the glycine receptor. Frog brain KBP is very similar (36% amino acid identity to the carboxyl half) to rat brain kainate receptor, suggesting that these two proteins evolved from a common ancestor. The function of KBP in frog brain remains a major question. Preliminary results showed that Xenopus laevis oocytes injected with KBP RNA did not produce a detectable electrophysiological response when perfused with kainate. These results suggest that additional subunits may be required to form a functional receptor or that KBP is not functionally related to a neurotransmitter receptor. Author(s): Wenthold, R J; Hampson, D R; Wada, K; Hunter, C; Oberdorfer, M D; Dechesne, C J Language: Source J Histochem Cytochem 1990 Dec;38(12):1717-23.

Issues raised by human cloning research: hearing before the Subcommittee on Oversight and Investigations of the Committee on Energy and Commerce, House of Representatives, One Hundred Seventh Congress, first session, March 28, 2001. Published/Created: Washington: U.S. G.P.O.: For sale by the Supt. of Docs., U.S. G.P.O. [Congressional Sales Office], 2001. Description: iii, 153 p.; 24 cm. ISBN: 0160657792 Notes: Distributed to some depository libraries in microfiche. Shipping list no.: 2001-0246-P. "Serial no. 107-5." Includes bibliographical references. Subjects: Human cloning--Research--United States. Human cloning--Moral and ethical aspects. Human reproductive technology--Moral and ethical aspects. LC Classification: KF27 .E5546 2001h Govt. Doc. No.: Y 4.C 73/8:107-5

Jefferis, David. Cloning: frontiers of genetic engineering / David Jefferis. Published/Created: New York: Crabtree, c1999. Description: 32 p.: ill. (chiefly col.); 29 cm. ISBN: 0778700488 (rlb) 0778700585 (paper) Summary: Discusses the science of genetics, the first successful cloning of a mammal, its implications, and its ethical aspects. Notes: Includes index. Subjects: Cloning--Juvenile literature. Genetic engineering--Juvenile literature. Medical genetics--Juvenile literature. Cloning. Genetic engineering. Series: Megatech LC Classification: QH442.2 .J44 1999 Dewey Class No.: 571.8/919 21

Kass, Leon. The ethics of human cloning / Leon R. Kass and James Q. Wilson. Published/Created: Washington, D.C.: AEI Press, 1998. Related Names: Wilson, James Q. Description: xxi, 101 p.; 19 cm. ISBN: 0844740500 (alk. paper) Notes: Includes bibliographical references. Subjects: Human cloning--Moral and ethical aspects. Human reproductive technology--Moral and ethical aspects. LC Classification: QH442.2 .K37 1998 Dewey Class No.: 174/.25 21

Kaye, Marilyn. And the two shall meet / Marilyn Kaye. Published/Created: New York: Bantam Books, c1999. Description: 182 p.; 20 cm. ISBN: 0553486861 (pbk.) Subjects: Cloning--Juvenile fiction. Identity--Juvenile fiction. Cloning--Fiction. Identity--Fiction. Series: Kaye, Marilyn. Replica; #6.

Kaye, Marilyn. Ice cold / Marilyn Kaye. Published/Created: New York: Bantam Books, c2000. Description: 152 p.; 20 cm. ISBN: 0553487116 (pbk.) Notes: "RL: 5.5"--T.p. verso. Subjects: Cloning--Juvenile fiction. Identity--Juvenile fiction. Cloning--Fiction. Identity--Fiction. Series: Kaye, Marilyn. Replica; #10. LC Classification: PZ7.K2127 Ic 2000 Dewey Class No.: [Fic] 21

Kaye, Marilyn. Like father, like son / Marilyn Kaye. Published/Created: New York: Bantam Books, 2001. Related Names: Copyright Paperback Collection (Library of Congress) Description: 133 p.; 20 cm. ISBN: 0553487485 (pbk.) Notes: RL 5.5. Subjects: Cloning--Juvenile fiction. Identity--Juvenile fiction. Series: Kaye, Marilyn. Replica; #20. LC Classification: CPB Box no. 1865 vol. 24

Kaye, Marilyn. Lucky thirteen / Marilyn Kaye. Published/Created: New York: Bantam Books, 2000. Related Names: Copyright Paperback Collection (Library of Congress) Description: 168 p.; 20 cm. ISBN: 0553487124 (pbk.) Subjects: Cloning--Fiction. Series: Kaye, Marilyn. Replica; #11. LC Classification: CPB Box no. 1794 vol. 21

Kaye, Marilyn. The beginning / Marilyn Kaye. Published/Created: New York: Bantam Books, c2000. Related Names: Copyright Paperback Collection (Library of Congress) Description: 149 p.; 19 cm. ISBN: 0553487159 : Notes: "A Bantam Skylark book"--T.p. verso. Subjects: Cloning--Juvenile fiction. Identity--Juvenile fiction. Series: Kaye, Marilyn. Replica; #14. LC Classification: CPB Box no. 1998 vol. 20

Kaye, Marilyn. Virtual Amy / Marilyn Kaye. Published/Created: New York: Bantam Books, 2001. Related Names: Copyright Paperback Collection (Library of Congress) Description: 136 p.; 18 cm. ISBN: 0553487493 (pbk.) : Subjects: Human cloning--Fiction. Candler, Amy (Fictitious character)--Fiction. Series: Kaye, Marilyn. Replica; #21. LC Classification: CPB Box no. 1963 vol. 9

Kerner, Charlotte. Blueprint / Charlotte Kerner; translated from the German by Elizabeth D. Crawford. Published/Created: Minneapolis: Lerner Publications Company, 2000. Related Names: Crawford, Elizabeth D. Description: 189 p.; 22 cm. ISBN: 0822500809 (lib. bdg.: alk. paper) Summary: Siri Sellin, one of the first human clones, writes a bitter memoir of her childhood as the daughter of a famous and self-absorbed composer. Subjects: Cloning--Fiction. Mothers and daughters--Fiction. Musicians--Fiction. Science fiction. LC Classification: PZ7.K45787 Bl 2000 Dewey Class No.: [Fic] 21

Kidd, J. S. (Jerry S.) Life lines: the story of the new genetics / J.S. Kidd and Renee A. Kidd. Published/Created: New York: Facts on File, c1999. Related Names: Kidd, Renee A. Description: vii, 152 p.: ill.; 24 cm. ISBN: 0816035865 (acid-free paper) Summary: Surveys the field of genetics, discussing genetic analysis, cloning, other new research and developments, and their ethical aspects. Notes: Includes bibliographical references (p. 146-147) and index. Subjects: Genetics--Juvenile literature. Genetics--Social aspects--Juvenile literature. Human genetics--Juvenile literature. Human genetics--Social aspects--Juvenile literature. Genetics. Series: Science and society (Facts on File, Inc.) LC Classification: QH437.5 .K53 1999 Dewey Class No.: 576.5 21

Kolata, Gina Bari, 1948- Clone: the road to Dolly, and the path ahead / Gina Kolata. Edition Information: 1st ed. Published/Created: New York: W. Morrow & Co., c1998. Description: 276 p.; 25 cm. ISBN: 0688156924 (acid-free paper) Notes: Includes bibliographical references (p. [249]-259) and index. Subjects: Cloning--Moral and ethical aspects. Cloning--History. Dolly (Sheep) LC Classification: QH442.2 .K65 1998 Dewey Class No.: 174/.957 21

Kunich, John C., 1953- The naked clone: how bans on cloning threaten our personal constitutional rights / John Charles Kunich. Published/Created: Westport, Conn.: Praeger, 2003. Description: p. cm. ISBN: 0275979644 (alk. paper) Notes: Includes bibliographical references and index. Subjects: Human cloning--Law and legislation--United States. Civil rights--United States. LC Classification: KF3831 .K86 2003 Dewey Class No.: 342.73/085 21

Large-scale applications of cloning technologies for agriculture: an industry perspective. Abstract: The present costs and efficiencies of producing cloned embryos, pregnancies and offspring using the simplified nuclear transfer techniques developed in the authors' laboratories are compared with those required for the large-scale application of such cloning technologies in cattle. The current costs in the laboratory of producing large numbers of genetically identical cloned embryos for transfer is around $15.00 per blastocyst, which is within the cost estimated to be commercially viable for cloned female dairy embryos for transfer. However, the pregnancy and calving rates from the transfer of such embryos are still well below that required for large-scale commercial application for which ongoing pregnancy rates of at least 50% per recipient will be required. The current pregnancy rate (30-40 days post-transfer) following the transfer of an average of three cloned embryos per recipient is 37%, and the calving rate 17%, representing high losses between pregnancy diagnosis and term. In the beef industry and in some dairy situations the final product (cloned bulls for natural mating) will have a much higher inherent value and different parameters will therefore apply. Recent developments in the technologies that are likely to increase the probabilities of large-scale application are discussed, including recycling nuclear transfer embryos, somatic cell cloning, new cryopreservation techniques and automated oocyte harvesting. Author(s): Lewis, I M; Peura, T T; Trounson, A O Language: eng Source: Reprod Fertil Dev 1998;10(7-8):677-81.

Laser chromosome microdissection and cloning of the genetic disease loci Abstract: We have developed an argon ion laser chromosome microdissection technique in conjunction with a single unique primer polymerase chain reaction (SUP-PCR) to directly amplify microdissected chromosomes. The 22-mer primer used in PCR, although unique in sequence, randomly primed and amplified any target DNA. These methods were applied to both the terminal region of the human chromosome 4p (4p 16) and Xq (Xq26-q28), and two chromosome region-specific DNA libraries were constructed. The resulting libraries contained approximately 1000 nonoverlapping DNA sequences with an average size of 230-350 bp, at an average spacing of 10-65 Kbp along the chromosomes of origin. Our new method is a simple and general approach for constructing a chromosome region-specific DNA library from a single metaphase spread. Author(s): Hadano, S; Ikeda, J E Language: jpn Source Nippon Rinsho 1993 Sep;51(9):2240-5.

Lasky, Kathryn. Star split / Kathryn Lasky. Edition Information: 1st ed. Published: New York: Hyperion Books for Children, 1999. Description: 203 p. 22 cm. ISBN: 0786804599 (trade: alk. paper) 0786824018 (lib.: alk. paper) Summary: In 3038, thirteen-year-old Darci uncovers an underground movement to save the human race from genetic enhancement technology. LC Classification: PZ7.L3274 St 1999 Dewey Class No.: [Fic] 21

Ledford, Jan Roadarmel. The cloning / Jan Roadarmel Ledford. Published/Created: Simpsonville, S.C.: Millennium III Publishers, c2002. Description: p. cm. ISBN: 0962522066 (alk. paper) Subjects: Women journalists--Cloning--Fiction. LC Classification: PS3612.E342 C57 2002 Dewey Class No.: 813/.6 21

Lester, Lane P. Cloning, miracle or menace? / Lane P. Lester, with James C. Hefley. Published/Created: Wheaton, Ill.: Tyndale House, c1980. Related Names: Hefley, James C. Description: 156 p.; 21 cm. ISBN: 0842302948 (pbk.) : Notes: Includes bibliographical references. Subjects: Cloning--Moral and ethical aspects--Social aspects. LC Classification: QH442.2 .L47 Dewey Class No.: 174/.2 19

Lester, Lane P. Human cloning: playing God or scientific progress? / Lane P. Lester with James C. Hefley. Published/Created: Grand Rapids, Mich.: Fleming H. Revell, c1998. Related Names: Hefley, James C. Lester, Lane P. Cloning. Description: 175 p.; 22 cm. ISBN: 0800756681 (pbk.) Notes: "Some chapters are taken largely from Cloning: Miracle or Menace? published in 1980 by Tyndale House"--T.p. verso. Includes bibliographical references (p. 171-175). Subjects: Human cloning--Moral and ethical aspects. LC Classification: QH442.2 .L474 1998 Dewey Class No.: 241/.66 21

Leukotriene B4 receptor. Cloning and intracellular signaling. Author(s):Yokomizo, T; Masuda, K; Kato, K; Toda, A; Izumi, T; Shimizu, T Language: eng Subject(s): Amino Acid Sequence; Animal; Chemotaxis, Leukocyte physiology; Cloning, Molecular methods; Molecular Sequence Data; Receptors, Leukotriene B4/chemistry ; signal transduction physiology; Structure-Activity Relationship Source Am J Respir Crit Care Med 2000 Feb;161(2 Pt 2):S51-5.

Levick, Stephen E., 1951- Clone being: exploring the psychological and social realities / Stephen E. Levick. Published/Created: Lanham, MD: Rowman & Littlefield, 2004. Description: p. cm. ISBN: 0742529894 (alk. paper) 0742529908 (pbk.: alk. paper) Contents: The identical twin model -- The assisted reproductive technologies and arrangements model -- The stepchild model -- The adoption model -- The parent-child resemblance model -- The child of the famous model -- The replacement child model -- The namesakes model -- The models integrated -- Wider social and cultural implications of cloning -- Intimacy, sex, and sexuality -- Implications for cloning ethics and policy. Notes: Includes bibliographical references and index. Subjects: Human cloning--Social aspects--Psychological aspects. LC Classification: QH442.2 .L485 2004 Dewey Class No.: 176 22

Liberty, identity, and human cloning. Author(s): Robertson, John A Language: eng Subject(s): Cloning, Organism; Coercion; Disclosure; Embryo; Embryo Research; Ethics; Eugenics; Family elations; Federal Government; Fees and Charges; Freedom; Genes; Genetic Determinism; Genetic Engineering; Genetic Enhancement; Genetics, Behavioral; Government; Government Regulation; Guidelines; Human Experimentation; Human Rights; Infertility; Informed Consent; Source Tex Law Rev 1998 May;76(6):1371-456.

Ligands for expression cloning and isolation of GABA(B) receptors. Abstract: Outlined is the rationale behind the syntheses of radioligands [125I]CGP64213 and [125I]CGP71872, which led to the identification of cloned GABA(B) receptors 1a and 1b 17 years after the first pharmacological characterisation of native GABA(B) receptors by Bowery et al. [Nature 283 (1980) 92-94]. More recently it was shown that the N-terminal extracellular domains of GABA(B) receptors 1a and 1b contain binding sites for agonists and antagonists [B. Malitschek et al., Mol. Pharmacol. 56 (1999) 448-454]. In order to isolate the extracellular domain(s) of GABA(B) receptors 1a (or 1b) and to purify and crystallise these proteins a third ligand [125I]CGP84963 was designed, which combines, in one molecule, a GABA(B) receptor binding part, an azidosalicylic acid as photoaffinity moiety and 2-iminobiotin, which binds to avidin in a reversible, pH-dependent fashion [W. Froestl et al., Neuropharmacology 38 (1999) 1641-1646]. Author(s): Froestl, W; Bettler, B; Bittiger, H; Heid, J; Kaupmann, K; Mickel, S J; Strub, D Language: eng Source Farmaco 2001 Jan-Feb;56(1-2):101-5.

Lipases: structure, mechanism, and genetic engineering: contributions to the CEC-GBF International Workshop, September 13 to 15, 1990, Braunschweig, Germany / edited by L. Alberghina, R.D. Schmid, R. Verger. Published/Created: Weinheim, Federal Republic of Germany; New York, NY, USA: VCH, c1991. Related Names: Alberghina, L. (Lilia) Schmid, Rolf, 1942- Verger, R. (Robert) Description: xvii, 440 p.: ill. (some col.); 24 cm. ISBN: 3527283323 (Weinheim: alk. paper) 156081165X (New York: alk. paper) Notes: Proceedings of the CEC-GBF Lipase Workshop. "BRIDGE." Includes bibliographical references and index. Subjects: Lipase--Genetic engineering--Protein engineering--Cloning, Molecular---congresses. Series: GBF monographs, 0930-4320; v. 16 LC Classification: QP609.L5 C43 1990 Dewey Class No.: 660/.634 20

Listen carefully: positional cloning of an audiogenic seizure mutation may yield Frings benefits. Abstract: Neurologists have long sought to understand what precipitates individual seizures in epileptic patients. Studies of reflex epilepsies seem well suited to this task. In this issue of Neuron, Skradski et al. describe a mutation in a novel gene underlying audiogenic seizures in the Frings mouse, providing a valuable resource for elucidating the pathophysiological mechanisms of this inherited form of reflex epilepsy. Author(s): Burgess, D L Language: eng Source Neuron 2001 Aug 30;31(4):507-8.

Liver plasma membrane ecto-ATPase. Purification, localization, cloning, and functions. Author(s): Lin, S H Language: eng Subject(s): Adenosinetriphosphatase genetics isolation & purification metabolism; Animal; Cell Membrane enzymology; Cloning, Molecular; Human; Liver enzymology; Rats; Transfection Source Ann N Y Acad Sci 1990;603:394-9; discussion 399-400.

Location cloning strategy for characterizing genetic defects in Huntington's disease and Alzheimer's disease. Abstract: The recognition that DNA polymorphisms are widespread in the human genome and can be used as high quality genetic markers has introduced a new strategy for approaching inherited disorders for which no protein defect has been identified. Genetic linkage analysis can establish the chromosomal position of the genetic defect, providing a potential opportunity for isolating the disease gene and characterizing its product in the absence of any knowledge of its biochemical function. The first step in this location cloning approach has been successful in mapping the Huntington's disease gene to chromosome 4, and has implicated chromosome 21 as the site of a defect in familial Alzheimer's disease. An intensive effort is under way to narrow the region containing the disease gene and identify the defect in each of these disorders. This review will present the success that has been achieved and the problems that remain and will assess the current status of the location cloning strategy with regard to Huntington's disease and familial Alzheimer's disease. Author(s): Gusella, J F Language: eng. Source FASEB J 1989 Jul;3(9):2036-41.

Long, Jeff. Year zero: a novel / Jeff Long. Published/Created: New York: Pocket Books, c2002. Description: 406 p.; 25 cm. ISBN: 0743406117 Subjects: Archaeological thefts--Fiction. End of the world--Fiction. Women scientists--Fiction. Anthropologists--Fiction. Messiah--Fiction. Cloning--Fiction. Plague--Fiction. Los Alamos (N.M.)--Fiction. LC Classification: PS3562.O4943 Y43 2002 Dewey Class No.: 813/.54 21

Long-range walking techniques in positional cloning strategies. Author(s): Stubbs, L Language: eng Subject(s): Animal; Chromosome Walking/methods; Cloning, Molecular/methods; Genetic Markers; Genomic Library; Human; Linkage (Genetics) Source Mamm Genome 1992;3(3):127-42.

Long-term B-lymphoid cultures from murine bone marrow establishment and cloning by using stromal cell line AC 6.21. Author(s): Whitlock, C A; Muller-Sieburg, C E Language: eng Subject(s): Animal; B-Lymphocytes/cytology; Bone Marrow Cells; Cell Culture/methods; Cell Line; Cell Separation; Mice; Stromal Cells/cytology Source Methods Mol Biol 1997;75:231-48.

Lucotte, Gérard. Introduction to molecular cloning techniques / by Gérard Lucotte and François Baneyx. Published/Created: New York: VCH Publishers, c1993. Related Names: Baneyx, François. Description: xvi, 298 p.: ill.; 25 cm. ISBN: 1560816139 Notes: Includes bibliographical references (p. 287-292) and index. Subjects: Molecular cloning--Laboratory manuals. LC Classification: QH442.2 .L8313 1993 Dewey Class No.: 574.87/328/0724 20

M13 cloning vehicles. Their contribution to DNA sequencing. Author(s): Messing, J Language: eng Subject(s): Bacteriophage M13/genetics; Bacteriophage phi X 174/genetics; Base Sequence; Cloning, Molecular/methods; DNA Replication; Escherichia coli/genetics; Genetic Vectors; History of Medicine, 20th Cent.; Molecular Sequence Data; Mutagenesis, Insertional; Sequence Analysis, DNA/history/methods Source Methods Mol Biol 1993;23:9-22.

Maienschein, Jane. Whose view of life?: embryos, cloning, and stem cells / Jane Maienschein. Published/Created: Cambridge, MA: Harvard University Press, 2003. Description: p. cm. ISBN: 0674011708 Notes: Includes bibliographical references and index. Subjects: Human embryo--Stem cells--Human cloning--Research--Political aspects--United States. LC Classification: QP277 .M356 2003 Dewey Class No.: 176 21

Mammalian cloning: advances and limitations. Abstract: For many years, researchers cloning mammals experienced little success, but recent advances have led to the successful cloning of several mammalian species. However, cloning by the transfer of nuclei from adult cells is still a hit-and-miss procedure, and it is not clear what technical and biological factors underlie this. Our understanding of the molecular basis of reprogramming remains extremely limited and affects experimental approaches towards increasing the success rate of cloning. Given the future practical benefits that cloning can offer, the time has come to address what should be done to resolve this problem. Author(s): Solter, D Language: eng Source Nat Rev Genet 2000 Dec;1(3):199-207.

Maniatis, Tom. Molecular cloning: a laboratory manual / T. Maniatis, E.F. Fritsch, J. Sambrook. Published/Created: Cold Spring Harbor, N.Y.: Cold Spring Harbor Laboratory, 1982. Related Names: Fritsch, E. F. Sambrook, Joseph. Description: x, 545 p.: ill.; 28 cm. ISBN: 0879691360 : Notes: Includes index. Bibliography: p. 507-520. LC Classification: QH442.2 .M26 1982 Dewey Class No.: 574.87/3224 19

Map-based cloning of quantitative trait loci: progress and prospects. Abstract: Map-based cloning has been considered problematic for isolating quantitative trait loci (QTLs) due to the confounding phenotypic effects of environment and other QTLs. However, five recent studies, all in plants, have succeeded in cloning QTLs using map-based methods. We review the important features of these studies and evaluate the prospects for broader application of the techniques. Successful map-based cloning requires that QTLs represent single genes that can be isolated in near-isogenic lines, and that genotypes can be unambiguously inferred by progeny testing. In plants or animals for which map-based cloning of genes with discrete phenotypes is feasible, the modified procedures required for QTLs should not be limiting in most cases. The choice between map-based cloning and alternative methods will depend on details of the species and traits being studied. Author(s): Remington, D L; Ungerer, M C; Purugganan, M D Language: eng Source Genet Res 2001 Dec;78(3):213-8.

MapInfo Professional [computer file]: a world of maps & data. Published/Created: Troy, N.Y.: MapInfo, c1995. Related Names: MapInfo (Firm) Description: 1 computer laser optical disc; 4 3/4 in. + 1 user's guide + 1 reference manual + 1 your turn guide. Computer File Info.: Computer data and program. Summary: Desktop mapping system that provides data visualization and analysis tools for graphic management of resources. Overlays information onto maps from dBase, Lotus 1-2-3, ASCII, or remote databases. Performs geographic searches, zooms, and address matches; shades or colors boundaries; and prints maps. New features for this ed. include automatic node tracing, map cloning, directory management, drag and drop, map clipping, and multi-file selection options. Notes: Title from disc label. System requirements: 486 PC; 8MB RAM; Windows 3.1, 95, or NT; hard disk with 19.5MB free space for standard installation (software requires 5.5MB; data requires 4MB); VGA monitor or better. Includes MapXpress version 1.0.1, a catalog on disc, and disk entitled Geographic Tracker with accompanying getting started guide. Subjects: United States--Maps--Software. LC Classification: G1200 Dewey Class No.: 025.06 12

Mapping and cloning hereditary deafness genes. Abstract: In the past two years, considerable progress has been made in the mapping and cloning of human deafness genes. Highlights are the chromosomal localization of at least five genes for autosomal forms of non-syndromic deafness and, more recently, the cloning of an X-linked deafness gene, DFN3, and the Usher syndrome type IB gene. This last gene encodes a myosin-like protein and was identified as the human homolog of the mouse shaker-1 gene. Author(s): Cremers, F P; Bitner-Glindzicz, M; Pembrey, M E; Ropers, H H Language: eng Source Curr Opin Genet Dev 1995 Jun;5(3):371-5.

McCaffrey, Tony. Emmanuel McClue and the mystery of the Shroud / by Tony McCaffrey. Published/Created: Worcester, Mass.: Ambassador Books, 2001. Description: p. cm. ISBN: 1929039085 Summary: Twelve-year-old Emmanuel McClue's parents are participating in an investigation of miracles that thrusts him into a world of excitement and danger. Subjects: Holy Shroud--Cloning--Racially mixed people--Fiction. Mystery and detective stories. LC Classification: PZ7.M122837 Em 2001 Dewey Class No.: [Fic] 21

McKinnell, Robert Gilmore. Cloning of frogs, mice, and other animals / Robert Gilmore McKinnell. Edition Information: Rev. ed. Published/Created: Minneapolis: University of Minnesota Press, c1985. Related Names: McKinnell, Robert Gilmore. Cloning. Description: ix, 127 p.: ill.; 24 cm. ISBN: 0816613605 : Notes: Rev. ed. of: Cloning. c1979. Includes index. Bibliography: p. 109-117. Subjects: Cloning. Cell nuclei--Transplantation. Embryology, Experimental. LC Classification: QH442.2 .M32 1985 Dewey Class No.: 596/.016 19

McKinnell, Robert Gilmore. Cloning: a biologist reports / Robert Gilmore McKinnell. Published/Created: Minneapolis: University of Minnesota Press, c1979. Description: viii, 130 p.: ill.; 24 cm. ISBN: 0816608830 : Notes: Includes index. Bibliography: p. 119-121. Subjects: Cloning. Cell nuclei--Transplantation. Embryology, Experimental. LC Classification: QH442.2 .M32 Dewey Class No.: 596/.01/6

McKinnell, Robert Gilmore. Cloning--nuclear transplantation in Amphibia: a critique of results obtained with the technique to which is added a discourse on the methods of the craft / Robert Gilmore McKinnell. Published/Created: Minneapolis: University of Minnesota Press, c1978. Description: xii, 319 p.: ill.; 24 cm. ISBN: 0816608318 Notes: Includes index. Bibliography: p. 287-310. Subjects: Amphibians--Cloning. Cell nuclei--Transplantation. Amphibians as laboratory animals. Embryology, Experimental--Technique. Amphibians--Embryology. LC Classification: QH442.2 .M33 1978 Dewey Class No.: 597/.6/0416

Medaka linkage map--from positional cloning and comparative genomics to genome evolution; Naruse, K; Kondo, S; Mitani, H; Shima, A; Fukamachi, S; Kondo, M; Mitani, H; Shima, A Language: jpn Source Tanpakushitsu Kakusan Koso 2000 Dec;45(17 Suppl):2844-52.

Medical science and bioethics: attack of the clones?: hearing before the Subcommittee on Criminal Justice, Drug Policy, and Human Resources of the Committee on Government Reform, House of Representatives, One Hundred Seventh Congress, second session, May 15, 2002. Published/Created: Washington: U.S. G.P.O., 2003. Description: iii, 514 p.; 24 cm. Notes: "Serial no. 107-194." Subjects: Human cloning--United States. LC Classification: KF27 .G64 2002e

Megakaryocyte biology and precursors: In Vitro cloning and cellular properties: proceedings of the Symposium on Megakaryocytes In Vitro held at the Centers [i.e. Center] for Disease Control, Public Health Service, U.S. Department of Health and Human Services, Atlanta, Georgia, U.S.A., May 1-2, 1980 / editors, Bruce Lee Evatt, Richard F. Levine, and Neil T. Williams. Published: New York, N.Y.: Elsevier/North Holland, c1981. Related Names: Evatt, Bruce L. Levine, Richard F. (Richard Frank) Williams, Neil T. (Neil Thomas) Description: xiv, 349 p.: ill.; 24 cm. ISBN: 0444005854 Notes: Includes bibliographies and index. Subjects: Megakaryocytes--Cloning--Congresses. LC Classification: QP97.5 .S94 1980 Dewey Class No.: 599.01/13 19

Melatonin receptors step into the light: cloning and classification of subtypes. Author(s): Reppert, S M; Weaver, D R; Godson, C Language: eng Subject(s): Amino Acid Sequence; Animal; Cloning, Molecular; Human; Molecular Sequence Data; Receptors, Cell Surface/chemistry/classification/metabolism; Receptors, Cytoplasmic and Nuclear/chemistry/classification/metabolism; Terminology Source Trends Pharmacol Sci 1996 Mar;17(3):100-2.

Menditto, Joseph. Genetic engineering, DNA, and cloning: a bibliography in the future of genetics / by Joseph Menditto and Debbie Kirsch. Published: Troy, N.Y.: Whitston Pub. Co., 1983. Related Names: Kirsch, Debbie. Description: xiii, 783 p.; 24 cm. ISBN: 0878752412 Notes: Includes index. Subjects: Genetic engineering--Recombinant DNA--Cloning--Bibliography. LC Classification: Z5322.G4 M46 1983 QH442 Dewey Class No.: 016.57487/3282 19

Merkley, Chad, 1990- Too many me's / story by Chad Merkley; illustrations by Barbara Johansen Newman. Published/Created: Austin, Tex.: Raintree Steck-Vaughn, c1999. Related Names: Newman, Barbara Johansen, ill. Description: 24 p.: col. ill.; 21 x 26 cm. ISBN: 0739800523 Summary: Eight-year-old boy genius Albert makes clones of himself to do the chores he does not want to do, but then he finds this solution not so simple as he thought. Subjects: Cloning--Fiction. Individuality--Fiction. Series: Publish-a-book LC Classification: PZ7.M5365 To 1999 Dewey Class No.: [Fic] 21

Metabolism and enzymology of nucleic acids including gene manipulations / edited by Ján Zelinka and Jozef Balan. Published/Created: New York: Plenum Press, c1988. Related Names: Zelinka, J. (Jan) Balan, J. (Jozef) Slovenská akadémia vied. Description: x, 312 p.: ill.; 26 cm. ISBN: 0306430894 Notes: "Proceedings of the Sixth International Symposium on Metabolism and Enzymology of Nucleic Acids Including Gene Manipulations, held June 7-11, 1987, at Smolenice Castle, Czechoslovakia"--T.p. verso. Sponsored by the Slovak Academy of Sciences. Includes bibliographical references. Subjects: Nucleic acids--Nucleuses--Methylation--Molecular cloning--Gene expression--Congresses. LC Classification: QP620 .I56 1987 Dewey Class No.: 574.87/3282 19

Metcalf, Donald. Clonal culture of hemopoietic cells: techniques and applications / Donald Metcalf. Published/Created: Amsterdam; New York: Elsevier; New York, NY, USA: Sole distributors for the USA and Canada, Elsevier Science Pub. Co., 1984. Related Names: Metcalf, Donald. Hemopoietic colony stimulating factors. Description: x, 167 p.: ill.; 24 cm. ISBN: 0444805656 (U.S.: pbk.) Notes: Companion v. to: The hemopoietic colony stimulating factors / Donald Metcalf. c1984. Includes index. Bibliography: p. [155]-163. Subjects: Hematopoietic stem cells--Growth. Cloning. Colony-stimulating factors (Physiology)--Analysis. Colony-stimulating factors (Physiology)--Purification. Clone Cells. Colony-Forming Units Assay. Colony-Stimulating Factor--isolation & purification. Hematopoietic Stem Cells--analysis. Tissue Culture--methods. LC Classification: QP92 .M483 1984 Suppl. Dewey Class No.: 611/.0184 19

Methods for finding genes. A major rate-limiting step in positional cloning. Abstract: Identification of transcribed sequences from within genomic regions has been a major rate-limiting step in the pursuit of genes involved in many human genetic diseases. Early efforts focused primarily on screening of cDNA libraries, identification of evolutionarily conserved sequences, and northern blot hybridization. In recent years, several innovative techniques for gene identification have been devised. These techniques expand the size of the genomic region capable of being scanned for genes, while also allowing detection of genes regardless of their expression patterns. This article reviews several new and older techniques and discusses the advantages and limitations of each. Author(s): Parrish, J E; Nelson, D L Language: eng Source Genet Anal Tech Appl 1993 Apr;10(2):29-41.

Methods for long term growth and cloning of T cells reactive with soluble antigens. Author(s): Sredni, B Language: eng Source J Immunol Methods 1982;49(1):R1-10.

Microbial beta-glucosidases: cloning, properties, and applications. Abstract: Beta-glucosidases constitute a major group among glycosylhydrolase enzymes. Out of the 82 families classified under glycosylhydrolase category, these belong to family 1 and family 3 and catalyze the selective cleavage of glucosidic bonds. This function is pivotal in many crucial biological pathways, such as degradation of structural and storage polysaccharides, cellular signaling, oncogenesis, host-pathogen interactions, as well as in a number of biotechnological applications. In recent years, interest in these enzymes has gained momentum owing to their biosynthetic abilities. The enzymes exhibit utility in syntheses of diverse oligosaccharides, glycoconjugates, alkyl- and aminoglucosides. Attempts are being made to understand the structure-function relationship of these versatile biocatalysts. Earlier reviews described the sources and properties of microbial beta-glucosidases, yeast beta-glucosidases, thermostable fungal beta-glucosidase, and the physiological functions, characteristics, and catalytic action of native beta-glucosidases from various plant, animal, and microbial sources. Recent efforts have been directed towards molecular cloning, sequencing, mutagenesis, and crystallography of the enzymes. The aim of the present article is to describe the sources and properties of recombinant beta-glucosidases, their classification schemes based on similarity at the structural and molecular levels, elucidation of structure-function relationships, directed evolution of existing enzymes toward enhanced thermostability, substrate range, biosynthetic properties, and applications. Author(s): Bhatia, Yukti; Mishra, Saroj; Bisaria, V S Language: eng Source Crit Rev Biotechnol 2002;22(4):375-407.

Molecular analysis of the fyn-complex: cloning of SKAP55 and SLAP-130, two novel adaptor proteins which associate with fyn and may participate in the regulation of T cell receptor-mediated signaling. Author(s): Schraven, B; Marie-Cardine, A; Koretzky, G Language: eng Source Immunol Lett 1997 Jun 1;57(1-3):165-9.

Molecular analysis of the Neurospora clock: cloning and characterization of the frequency and period-4 genes. Abstract: Genetic analysis of Neurospora crassa has identified many mutants that affect the biological clock. In this article we review the cloning of two of these genes, frq and prd-4. Both genes were isolated using a chromosome walk technique. Subcloning experiments and subsequent Northern analysis of frq implicate the importance of two transcripts that emanate from this locus. In preliminary data, no protein-coding region is evident in the smaller transcript; the larger transcript contains a 962-amino acid open reading frame. The open reading frame shows limited homology to per, a clock gene identified in Drosophila. Sequence analysis of all existing frq alleles suggests that the defect in each case lies within the open reading frame. Successful cloning of the prd-4 gene required walking a distance of greater than 40 kb. A physical map of this region has been constructed using restriction analysis. The dominance-recessive relationship of prd-4 and prd-4+ was established by examining the period lengths of strains harboring a wide range of prd-4/prd-4+ nuclear ratios. Author(s): Aronson, B D; Johnson, K A; Liu, Q; Dunlap, J C Language: eng Source Chronobiol Int 1992 Jun;9(3):231-9.

Molecular analysis of vitamin A formation: cloning and characterization of beta-carotene 15,15'-dioxygenases. Abstract: Beta-carotene 15,15'-dioxygenase cleaves beta-carotene into two molecules of retinal and is the key enzyme in the metabolism of carotene to vitamin A. Although the enzyme has been known for more than 40 years, all attempts to purify the protein to homogeneity or to clone its gene have failed until recently, when the successful cloning and sequencing of cDNAs encoding enzymes with beta-carotene 15,15'-dioxygenase activity from Drosophila (J. von Lintig and K. Vogt, 2000, J. Biol. Chem. 275, 11915-11920) and chicken (A. Wyss et al., 2000, Biochem. Biophys. Res. Commun. 271, 334-336) were reported. Very soon it became clear, that we have cloned two members of a new family of carotenoid cleaving enzymes. Overall homologies are very high, certain amino acid stretches almost identical. Thus, beta-carotene 15,15'-dioxygenase can be considered as evolutionarily well conserved. These findings open up wide perspectives for further analysis of this important biosynthetic pathway, concerning basic and medical research as well as biotechnological aspects related to vitamin A supply, which are discussed here. Author(s): von Lintig, J; Wyss, A Language: eng Source Arch Biochem Biophys 2001 Jan 1;385(1):47-52.

Molecular biological methods for Bacillus / edited by Colin R. Harwood and Simon M. Cutting; with contributions by R. Chambert ... [et al.]. Published/Created: Chichester; New York: Wiley, c1990. Related Names: Harwood, Colin R. Cutting, Simon M. Description: xxxv, 581 p.: ill.; 24 cm.

ISBN: 0471923931 : Notes: "A Wiley-Interscience publication." Folded ill. inserted. Includes bibliographical references and index. Subjects: Bacillus (Bacteria)--Genetics. Bacillus (Bacteria)--Biotechnology. Bacillus--genetics. Biotechnology. Cloning, Molecular. Series: Modern microbiological methods LC Classification: QR82.B3 M63 1990 Dewey Class No.: 589.9/5 20

Molecular biology and pathology of cystic fibrosis. Two years after cloning of the gene; Goossens, M Language: fre Source Arch Fr Pediatr 1992 Apr;49(4):329-31.

Molecular biology in medicine. III. DNA cloning; Plata, C; Gamba, G Language: spaSubject(s): Base Sequence; Cloning, Molecular/methods; DNA, Complementary; DNA, Recombinant; Genes; Genetic Diseases, Inborn; Genetic Vectors; Human; Molecular Sequence Data; Polymerase Chain Reaction; Transformation. Source Rev Invest Clin 1996 May-Jun;48(3):239-44.

Molecular biology of blood groups: cloning the Kell gene. Author(s): Marsh, W L Language: eng Subject(s): Cloning, Molecular; DNA genetics; Gene Library; Kell Blood-Group System genetics; Neprilysin genetics; Source Transfusion 1992 Feb;32(2):98-101.

Molecular cloning and analysis of Staphylococcus aureus chromosomal aminoglycoside resistance genes. Abstract: Most of the aminoglycoside resistant Staphylococcus aureus strains isolated in France are resistant to all the antibiotics belonging to this family. Two amino-glycoside-modifying enzymes were detected in the wild-type strains studied: an APH3'III and an AAC6'-APH2". These strains also carry two types of streptomycin resistance: high-level resistance due to chromosomal mutation(s) affecting ribosome affinity and low-level resistance, the mechanism of which was not characterized. All the aminoglycoside resistance genes were located on the chromosome. DNA fragments of 1.5 and 1.95 kb carrying the aphA and aacA genes, respectively, were isolated, by cloning, from the cellular DNA of a clinical isolate. When these genes were introduced into Escherichia coli and Bacillus subtilis strains, the enzymes synthesized were indistinguishable from those produced by the S. aureus strains. When the cellular DNAs of wild-type and resistant strains were hybridized with the cloned fragments, sequences homologous to the fragment carrying the aphA gene were found to be located at the same chromosomal site, while those hybridizing with the fragment carrying the aacA gene were at different chromosomal sites. Author(s): el Solh, N; Moreau, N; Ehrlich, S D Language: eng Source Plasmid 1986 Mar;15(2):104-18.

Molecular cloning and analysis of the protein modules of aggrecans. Abstract: The large aggregating chondroitin sulfate proteoglycan of cartilage, aggrecan, has served as a prototype of proteoglycan structure. Molecular cloning has elucidated its primary structure and revealed both known and unknown domains. To date the complete structures of chicken, rat and human aggrecans have been deduced, while partial sequences have been reported for bovine aggrecan. A related proteoglycan, human versican, has also been cloned and sequenced. Both aggrecan and versican have two lectin domains, one at the amino-terminus which binds hyaluronic acid and one at the carboxyl-terminus whose physiological ligand is unknown. Both lectins have homologous counterparts in other types of proteins. Within the aggrecans the keratan sulfate domain may be variably present and also has a prominent repeat in some species. The chondroitin sulfate domain has three distinct regions which vary in their prominence in different species. The complex molecular structure of aggrecans is consistent with the concept of exon shuffling and aggrecans serve as suitable prototypes for comprehending the evolution of multi-domain proteins. Author(s): Upholt, W B; Chandrasekaran, L; Tanzer, M L Language: eng Source EXS 1994;70:37-52.

Molecular cloning and analysis of the protein modules of aggrecans. Abstract: The large aggregating chondroitin sulfate proteoglycan of cartilage, aggrecan, has served as a prototype of proteoglycan structure. Molecular cloning has elucidated its primary structure and revealed both known and unknown domains. To date the complete structures of chicken, rat and human aggrecans have been deduced, while partial sequences have been reported for bovine aggrecan. A related proteoglycan, human versican, has also been cloned and sequenced. Both aggrecan and versican have two lectin domains, one at the amino-terminus which binds hyaluronic acid and one at the carboxyl-terminus whose physiological ligand is unknown. Both lectins have homologous counterparts in other types of proteins. Within the aggrecans the keratan sulfate domain may be variably present and also has a prominent repeat in some species. The chondroitin sulfate domain has three distinct regions which vary in their prominence in different species. The complex molecular structure of aggrecans is consistent with the concept of exon shuffling and aggrecans serve as suitable prototypes for comprehending the evolution of multi-domain proteins. Author(s): Upholt, W B; Chandrasekaran, L; Tanzer, M L Language: eng Subject(s): Amino Acid Sequence; Animal; Cloning, Molecular; DNA/genetics; Human; Molecular Sequence Data; Proteochondroitin Sulfates/chemistry/genetics; Proteoglycans/biosynthesis/chemistry/genetics; Sequence Homology, Amino Acid Source Experientia 1993 May 15;49(5):384-92.

Molecular cloning and characterization of a gene coding for human cardiac myosin heavy-chain Abstract: In vertebrates, the myosin heavy chain (MHC) genes are very closely related at the nucleotide sequence level. Amino acid sequence comparison of the human cardiac alpha- and beta-MHC have demonstrated that there are, at least, 7 isoform-specific divergent regions, including important binding protein-related sites such as ATP, actin and myosin light chain. It has been reported that in the rat, there are 8 isoform-specific divergent regions. The 7th divergent area which is thought to mediate thick filament in the light meromyosin region in the rat is not apparent in the human. Also, recently, 9 kinds of human genomic MHC genes have been investigated. The 4 human skeletal MHC genes are clustered at chromosome region 17 pter-p11, 17p13. The cardiac alpha- and beta-MHC genes are in tandem and mapped to chromosome region 14q11.2, the smooth muscle MHC gene is mapped to 16q11.2 and two nonmuscle MHC genes are mapped to chromosomes 17 and 22. We are studying the relationship between hypertrophic cardiomyopathy and the cardiac MHC gene, and between other MHC genes and other cardiac diseases. Author(s): Matsuoka, R Language: jpn Subject(s): Animal; Chromosome Mapping; English Abstract; Human; Myosins/genetics; Rats Source Nippon Rinsho 1993 Jun;51(6):1441-7.

Molecular cloning and characterization of a parathyroid hormone/parathyroid hormone-related peptide receptor: a member of an ancient family of G protein-coupled receptors. Abstract: The parathyroid hormone (PTH)/parathyroid hormone-related peptide (PTHrP) receptor belongs to a newly discovered family of G protein-coupled receptors. Members of this family, which have been isolated from mammals, include the receptors for PTH/PTHrP, calcitonin, secretin, growth hormone-releasing hormone, vasoactive intestinal polypeptide (types 1 and 2), gastric-inhibitory polypeptide, glucagon-like peptide 1, glucagon, corticotropin-releasing factor, and the pituitary adenylate cyclase-activating peptide. Very recently, a receptor with remarkable homology to these mammalian receptors was isolated from the insect Manduca sexta, which indicates considerable conservation of these related proteins during evolution. Thus far the cognate ligands for these receptors are 27- to 46-amino-acid residues in length. Members of this novel receptor family are characterized by seven membrane-spanning domains and at least two conserved sites for N-linked glycosylation. Furthermore, 48-amino-acid residues, including eight extracellular cysteines, are identical in all receptors, and many other residues are highly conserved. The PTH/PTHrP receptor is expressed in a large variety of fetal and adult tissues, binds two ligands (PTH and PTHrP) with high affinity, and activates at least two second-messenger systems (adenylate cyclase and phospholipase C). Author(s): Juppner, H Language: eng Source Curr Opin Nephrol Hypertens 1994 Jul;3(4):371-8.

Molecular cloning and characterization of complementary and genomic DNA clones for mouse C4 and Slp. Author(s): Nonaka, M; Nakayama, K; Yeul, Y D; Shimizu, A; Takahashi, M Language: eng Source Immunol Rev 1985 Oct;87:81-99.

Molecular cloning and characterization of neurotransmitter transporters. Author(s): Blakely, R D Language: eng Subject(s): Animal; Carrier Proteins/metabolism; Cloning, Molecular; Human; Neurotransmitters/metabolism Source NIDA Res Monogr 1992;126:66-83.

Molecular cloning and characterization of the gene coding for red cell membrane skeletal protein 4.1. Abstract: The application of molecular biology to the study of the human erythrocyte membrane is presented in this report. We describe the strategy employed to clone DNA sequences encoding the membrane skeletal protein, protein 4.1. We demonstrate how these sequences may be utilized to deduce detailed structural and functional information about the 4.1 polypeptide. We also illustrate the use of cloned 4.1 DNA sequences as probes to explore the structural organization of the protein 4.1 gene in normal individuals and in patients with dysfunctional erythrocyte membranes. Ultimately it will be possible to generate a molecular description both of the structural proteins which constitute the membrane skeleton, and of the genetic mechanisms regulating their expression in erythroid and nonerythroid cells. Author(s): Conboy, J Language: eng Source Biorheology 1987;24(6):673-87.

Molecular cloning and characterization of the T-cell mono(ADP-ribosyl)transferase RT6. Relationships to other mADPRTs and possible functions. Author(s): Thiele, H G; Haag, F; Koch-Nolte, F Language: eng Subject(s): ADP Ribose Transferases; Antigens, Cloning, Molecular; Isoantigens; Mice; Primates; Rats; Rats, Inbred BN; T-Lymphocytes. Source Adv Exp Med Biol 1997;419:109-20.

Molecular cloning and developmental pattern of expression of MSJ-1, a new male germ cell-specific DNAJ homologue. Abstract: A cDNA encoding a new member of the DnaJ protein family has been isolated by screening a mouse testicular expression library. The predicted protein, named MSJ-1, is 242 amino acid residues-long, containing the fingerprinting J domain in the NH2 terminus. A wide tissutal Northern blot analysis reveals that MSJ-1 is expressed only in the testis, while in situ hybridization analyses demonstrate that the mRNA is first transcribed in spermatids. The antiserum developed against a MSJ-1/GST fusion protein recognizes a protein of 30 kDa in germ cell protein extracts. Author(s): Berruti, G; Perego, L; Martegani, E Language: eng Source Adv Exp Med Biol 1998;444:145-50; discussion 151.

Molecular cloning and expression of angiotensin II type 2 receptor gene. Author(s): Ichiki, T; Kambayashi, Y; Inagami, T Language: eng Subject(s): Amino Acid Sequence; Animal; Cloning, Molecular; GTP-Binding Proteins; Gene Expression; Human; Molecular Sequence Data; RNA, Messenger; Receptors, Angiotensin/chemistry/genetics/metabolism; Signal Transduction Source Adv Exp Med Biol 1996;396:145-52.

Molecular cloning and expression of carcinoembryonic antigen gene family members. Abstract: The carcinoembryonic antigen (CEA) gene family consists of approximately 30 genes, of which 17 are transcriptionally active. These genes are tightly clustered on the long arm of chromosome 19. Two main subgroups are distinguishable, one encoding CEA and its classical non-specific cross-reacting antigens (NCAs) and the second encoding the pregnancy-specific glycoproteins (PSGs). The CEA family belongs to the immunoglobulin superfamily. CEA subgroup members are either membrane bound through a glycosyl phosphatidylinositol anchor, or are integral membrane proteins, whereas PSG molecules are secreted from the cell. Despite high sequence similarity, CEA family members reveal variable expression patterns, indicating different functions. Apart from its high expression in various adenocarcinomas, CEA is found in epithelia lining the gastrointestinal tract, in eccrine sweat glands, lung epithelia and testes. The NCAs are often coexpressed with CEA, but are also found in granulocytes. PSGs are coexpressed in syncytiotrophoblast cells of the placenta, submandibular salivary glands and fetal liver, as well as in choriocarcinomas and breast tumours. Author(s): Thompson, J A Language: eng Source Tumour Biol 1995;16(1):10-6.

Molecular cloning and expression of rat V1a and V2 arginine vasopressin receptors. Abstract: Vasopressin, one of the first characterized neuropeptides, has a wide spectrum of biological action, acting on distinct tissues. Indeed, it is involved in water retention, glucose metabolism, blood pressure and its implication in the CNS has also been described. This diversity of effects on mammalian tissues is mediated by distinct G protein-coupled receptors, acting via distinct second messenger pathways. This receptor family has been subtyped by pharmacological studies, as V1a receptor whose action is mediated by intracellular calcium mobilization, and V2 receptor which is linked to adenylyl cyclase. Since so many essential functions were ensured by vasopressin, molecular characterization of its receptors became soon a great challenge. This prompted us to isolate the cDNA of AVP V1a receptor as the first member of this family, by expression cloning. Intracellular calcium mobilization was therefore assayed after rat liver mRNA injection into Xenopus oocytes. A single clone, encoding a functional AVP receptor corresponding to the V1a subtype was finally characterized as a G protein-coupled receptor. Furthermore, we used homology cloning strategy in order to clone the AVP V2 subtype from a rat kidney cDNA library. A putative receptor clone was finally characterized as the rat V2 receptor cDNA by binding and cAMP increase experiments, on transfected cells. Author(s): Morel, A; Lolait, S J; Brownstein, M J Language: eng Source Regul Pept 1993 Apr 29;45(1-2):53-9.

Molecular cloning and functional analysis of sialyltransferases. Abstract: To elucidate the regulatory mechanism for carbohydrate expression and to understand the meaning of the carbohydrate-structural diversity, we started to clone sialyltransferase (ST) genes based on two different strategies, i.e. expression and homology cloning. So far, 13 STs have been cloned in our laboratory, 7 of which turned out to be new ones. The primary enzyme structures deduced from all the cloned ST genes suggest a putative domain structure with a type II transmembrane topology. There are no significant amino acid sequence similarities among these cloned STs, except for in two sialyl motifs, L and S, which are proposed to be the CMP-sialic acid recognition and/or catalytic sites. Northern blot analysis revealed the developmental stage-dependent and/or tissue-specific expression of most of the cloned STs. The cloned STs are classified into four families according to the carbohydrate linkages they synthesize, i.e. the ST3Gal-, ST6Gal-, ST6GalNAc-, and ST8Sia-families. Generally, enzymes in these families exhibit strong activity toward certain acceptor groups but show very weak activity toward other acceptor groups, and the substrate specificities of the enzymes overlap one another, as indicated by in vitro experiments. Enzymes in the ST3Gal-family are expressed mainly in a tissue-specific manner. However, those in the ST6GalNAc- and ST8Sia-families are expressed in a tissue-as well as developmental stage-specific manner. In vivo conditions are supposed to be more complex. Therefore, it is quite important to examine their substrate specificities in vivo and the mechanism of their expression to elucidate the physiological role of each enzyme and the meaning of the diversity in carbohydrate structure. Using cloned cDNAs and expressed enzymes, we have been studying how sialylcarbohydrate expression is regulated and what the functions of sialylcarbohydrate chains are. Recently, we found that transfection of the GD3 synthase, an alpha 2,8-ST (ST8Sia I), gene triggers cholinergic neuritogenesis in Neuro2a cells through the de novo expression of GD3, suggesting that the GD3 synthase gene behaves as a neural differentiation inducer. Author(s): Tsuji, S Language: eng Source J Biochem (Tokyo) 1996 Jul;120(1):1-13.

Molecular cloning and functional expression of the human heart inwardly-rectifying potassium channel. Author(s): Takeda, J; Koumi, S; Sato, R; Hayakawa, H Language: eng Subject(s): Animal; Base Sequence; Cloning, Molecular; Electrophysiology; Human; Molecular Sequence Data; Myocardium/metabolism; Potassium Channels/genetics/metabolism/physiology Source Nippon Ika Daigaku Zasshi 1998 Feb;65(1):73-5.

Molecular cloning and functions of rat liver hydroxysteroid sulfotransferases catalysing covalent binding of carcinogenic polycyclic arylmethanols to DNA. Abstract: Three sulfotransferases (STs) catalysing the metabolic activation of potent carcinogenic polycyclic arylmethanols were purified from female Sprague-Dawley (SD) rat liver cytosol without loss of their enzyme activities in the presence of Tween 20 used for preventing the enzymes from aggregation during purification and identified as hydroxysteroid sulfotransferases (HSTs). All the purified HSTs, STa, STb, and STc, with different electric charges had an apparently equal size of subunit (30.5 kDa) and cross-reacted with polyclonal antibody raised against STa. Our study on molecular cloning of cDNA libraries from two female SD rat livers indicated that both contained cDNA inserts coding for 5 different HST subunits, consisting of 284-285 amino acid residues (M(r), 33,084-33,535) and sharing strong amino acid sequence identity (> 83%). Of the 5 HST subunits, two had an identical amino acid sequence except for only one amino acid residue, and the other two contained only 6 amino acid substitutions in their sequences. Author(s): Watabe, T; Ogura, K; Satsukawa, M; Okuda, H; Hiratsuka, A Language: eng Source Chem Biol Interact 1994 Jun;92(1-3):87-105.

Molecular cloning and gene regulation in bacilli / edited by A.T. Ganesan, Shing Chang, James A. Hoch. Published/Created: New York: Academic Press, 1982. Related Names: Ganesan, A. T. Chang, Shing. Hoch, James A. Cetus Conference on Genetics (1981: Stanford University) Description: xxii, 359 p.: ill.; 24 cm. ISBN: 0122741501 Notes. "Articles based on talks given at the Cetus Conference on Genetics held at Stanford University, Stanford, California, June 22-24, 1981"--Pref. Includes bibliographies. Subjects: Bacillus (Bacteria)--Bacillus subtilis--Bacterial genetics--Molecular cloning--Genetic regulation--Genetics--Congresses. LC Classification: QR82.B3 M65 1982 Dewey Class No.: 589.9/5 19

Molecular cloning and in vitro expression of C. elegans and parasitic nematode ionotropic receptors. Abstract: The free living nematode, C. elegans is understood at a level of detail equalled by few other organisms, and much of the cell biology and sequence information is proving of considerable utility in the study of parasitic nematodes. Already, C. elegans provides a convenient vehicle for investigating anthelmintic drug action and resistance mechanisms. Among the ionotropic receptors, with their important roles in the behaviour and development of the organism, are targets for anthelmintics. The subunits of nicotinic acetylcholine receptors of C. elegans form a large and diverse multigene family. Members of this family are among the 11 genes associated with resistance to the anthelmintic drug levamisole. Author(s): Fleming, J T; Baylis, H A; Sattelle, D B; Lewis, J A Language: Source Parasitology 1996;113 Suppl:S175-90.

Molecular cloning and molecular biological studies of human thymidylate synthase gene Abstract: Thymidylate synthase (TS) provides the only de novo source of thymidylate for deoxyribonucleic acid (DNA) synthesis and is a key target of cancer chemotherapeutic agents. Expression of its activity is strongly dependent on the cell cycle. Therefore, it is important for understanding the abnormal growth of cancer cells as well as for cancer chemotherapy to elucidate the structure of a TS gene and the regulatory mechanism of expression of its gene at the molecular level. From this point of view, genomic DNA segments partially coding for human TS were cloned from the mouse cell transformant obtained by gene transfer. By using them as a probe, functional cDNA and genomic DNA (approx. 18 kb) clones for human TS were isolated, and the entire nucleotide sequence of these cloned DNAs including all the introns determined. On the basis of the sequence data, we revealed an amino acid sequence of human TS, and organization and several structural features of its gene were revealed. Then, multiple transcription initiation sites and transcriptional regulatory elements of the TS gene were identified. Furthermore, the importance of the post-transcriptional regulation in the cell-cycle dependent expression of the gene was shown. The results obtained so far have opened the way to elucidate the molecular mechanism regulating the TS gene expression entirely. Author(s): Takeishi, K Language: jpn Source Yakugaku Zasshi 1990 Dec;110(12):891-907.

Molecular cloning and posttranslational processing of human preprothyrotropin-releasing hormone Abstract: Human PreproTRH gene and its hypothalamic cDNA were cloned. The human preproTRH protein contains six copies of TRH progenitor sequence flanked by paired basic amino acids. Radioimmunoassays with specific antibodies raised against five human preproTRH cryptic peptides and chromatographic analyses identified that the posttranslational processing of preproTRH in hypothalami and placentae could be different. Moreover, transfection assays using human preproTRH promoter-luciferase chimeric-plasmids demonstrated that human preproTRH promoter activities were regulated negatively by T3 in GH3 cells. Cloning of the human preproTRH gene would facilitate the elucidation of the molecular mechanism of actions of TRH and TRH cryptic peptides in human hypothalamo-pituitary-thyroid axis and central nervous system. Author(s): Satoh, T; Mori, M Language: jpn Source Nippon Rinsho 1994 Apr;52(4):995-1000.

Molecular cloning and sequence analysis of full-length cDNA coding for mouse contrapsin. Abstract: Four overlapping cDNA clones encoding contrapsin were isolated from a mouse liver cDNA library constructed in the expression vector, lambda gt11. M13 vector sequence analysis revealed that contrapsin cDNA contained an open reading frame of 1,254 bases encoding 418 amino acids. The N-terminal amino acid sequence of the isolated contrapsin matched residues 30 to 48 of the sequence deduced on nucleotide analysis. One clone, which had the longest 3' untranslated region, contained two sets of tandem polyadenylation signals, AATACA and AATAAA, which were located 497 bases apart, while the remaining three clones terminated at the first signal. Author(s): Suzuki, Y; Yamamoto, K; Sinohara, H Language: eng Source J Biochem (Tokyo) 1990 Sep;108(3):344-6.

Molecular cloning for gene expression of nerve activity; Kojima, N; Inokuchi, K Language: jpn Subject(s): Cloning, Molecular methods; Gene Expression Regulation genetics; Genes, Immediate-Early genetics; Neurons physiology Source Nippon Seirigaku Zasshi 1998;60(7-8):301-17.

Molecular cloning in heterologous systems. Author(s): Esser, K; Lang-Hinrichs, C Language: eng Subject(s): Animal; Cloning, Molecular/methods; Comparative Study; DNA, Recombinant; Eukaryotic Cells; Genetic Vectors; Prokaryotic Cells Source Adv Biochem Eng Biotechnol 1983;26:143-73.

Molecular cloning in Streptomyces; Suarez, J E Language: spa; Bacteriophages/genetics; Cloning, Molecular; Genetic Vectors; Mutation; Plasmids; Streptococcus genetics; Transformation, Genetic Source Microbiol Esp 1982;35:53-76.

Molecular cloning of a human oxytocin receptor. Author(s): Kimura, T; Azuma, C, Takemura, M; Inoue, T; Kikuchi, T; Kubota, Y; Ogita, K; Saji, F; Tanizawa, O Language: eng Subject(s): Amino Acid Sequence; Cloning, Molecular; Female; Human; Molecular Sequence Data; Myometrium physiology; Receptors, Oxytocin; Receptors, Vasopressin genetics; Xenopus Source Regul Pept 1993 Apr 29;45(1-2):73-7.

Molecular cloning of a second human 15S-lipoxygenase and its murine homologue, an 8S-lipoxygenase. Their relationship to other mammalian lipoxygenases. Author(s): Brash, A R; Jisaka, M; Boeglin, W E; Chang, M S Language: eng Source Adv Exp Med Biol 1999;447:29-36.

Molecular cloning of adenosine A1 and A2 receptors. Author(s): Linden, J; Tucker, A L; Lynch, K R Language: eng Subject(s): Amino Acid Sequence; Animal; Cloning, Molecular; Molecular Sequence Data; Protein Conformation; Receptors, Purinergic/genetics; Sequence Homology, Nucleic Acid Source Trends Pharmacol Sci 1991 Sep;12(9):326-8.

Molecular cloning of ADH7 encoding human class IV (sigma) ADH and its clinical significance; Yokoyama, H; Ishii, H Language: jpn Subject(s): Alcohol Dehydrogenase/genetics/physiology; Animal; Base Sequence; Cloning, Molecular; Gene Expression Regulation, Enzymologic; Human; Isoenzymes/genetics/physiology; Molecular Sequence Data; Racial Stocks; Stomach/enzymology; Stomach Neoplasms/etiology; Vitamin A/metabolism Source Tanpakushitsu Kakusan Koso 1997 Jan;42(1):42-9.

Molecular cloning of Aspergillus fumigatus allergens and their role in allergic bronchopulmonary aspergillosis.. Author(s):Crameri, Reto Language: eng Source Chem Immunol 2002;81:73-93.

Molecular cloning of AT1 angiotensin receptors. Abstract: Angiotensin II is the principal effector molecule of the renin-angiotensin system. Its effects are mediated by cell surface proteins termed AT receptors. On the basis of radioligand binding studies, these have been pharmacologically subdivided into two classes, termed AT1 receptors and AT2 binding sites (Chiu AT, et al, Biochem Biophys Res Commun 1989;165:196-203). AT1 receptors appear to mediate the major cardiovascular effects of angiotensin II, whereas no known physiological properties appear to be coupled to AT2 binding sites (Wong PC, et al, J Pharmacol Exp Ther 1990;255:584-592). To gain further insight into the function of AT1 receptors we have isolated rat cDNA's and genes encoding two distinct but highly similar isoforms of AT1 receptors, termed AT1a and AT1b receptors. Two cDNA's encoding the vascular AT1a receptor were isolated by an expression cloning strategy from a cDNA library prepared from vascular smooth muscle cells. The properties of the clones isolated by this approach are consistent with known pharmacological, biochemical signaling, and tissue distribution properties of AT1 receptors. Using this cDNA as a probe, a second isoform of rat AT1 receptor was isolated from a genomic library. This receptor, termed the AT1b receptor, is 95% identical in amino acid sequence and is pharmacologically indistinguishable from the AT1a receptor. However, the tissue-specific expression pattern of the AT1b gene differs significantly from that for the AT1a receptor. Author(s): Murphy, T J; Takeuchi, K; Alexander, R W Language: eng Source Am J Hypertens 1992 Dec;5(12 Pt 2):236S-242S.

Molecular cloning of bacterial antigens and virulence determinants. Author(s): Macrina, F L Language: eng Source Annu Rev Microbiol 1984;38:193-219.

Molecular cloning of BRCA1: a gene for early onset familial breast and ovarian cancer. Abstract: Molecular analyses allow one to determine genetic lesions occurring early in the development of tumors. With positional cloning approaches we are searching for a gene involved in the development of early onset familial breast and ovarian cancer that maps to human chromosome 17q21 and is termed BRCA1. This involves localizing the region genetically within families with multiply affected members, capturing the region identified by genetic analyses in YACs (yeast artificial chromosomes), converting those YACs to smaller manipulable pieces (such as cosmids), and searching for genes via a variety of approaches such as direct screening of cDNA libraries with genomic clones, direct selection by hybridization, "exon trapping", and CpG island rescue. Once identified, candidate genes will be screened for mutations in affected family members in whom breast cancer segregates with the locus on 17q21. The frequency of this gene has been calculated to be 0.0033; from this the incidence of carriers, i.e. those carrying such a predisposition, is one in 150 women. The isolation of BRCA1 and the elucidation of the mutations resulting in breast and ovarian cancer predisposition will allow identification of women who have inherited germ-line mutations in BRCA1. In families known to harbor a germ-line BRCA1 mutation, diagnosis of affected members will be rapid. It is possible that one will also be able to detect alterations of the second copy of this gene early in tumor development in individuals carrying a germ-line mutation. It is not yet known how frequently somatic BRCA1 mutations predispose to breast and ovarian carcinoma in the general female population. If, as in other genetic diseases, new germ-line mutations occur in some women and thus contribute to the development of breast cancer, it may be feasible to screen women in the general population for predisposing mutations. In addition, if acquired genetic mutations of the BRCA1 gene are involved as early events in the development of non-familial forms of the disease, early detection of possible breast carcinoma may become feasible in biopsy of breast tissue. Author(s): Bowcock, A M Language: eng Source Breast Cancer Res Treat 1993 Nov;28(2):121-35.

Molecular cloning of Candida albicans phospholipase D Abstract: Phospholipase D (PLD) catalyses the hydrolysis of phosphatidylcholine, a major substrate, to phosphatidic acid and choline, and its activity is regulated by a variety of hormones, growth factors, and other extracellular signals in mammalian cells. Thus, it is now recognized as a signal transducing enzyme such as phosphatidylinositol-specific phospholipase C, adenylate cyclase, or protein tyrosine kinases. Furthermore, recent findings that regulation by members of the ADP-ribosylation factor (ARF) and Rho families of monomeric GTP-binding protein suggest roles of PLD in intracellular vesicle traffi-cking, morphological changes, and mitogenic signaling process. In Saccharomyces cerevisiae, PLD gene has been cloned and revealed to be essential for meiosis. In contrast, little is known about PLD in Candida albicans. As a first step to understand possible physiological roles of PLD in C. albicans, we cloned a PLD gene from a C. albicans genomic DNA library. Deduced amino acid sequence analysis showed the structural similarity to mammalian, yeast, and plant PLDs. It was also suggested employing RT-PCR (reverse transcriptase polymerase chain reaction) that an isozyme of C. albicans PLD was present. Author(s): Kanoh, H; Nakashima, S; Kitajima, Y; Nozawa, Y Language: jpn Source Nippon Ishinkin Gakkai Zasshi 1998;39(2):79-83.

Molecular cloning of cDNAs encoding vitamin D3 25-hydroxylase and 25-hydroxyvitamin D3 24-hydroxylase. Author(s): Okuda, K; Ohyama, Y; Usui, E; Noshiro, M Language: eng Source J Nutr Sci Vitaminol (Tokyo) 1992;Spec No:100-3.

Molecular cloning of chloroplasts genes--chloroplast DNA replication origin and transcriptional promoters; Ohyama, K Language: jpn Source Tanpakushitsu Kakusan Koso 1985 Mar;30(3):184-96.

Molecular cloning of CL-P1 gene; Wakamiya, N; Suzuki, Y Language: jpn Subject(s): Animal; Carrier Proteins genetics physiology; Cloning, Molecular; Collectins; Endothelium, Vascular metabolism; Human; Receptors, Immunologic Source Seikagaku 2001 Mar;73(3):205-8.

Molecular cloning of developmentally regulated neonatal rat submandibular gland proteins. Abstract: At birth, the rat submandibular gland (SMG) contains two transient secretory cell types that produce several characteristic salivary proteins. Proteins SMG-A, B1, and B2 (23.5, 26 and 27.5 kDa) are products of the neonatal type III cells, but not the adult acinar cells. Protein C (89 kDa), a major product of the neonatal type I cells, is either absent or present at greatly diminished levels in the secretory cells of the adult gland. The decrease in biosynthesis of these neonatal salivary proteins occurs concomitantly with the increase in levels of characteristic adult SMG products. In order to understand these developmentally regulated changes in SMG salivary protein gene expression, we have initiated the molecular cloning and characterization of neonatal submandibular gland proteins from a 5-d-old rat submandibular gland cDNA library. Clones encoding SMG-A were isolated by homology to the mouse parotid secretory protein (PSP). SMG-A was shown to be derived from a salivary protein multigene family that also includes PSP. Cloning and characterization of additional neonatal rat submandibular gland proteins was initiated by screening the 5-d-old rat submandibular gland cDNA library with first strand cDNA prepared from 1-d-old rat submandibular glands. Clones corresponding to a highly abundant 3 kb transcript present in the neonatal rat SMG, but not in adult submandibular, sublingual, or parotid gland have been identified. The size, abundance, and organ specificity of this transcript suggest that it may encode protein C.(ABSTRACT TRUNCATED AT 250 WORDS); Mirels, L; Girard, L R Language: eng Source Crit Rev Oral Biol Med 1993;4(3-4):525-30.

Molecular cloning of differentially expressed genes during plant somatic embryogenesis; Kayano, T; Momiyama, T; Tabei, Y; Takaiwa, F; Nishio, T Language: jpn Subject(s): Amino Acid Sequence; Cloning, Molecular; Genes, Plant; Molecular Sequence Data; Plants embryology genetics; Polymerase Chain Reaction methods; Sequence Analysis, DNA; Transcription Factors genetics Source Tanpakushitsu Kakusan Koso 1996 Apr;41(5):721-6.

Molecular cloning of DNA. An introduction into techniques and problems. Author(s): Vosberg, H P Language: eng Subject(s): Bacteriophages genetics; Coliphages genetics; DNA Restriction Enzymes; DNA, Viral genetics; National Institutes of Health (U.S.); Plasmids; Recombination, Genetic; Research standards; Risk; Transfection; Transformation, Genetic; United States Source Hum Genet 1977 Dec 29;40(1):1-72.

Molecular cloning of eukaryotic glycoprotein and glycolipid glycosyltransferases: a survey. Abstract: The rapidity with which molecular sequence data are gathered continues to grow. The result is that, for many workers, it is increasingly difficult to keep abreast of the current state of play of molecular cloning, even for those genes that encode proteins of special interest. The clear success of the various worldwide genome projects has made this even more apparent, and by the end of 1996 the complete determination of the nucleotide sequences of the genomes of two eukaryotes, Saccharomyces cerevisiae and Caenorhabditis elegans, will have either been completed or will be nearing completion. This article is an attempt to provide, in an easily accessible format, a compilation of genes and cDNAs that have been sequenced and deposited in GenBank that encode transferase enzymes involved in eukaryotic glycoprotein or glycolipid biosynthesis. The full sequence information can be easily retrieved from a databank, e.g. GenBank, using the relevant accession number(s). Author(s): Field, M C; Wainwright, L J Language: eng Glycobiology 1995 Jul;5(5):463-72.

Molecular cloning of genes encoding gram-positive virulence factors. Author(s): Gilmore, M S Language: eng Source Curr Top Microbiol Immunol 1985;118:219-234.

Molecular cloning of genes encoding transcription factors with the use of recognition site probes. Author(s): Singh, H Language: eng Source Genet Eng (N Y) 1990;12:317-30.

Molecular cloning of hormone genes / edited by Joel F. Habener. Published/Created: Clifton, N.J.: Humana Press, c1987. Related Names: Habener, Joel F. Description: xvii, 441 p.: ill.; 24 cm. ISBN: 0896030911 Notes: Includes bibliographies and index. Subjects: Molecular cloning. Peptide hormones. Series: Molecular biology and biophysics LC Classification: QH442.2 .M65 1987 Dewey Class No.: 574.19/256 19

Molecular cloning of interferon receptors: a short review. Abstract: Interferons (IFNs) exert their pleiotropic effects through at least two different receptors. IFNs alpha/beta (type I IFNs) crossreact with presumably common receptors, while IFN gamma (type II IFN) initiates its biological effects through its own specific receptor system. Transmembrane proteins corresponding to the ligand binding chains of these receptors have recently been cloned. But transfection experiments revealed that additional components are probably involved in high affinity binding of some type I IFN subtypes and that a species specific putative signal transducer is required to confer biological responsiveness to IFN gamma. Although some elements involved in transcriptional regulation of interferon inducible genes have been identified, the pathways involved in these regulatory processes are not understood as yet. The current knowledge on the structure of IFN receptors is reviewed in the context of some recent advances on other cytokine receptors. Author(s): Aguet, M Language: eng ource Br J Haematol 1991 Oct;79 Suppl 1:6-8.

Molecular cloning of ion transporters: potential clinical implications. Abstract: The ion transporters are a group of integral membrane proteins which, like enzymes, can be regulated at multiple levels. Understanding of function and regulation of these proteins has been greatly facilitated by the knowledge of their molecular structures. In this article, we discuss the various approaches used in the cloning of these proteins. We then focus on genetic disorders known to be caused by abnormal structure of the transporter protein including cystic fibrosis, generalized myotonia and myotonia congenita, hyperkalemic periodic paralysis, hereditary ovalocytosis, hereditary spherocytosis and glucose malabsorption. Renal disorders thought, but not proved, to be due to abnormal transporter structure/function are briefly mentioned. Author(s): Cano, A; Alpern, R J Language: eng Source Semin Nephrol 1995 Jan;15(1):2-8.

Molecular cloning of lactose genes in dairy lactic streptococci: the phospho-beta-galactosidase and beta-galactosidase genes and their expression products. Abstract: The mesophilic (S. lactis and S. cremoris) and thermophilic (S. thermophilus) dairy lactic streptococci, which are used in industrial dairy fermentations, contain two different lactose hydrolysing enzymes, a phospho-beta-galactosidase and a beta-galactosidase. The central role of these enzymes in the pathways used for lactose transport and degradation is discussed along with their properties and distributions in lactic streptococci. In addition, recent results on the cloning, expression and sequence organization of the genes for the mesophilic phospho-beta-galactosidase and thermophilic beta-galactosidase are reviewed. Original data are presented concerning heterologous gene expression in the study of lactose hydrolysis in lactic streptococci. These include 1) the purification of the S. lactis phospho-beta-galactosidase from an overproducing Escherichia coli, and 2) the expression of the E. coli beta-galactosidase (lacZ) gene in S. lactis employing a lactic streptococcal expression vector. Author(s): De Vos, W M; Simons, G Language: eng Source Biochimie 1988 Apr;70(4):461-73.

Molecular cloning of liver/bone/kidney-type alkaline phosphatase complementary and genomic DNA: analyses of its deficiency, infantile hypophosphatasia Abstract: Alkaline phosphatase is an enzyme present in nearly all living organisms. The liver/bone/kidney-type isozyme (ALPL) is expressed in the liver, bone, kidney and in most other tissues. We have isolated the ALPL cDNA and its gene and indicated that the gene is divided into two leader exons (exon 1B and 1L) and 11 coding exons and the liver- and bone-specific transcriptions are regulated by their own promoters. The defect of ALPL results in infantile hypophosphatasia, a disorder characterized by defective bone mineralization and subnormal activity of circulating alkaline phosphatase. Prenatal diagnoses of the disease were successfully carried out. Mutation analysis of the family member is in progress. Author(s): Kishi, F Language: jpn Source Nippon Rinsho 1993 Feb;51(2):488-94.

Molecular cloning of neuronal calmodulin mRNA species. Author(s): Brown, I R; Gurd, J W; Ni, B Language: eng Subject(s): brain physiology; calmodulin genetics; DNA genetics; gene library; neurons physiology; RNA, messenger genetics; rats; species specificity Source Biochem Soc Trans 1992 May;20(2):382-5.

Molecular cloning of neuropathy target esterase (NTE). Abstract: Covalent modification of NTE, a neuronal protein with serine esterase activity, by certain organo-phosphates (OP) initiates degeneration of long axons in the peripheral and central nervous system. Simple inhibition of NTE esterase activity does not initiate neuropathy; the latter requires aging of the OP bound to the catalytic serine residue so that a negatively-charged species is left attached to the active site. This may indicate that a non-esterase function of NTE is important for axonal maintenance. We have recently cloned NTE and shown that it is unrelated to any known serine hydrolases but contains a novel C-terminal domain which is conserved from bacteria to man. Furthermore, the catalytic serine is located within this domain at the centre of a helical hydrophobic segment of the polypeptide's secondary structure. The integrity of NTE would be severely compromised by the presence of a negatively-charged organophosphate moiety at this site. Implications for possible higher-order structures and functions for NTE are discussed. Author(s): Glynn, P; Read, D J; Lush, M J; Li, Y; Atkins, J Language: eng Source Chem Biol Interact 1999 May 14;119-120:513-7.

Molecular cloning of pp120/ECTO-ATPase, an endogenous substrate of the insulin receptor kinase. Author(s): Najjar, S M; Philippe, N; Taylor, S I; Accili, D Language: eng Source Adv Exp Med Biol 1993;343:169-74.

Molecular cloning of protozoan myosin I heavy chain genes. Abstract: Myosins I are ubiquitous, nonfilamentous, actin-based mechanoenzymes originally discovered in protozoa. The extensive in vitro biochemical studies of purified protozoan myosins I are now being complemented with in vivo studies using cloned myosin I heavy chain genes and gene targeting techniques. Here we review briefly the systems and methods being used in these efforts to dissect protozoan myosin I structure and function using molecular genetic approaches. Author(s): Hammer, J A 3rd; Jung, G Language: eng Source J Cell Sci Suppl 1991;14:37-40.

Molecular cloning of recombinant DNA: proceedings of the Miami winter symposia, January 1977 / edited by W. A. Scott, R. Werner; sponsored by the Department of Biochemistry, University of Miami, School of Medicine, Miami, Florida. Published/Created: New York: Academic Press, 1977. Related Names: Scott, Walter A. Werner, Rudolf, 1934- Description: xii, 249 p.: ill.; 24 cm. ISBN: 0126342504 Notes: Includes bibliographical references. Subjects: Molecular cloning--Congresses. Genetic recombination--Congresses. DNA--Congresses. Series: Miami winter symposia; v. 13 LC Classification: QH442 .M64 Dewey Class No.: 574.8/732

Molecular cloning of ricin. Abstract: A variety of strategies have been used to obtain cDNA and genomic clones encoding ricin. Since their isolation these sequences have been manipulated to allow expression of A chain (19) and A chain mutants (15,20,34), B chain (14,21-23) and proricin (24). Utilizing structural information (35), precise changes have been introduced into both A and B chains with the aim of probing catalytic and sugar-binding residues, respectively. In the longer term, such manipulations, coupled with successful expression and purification schemes, will allow the delineation of functional residues and domains, ensuring that ricin remains the prototype plant toxin with which to study cellular intoxication and ribosome inactivation and to utilize in pharmaceutical product development. Author(s): Roberts, L M; Tregear, J W; Lord, J M Language: eng Source Targeted Diagn Ther 1992;7:81-97.

Molecular cloning of schistosome genes. Author(s): Taylor, D W; Cordingley, J S; Dunne, D W; Johnson, K S; Haddow, W J; Hormaeche, C E; Nene, V; Butterworth, A E Language: Source Parasitology 1986;92 Suppl:S73-81.

Molecular cloning of sequence-specific DNA binding proteins using recognition site probes. Abstract: Genes encoding sequence-specific DNA binding proteins can be isolated by screening lambda gt11 expression libraries with recognition site DNAs. This strategy is derived from that developed for the isolation of genes using antibody probes. Many different genes encoding transcriptional regulatory proteins have been cloned using this strategy. The DNA binding domains of these regulatory proteins contain different structural motifs including the helix-turn-helix, the "zinc finger" and the "leucine zipper". Various aspects of the screening strategy are evaluated and a detailed protocol is provided. In addition to binding site DNAs, protein and nucleotide probes have been successfully used to screen expression libraries. Therefore ligand based expression screening may be quite general in scope. Author(s): Singh, H; Clerc, R G; LeBowitz, J H Language: eng Source Biotechniques 1989 Mar;7(3):252-61.

Molecular cloning of the genes in genetic chorioretinal diseases--positional cloning and the candidate gene approach Abstract: Two different molecular biological approaches to the disease-causing genes of genetic eye diseases are described. In gyrate atrophy of the chroid and retina where the biochemical defect was identified as inactivation of ornithine aminotransferase, the gene was cloned by using antibody for the enzyme. In most genetic eye diseases, however, the biochemical defects are unknown. Positional cloning and/or the candidate gene approach are used to identify the disease-causing genes for these diseases. The genes of chroideremia and Norrie disease were cloned by positional cloning. Several genes expressed in the photoreceptor cells have been identified recently and may be the genes causing progressive degeneration of the retina and choroid. Rhodopsin, peripherin (RDS), rom-1, and beta subunit-cGMP phosphodiesterase are identified as the disease-causing genes for retinitis pigmentosa by the candidate gene approach. Author(s): Mashima, Y; Saga, M; Oguchi, Y Language: jpn Source Nippon Ganka Gakkai Zasshi 1993 Nov;97(11):1253-64.

Molecular cloning of the human AT2 receptor. Author(s): Martin, M M; Su, B; Elton, T S Language: eng Source Adv Exp Med Biol 1996;396:153-65.

Molecular cloning of the selenocysteine-containing enzyme type I iodothyronine deiodinase. Abstract: It has proved impossible to purify the three types of iodothyronine deiodinase to homogeneity by use of standard physiochemical techniques. However, using expression cloning with Xenopus oocytes, we were successful in isolating a cDNA coding for the rat type I iodothyronine deiodinase. An extremely puzzling aspect of this cDNA was an in-frame thymine-guanine-adenine (TGA) codon at position 126. By mutagenesis studies we showed that this coded for selenocysteine, consistent with the previously described requirement for selenium for hepatic and renal thyronine deiodination. The cloned enzyme can be expressed transiently in JEG, a human cell line, or COS, a monkey cell line, by gene-transfer techniques, and the expressed protein has the appropriate biochemical characteristics of type I iodothyronine deiodinase. These include a newly identified extreme sensitivity to competitive inhibition of reverse-3,5,3'-triiodothyronine (rT3) deiodination by gold. The identification of this elusive protein should prove helpful in understanding the process of thyroid hormone activation. Author(s): Berry, M J; Larsen, P R Language: eng Subject(s): Animal; Cloning, Molecular; Gene Expression; Human; Iodide Peroxidase/analysis/chemistry/genetics/isolation & purification/metabolism; Selenocysteine/analysis; Transfection Source Am J Clin Nutr 1993 Feb;57(2 Suppl):249S-255S.

Molecular cloning of the von Hippel-Lindau tumor suppressor gene and its role in renal carcinoma. Author(s): Gnarra, J R; Duan, D R; Weng, Y; Humphrey, J S; Chen, D Y; Lee, S; Pause, A; Dudley, C F; Latif, F; Kuzmin, I; Schmidt, L; Duh, F M; Stackhouse, T; Chen, F; Kishida, T; Wei, M H; Lerman, M I; Zbar, B; Klausner, R D; Linehan, W M Language: eng Source Biochim Biophys Acta 1996 Mar 18;1242(3):201-10.

Molecular cloning vectors for Aspergillus and Neurospora. Author(s): Cullen, D; Gray, G L; Berka, R M Language: eng Subject(s): Aspergillus nidulans/genetics; Cloning, Molecular/methods; DNA Replication; Escherichia coli/genetics; Genetic Markers; Genetic Vectors; Neurospora/genetics, Neurospora crassa/genetics; Saccharomyces cerevisiae/genetics; Transformation, Genetic Source Biotechnology 1988;10:419-33.

Molecular cloning vectors of Saccharomyces: generalized cloning vectors. Author(s): West, R W Jr Language: eng Source Biotechnology 1988;10:387-404.

Molecular cloning, expression and assembly of multimeric von Willebrand factor. Abstract: Recently, substantial progress has been made in our knowledge of the domains involved in correlating structure and function of vWF, as well as in the biosynthesis and assembly of multimeric vWF. These studies were greatly supported by the development of three new techniques. (1) In vitro culturing of (human) vascular endothelial cells has allowed studies on the subcellular localization for dimerization of pro-vWF subunits, multimerization, carbohydrate and proteolytic processing. Moreover, this approach has provided insight into the complex intracellular routing of vWF that proceeds by either one of two pathways. During the constitutive pathway, vWF is packaged in secretory vesicles that are rapidly secreted both at the luminal and at the basolateral side. The regulated pathway includes storage of vWF molecules in specialized organelles, i.e. the Weibel-Palade bodies. (2) The application of proteases to dissect purified multimeric vWF and to assign function(s) to defined proteolytic fragments of vWF. Monoclonal antibodies, raised against native vWF, that block specific functions and bind to fragments are subsequently employed to correlate structure and function. More precise localizations of functions are now feasible, using overlapping synthetic peptides derived from the primary amino acid sequence of (pro)-vWF and antibodies raised against such peptides. This approach has permitted a fine mapping of the interaction site of vWF with the platelet receptor glycoprotein (GP) IIB/IIIA. (3) Molecular cloning of full-length vWF cDNA and the development of eukaryotic expression systems have substantially increased the possibilities to investigate structures on pro-vWF involved in the biosynthesis and the assembly of multimers. In particular, the construction of point and deletion mutants of vWF, employing vWF cDNA, and subsequent expression in heterologous cells have demonstrated that proteolytic processing of pro-vWF, between arginine (Arg-763) and serine (Ser-764), to generate free propolypeptide and mature vWF is not required for multimerization. Finally, the propolypeptide has an obligatory role for the formation of multimers, enabling interdimer disulphide bonding of free sulphhydryl groups located within the mature vWF part of pro-vWF. Author(s): Pannekoek, H; Voorberg, J Language: eng Source Baillieres Clin Haematol 1989 Oct;2(4):879-96.

Molecular cloning, expression and epitope mapping of nuclear antigen Abstract: The antinuclear antibody is a major characteristic phenomenon in systemic type of autoimmune diseases. There is a good correlation between symptoms or diseases and the types of antinuclear antibodies in these patients. Therefore, the study of the mechanisms of antinuclear antibody production is thought to be important in the elucidation of the pathogenesis of such autoimmune diseases. The conventional methods to detect antinuclear antibodies have been useful to date in clinical studies. However, these methods are not sufficient for more precise investigations. Thus, we have applied the recently developed molecular biology to this field. cDNAs which encode nuclear antigens have been cloned and their protein products were expressed as recombinant fusion protein in E. coli. By this procedure, we could establish simple and reproducible method to detects antinuclear antibodies. Furthermore, epitope mapping of nuclear antigens could be performed using deletion mutants generated by enzymatic manipulations of the cDNAs. These epitope studies are facilitating a more precise understanding of the mechanisms of antinuclear antibody production. Author(s): Yamamoto, K Language: jpn Source Rinsho Byori 1994 May;42(5):460-6.

Molecular cloning, expression, and uses of mammalian glycosyltransferases. Abstract: Lineage specific expression patterns of mammalian oligosaccharides are regulated largely by glycosyltransferases. Several distinct glycosyltransferase genes have recently been isolated by gene transfer and traditional cloning approaches. These each maintain a type II transmembrane topology, that yields a COOH-terminal, Golgi-resident catalytic domain. However, significant primary sequence similarities between cloned glycosyltransferases have been found only between enzymes with substantial functional similarity. These cloned DNA segments, and their cognate enzymes, have been used to determine glycosyltransferase expression patterns, characterize the molecular basis for mutant glycosyltransferase alleles, and identify oligosaccharide ligands for members of the Selectin family of cell adhesion molecules. They have also shown promise for use in the preparative synthesis of defined oligosaccharide molecules, and for transgenic animal approaches designed to explore the function(s) of oligosaccharides in mammalian organisms. Author(s): Lowe, J B Language: eng Source Semin Cell Biol 1991 Oct;2(5):289-307.

Molecular cloning, structural characterization, and analysis of transcription of the melanoma oncogene of xiphophorus. Author(s): Schartl, M; Adam, D Language: eng Source Pigment Cell Res 1992;Suppl 2:173-80.

Molecular cloning. Author(s): Hamer, D H; Thomas, C A Jr Language: eng Subject(s): Base Sequence; Clone Cells/metabolism; DNA Restriction Enzymes; DNA, Recombinant/analysis/metabolism; Drug Resistance, Microbial; Electrophoresis, Agar Gel; Escherichia coli/metabolism; In Vitro; Methods; Mutation; Plasmids; Polynucleotide Ligases; Suppression, Genetic; Transformation, Genetic Source Adv Pathobiol 1977;(6):306-19.

Molecular cloning. Author(s): Kobayashi, Y; Kawamura, F Language: eng Subject(s): Bacillus subtilis/genetics; Bacteriophages/genetics; Cloning, Molecular/methods; Genetic Vectors; Plasmids Source Biotechnology 1992;22:123-41.

Molecular control of melanogenesis in malignant melanoma: functional assessment of tyrosinase and lamp gene families by UV exposure and gene co-transfection, and cloning of a cDNA encoding calnexin, a possible melanogenesis "chaperone". Abstract: Melanogenesis is a cascade of events significantly controlled by regulatory genes which are associated with the melanosomal membrane. This report introduces our current research efforts dealing with (a) the gene and protein expressions of tyrosinase and Lamp (lysosome-associated membrane protein) families by human melanoma cells after repeated exposures to UV light, (b) the coordinated alterations in the expression of the Lamp family gene and its encoding product after transfection of two genes of the tyrosinase family in human melanoma cells and (c) cloning and sequencing of a Ca(2+)-binding phosphoprotein, calnexin, which could be a candidate as a chaperone for sorting and maturation of tyrosinase and Lamp family glycoproteins in melanogenesis cascade. Our UV exposure study, as well as gene transfection and antisense hybridization experiments, has clearly indicated a marked and coordinated interaction of the Lamp-1 gene with the tyrosinase and TRP-1 genes in this process. We propose that melanogenesis is controlled at least by two major gene family products, i.e., (a) the tyrosinase family of tyrosinase, TRP-1 and TRP-2, and the Lamp family of Lamp-1, Lamp-2 and Lamp-3. These two gene families probably derived from primordial melanogenesis-associated genes which are common or closely related to each other. Author(s): Jimbow, K; Hara, H; Vinayagamoorthy, T; Luo, D; Dakour, J; Yamada, K; Dixon, W; Chen, H Language: eng Source J Dermatol 1994 Nov;21(11):894-906.

Molecular genetic analysis of the V kappa Ser group associated with two mouse light chain genetic markers. Complementary DNA cloning and southern hybridization analysis. Abstract: Previous studies (21) have shown that two mouse kappa light (L) chain variable (V) region polymorphisms, the IB-peptide and Efla markers, reflect expression of a characteristic group of V kappa regions, called V kappa Ser, by some inbred strains and not others. Expression of V kappa Ser is controlled by a locus on chromosome 6, the chromosome that contains the kappa locus. To further characterize this V kappa group and begin to analyze the basis for its strain-specific expression, full-length complementary DNA (cDNA) copies were produced of L chain mRNA from the M75 myeloma that had been induced in the C.C58 strain of mice, and which produces a V kappa Ser L chain. The C.C58 strain is congenic with

BALB/cAn, differing in the region of chromosome 6 that controls expression of the V kappa polymorphisms and the Lyt-2 and Lyt-3 T cell alloantigens. The complete nucleotide sequence of this cloned cDNA was determined and compared with the nucleotide sequences the most closely related BALB/c myeloma L chains known. Results indicated significant differences throughout the variable region, but particularly toward the 5' portion of the sequence. A probe corresponding to 200 bp of the 5' end of the cloned V kappa Ser cDNA was used in Southern hybridizations of restriction digests of liver DNA from a number of inbred, recombinant, and recombinant inbred strains. Under stringent hybridization conditions, one strongly-hybridizing fragment was observed in Bam HI, Hind III, and Eco RI digests, and based on the size of the fragments, strains could be organized into two groups. The presence of strongly hybridizing Bam HI, Hind III, and Eco RI fragments of 3.2, 2.8, and 2.1 kb, respectively, was found to correlate completely with expression by the strain of the IB-peptide and Efla markers. All nonexpressor strains yielded hybridizing fragments of 7.8, 8.4, and 2.8 kb, respectively. Possible explanations for strain-specific expression of V kappa Ser-associated phenotypic markers are discussed. Author(s): Goldrick, M M; Boyd, R T; Ponath, P D; Lou, S Y; Gottlieb, P D Language: eng Source J Exp Med 1985 Aug 1;162(2):713-28.

Molecular genetic approach for positional cloning of X-chromosome-linked diseases; Onodera, O; Tanabe, H; Tuji, S; Kaneko, K; Kobayashi, H Language: jpn Subject(s): Cloning, Molecular; DNA/genetics; DNA Probes; Genome, Human; Human; Linkage (Genetics); Mutation; Neuromuscular Diseases/genetics; Repetitive Sequences, Nucleic Acid; Transcription, Genetic; X Chromosome Source Tanpakushitsu Kakusan Koso 1993 Feb;38(3):354-60.

Molecular genetics of phosphorylase kinase: cDNA cloning, chromosomal mapping and isoform structure. Abstract: A deficiency in phosphorylase kinase is responsible for several forms of glycogen storage disease which differ in heredity and affected tissues. This is so because phosphorylase kinase consists of four different subunits and has multiple tissue-specific isoforms. To elucidate the molecular basis of phosphorylase kinase deficiencies, the cDNAs encoding the subunits alpha and beta were cloned and sequenced. Each subunit was shown to be encoded by a single gene. The alpha subunit gene was mapped to chromosome Xq12-q13 and the beta subunit gene to chromosome 16q12-q13. Isoform cDNAs reveal differential mRNA splicing. Thus, the stage is set for the molecular characterization of the genes and their deficiency mutations. Author(s): Kilimann, M W Language: eng Source J Inherit Metab Dis 1990;13(4):435-41.

Molecular neurobiology: recombinant DNA approaches / edited by Steve Heinemann and James Patrick. Published/Created: New York: Plenum Press, c1987. Related Names: Heinemann, Steve. Patrick, James, 1942- Description: xvii, 297 p.: ill.; 24 cm. ISBN: 0306424401 Notes: Includes bibliographies and index. Subjects: Molecular neurobiology. Recombinant DNA. Cloning, Molecular. DNA, Recombinant. Gene Expression Regulation. Molecular Biology. Nervous System. Neurobiology. Series: Current topics in neurobiology LC Classification: QP356.2 .M65 1987 Dewey Class No.: 599/.0188 19

Monozygotic twins: reproductive anomaly or successful natural cloning?; Blickstein, I Language: heb; Animal; Cloning, Organism; Female; Human; Male; Reproduction; Twins, Monozygotic/genetics Source Harefuah 1998 Sep;135(5-6):213-7.

Moody, Skye Kathleen. Habitat / Skye Kathleen Moody. Edition Information: 1st ed. Published/Created: New York: St. Martin's Minotaur, 1999. Description: 274 p.: ill.; 22 cm. ISBN: 031220390X Subjects: Diamond, Venus (Fictitious character)--Fiction. Women environmentalists--Fiction. Government investigators--Fiction. Endangered species--Fiction. Cloning--Fiction. Washington (State)--Fiction. LC Classification: PS3563.O5538 H3 1999 Dewey Class No.: 813/.54 21

Moral aspects of human cloning and relations between humans Abstract: It is usually assumed that there is a strong logical relationship between traditional worldviews and attitudes towards human cloning. I argue that the association is contingent, which can be seen in the analysis of arguments of such leading bioethicists as G. J. Annas, H. Jonas, and J. A. Robertson. Traditional worldviews lack conceptual resources that would enable them to provide normative conclusions about cloning. It is therefore mistake to look to these traditional worldviews in search for moral instruction. Instead, one should appeal to knowledge of psychological and social aspects of human life, and in particular to the process of emotional ties which form relations between humans. Author(s): Lukow, P Language: Source Med Wieku Rozwoj 2001;5(1 Suppl 1):103-11.

More about interferon cloning. Author(s): Derynck, R Language: eng Subject(s): Amino Acid Sequence; Animal; Base Sequence; Cloning, Molecular; Comparative Study; DNA/genetics; Escherichia coli/genetics/metabolism; Genes, Structural; Human; Interferon Type II/biosynthesis/genetics; Saccharomyces cerevisiae/genetics Source Interferon 1983;5:181-203.

Morgan, Sally. Body doubles: cloning plants and animals / Sally Morgan. Published/Created: Chicago: Heinemann Library, c2002. Description: 64 p.: col. ill.; 24 cm. ISBN: 1588106985 Contents: The path to Dolly -- Natural cloning -- Chromosomes and DNA -- Cloning in the lab -- Cloning in action -- Spare-part surgery -- Human cloning -- Future developments. Notes: Includes bibliographical references (p. 63) and index. Subjects: Cloning--Juvenile literature. Cloning. Series: Science at the edge LC Classification: QH442.2 .M67 2002 Dewey Class No.: 660.6/5 21

Mouse glutaredoxin - cDNA cloning, high level expression in E. coli and its possible implication in redox regulation of the DNA binding activity in transcription factor PEBP2. Abstract: We have isolated a cDNA encoding glutaredoxin (GRX) from a mouse splenic cDNA library. This cDNA encoded a protein of 107 amino acids with a calculated molecular weight of 11.9 kDa. The deduced amino acid sequence of glutaredoxin in mouse was highly homologous with that in other mammals (81-89%), containing a putative active sequence of -Cys-Pro-Try-Cys-. Recombinant mouse glutaredoxin expressed in E. coli showed glutathione-disulfide oxidoreductase activity with beta-hydroxyethyl disulfide as its substrate, whereas mutant glutaredoxin (Cys 22, Cys 25 to Ser) showed no activity. In electrophoretic mobility shift assay, we proved that wild type GRX, not mutant one, recovered the DNA-binding activity of a transcription factor, PEBP2, oxidized by diamide. This showed that GRX may be involved in the redox regulation of the DNA-binding activity of PEBP2 as is the case with thioredoxin. Author(s): Nakamura, T; Ohno, T; Hirota, K; Nishiyama, A; Nakamura, H; Wada, H; Yodoi, J Language: eng Source Free Radic Res 1999 Oct;31(4):357-65.

Mucin cDNA cloning. Author(s):Aubert, J P; Porchet, N Language: eng Subject(s): Cloning, Molecular/methods; DNA, Complementary; Human; Mucins/genetics Source Methods Mol Biol 2000;125:297-303.

Multiple serotonin receptor subtypes: molecular cloning and functional expression. Abstract: The field of serotonin (5HT) receptor pharmacology has been at least as dramatically altered by the advent of molecular neurobiology and recombinant DNA techniques as has any other neurotransmitter receptor field. The principal reason for this is the unforeseen multitude of genes expressing different types of 5HT receptors. Classic pharmacological studies as well as radioligand binding studies convinced workers in the field that there were multiple 5HT receptors. The extent of that multiplicity was not generally foreseen. At the time of this writing, no less than 13 5HT receptor genes have been cloned and functionally expressed. In addition, a fourteenth receptor has been well studied and will undoubtedly be cloned in the near future. These novel 5HT receptor clones are appearing at an astonishing rate. All 13 receptors have been cloned between the years 1987 to 1993, with more than half of the clones appearing in the literature in the last 18 months. Furthermore, there is no indication that all of the 5HT receptors present in the mammalian genome have been cloned. In fact, there are classic pharmacological data as well as radioligand binding data that implicate the existence of additional 5HT receptor subtypes not yet fully defined or cloned. Bringing order to this rapid outpouring of information at this time is a very difficult task. Not only is there a great multiplicity of receptor subtypes, each with a unique pharmacology and distribution, but there are species variants of 5HT receptors. In order to provide a concise and timely review, this article focuses on the strategies used to clone and express multiple 5HT receptors. Unique properties of the various receptors are emphasized. Author(s): Teitler, M; Herrick-Davis, K Language: eng Source Crit Rev Neurobiol 1994;8(3):175-88.

Murine natural killer cell cloning from fetal thymic organ cultures. Author(s):Ballas, Z K; Rasmussen, W L Language: eng Subject(s): Animal; Clone Cells; Human; Killer Cells, Natural/cytology; Mice; Organ Culture; Thymus Gland/cytology/embryology Source Methods Mol Biol 2000;121:39-47.

Muscle coding sequences and their regulation during myogenesis: cloning of muscle actin cDNA probes. Abstract: For a number of years our group has been mainly interested in the regulation of muscle gene expression during myogenesis. Using primary cultures and cell lines we have tried to find out whether the coding sequences for muscle proteins are already present in an unexpressed form or if there is a transcriptional switch at the onset of differentiation. Metabolic studies on pulse-labelled RNA, together with translation and molecular hybridization experiments have given a certain number of indications. More recently the development of genetic engineering techniques has made it possible to answer these questions directly with probes which are complementary to specific muscle coding sequences. We have identified a plasmid which contains a coding sequence for muscle actin. Other recombinant plasmids are being characterized. Such plasmids, used as probes, will permit us to study the organization and expression of the genes coding for the contractile proteins in muscle cells. Author(s): Minty, A; Caravatti, M; Robert, B; Cohen, A; Daubas, P; Weydert, A; Gros, F; Buckingham, M Language: eng Source Reprod Nutr Dev 1981;21(2):247-55.

Na+:HCO(3-) cotransporters (NBC): cloning and characterization. Abstract: The sodium bicarbonate cotransporter (NBC1) is essential for bicarbonate transport across plasma membranes in epithelial and nonepithelial cells. The direction of the $NaHCO_3$ movement in secretory epithelia is opposite to that in reabsorptive epithelia. In secretory epithelia (such as pancreatic duct cells) NBC is responsible for the transport of bicarbonate from blood to the cell for eventual secretion at the apical membrane. In reabsorptive epithelia (such as kidney proximal tubule cells) NBC is responsible for the reabsorption of bicarbonate from cell to the blood. In nonepithelial cells this transporter is mainly involved with cell pH regulation. Recent molecular cloning experiments have identified the existence of four NBC isoforms (NBC1, 2, 3 and 4) and two NBC-related proteins AE4 and NCBE (Anion Exchanger 4 and Na-dependent Chloride-Bicarbonate Exchanger). All but

AE4 are presumed to mediate the cotransport of Na+ and HCO(3-) under normal conditions and may be functionally altered in certain pathologic states. NBC1 shows a limited tissue expression pattern, is electrogenic and plays an important role in bicarbonate reabsorption in kidney proximal tubule. In addition to the kidney, NBC1 is expressed in pancreatic duct cells, is activated by cystic fibrosis transmembrane conductance regulator (CFTR) and plays an important role in HCO3- secretion. NBC2 and NBC3 have a wider tissue distribution than NBC1, are electroneutral, and are involved with cell pH regulation. The characterization of NBC4 is incomplete. The NBC-related protein called NCBE mediates Na-dependent, Cl-/Bicarbonate Exchange. The purpose of this review is to summarize recent advances on the cloning of NBC isoforms and related proteins and their role and regulation in physiologic and pathologic states. Author(s): Soleimani, M; Burnham, C E Language: eng Source J Membr Biol 2001 Sep 15;183(2):71-84.

Nardo, Don, 1947- Cloning / by Don Nardo. Published/Created: San Diego, Calif.: Kidhaven Press, c2003. Description: 48 p.: col. ill.; 23 cm. ISBN: 0737714034 (hbk.: alk. paper) Summary: Discusses the history of cloning, its possible applications, and some of the controversy surrounding it. Contents: The long road to Dolly -- Cloning endangered species -- Medical benefits of animal cloning -- The question of human cloning. Notes: Includes bibliographical references (p. 44-45) and index. Subjects: Cloning--Juvenile literature. Series: The Kidhaven science library LC Classification: QH442.2 .N372 2003 Dewey Class No.: 660.6/5 21

Nardo, Don, 1947- Cloning / by Don Nardo. Published/Created: San Diego, Calif.: Blackbirch Press, c2003. Description: 48 p.: col. ill.; 24 cm. ISBN: 1567117821 Summary: Discusses the history of the concept of cloning and the pros and cons of cloning animals and humans. Contents: Cloning is nothing new -- The potential benefits of animal cloning -- The promise of human cloning. Notes: Includes bibliographical references and index. Subjects: Cloning--Juvenile literature. Series: Science on the edge LC Classification: QH442.2 .N37 2003 Dewey Class No.: 660.6/5 21

Nardo, Don, 1947- Cloning / Don Nardo. Published/Created: San Diego, Calif.: Lucent Books, c2002. Description: 128 p.: ill.; 24 cm. ISBN: 1560069279 (hbk.: alk. paper) Contents: Before Dolly: cloning in nature, agriculture, and the imagination A big splash for a little lamb: the creation of Dolly -- Of cows, pigs, and rats: potential medical benefits of animal cloning -- Diversity and dinosaur DNA: cloning endangered and extinct species -- The advent of human cloning: who will be cloned and why? -- Playing God?: ethical and moral concerns about human cloning -- Should and will human cloning be banned? Notes: Includes bibliographical references (p. 114-121) and index. Subjects: Cloning--Juvenile literature. Series: Great medical discoveries LC Classification: QH442.2 .N37 2002 Dewey Class No.: 660.6/5 21

Neurofibromatosis type 1. Beyond positional cloning. Abstract: Recent advances in molecular genetics have enabled researchers to more rapidly identify human disease genes. The identification of these genes by positional cloning has opened the door to a better understanding of such diseases through a more complete appreciation of the molecular biologic processes that underlie them. In this review, the approaches taken to dissect the function of the gene for von Recklinghausen neurofibromatosis are presented. These general approaches involve identification of the protein product, determination of its relation to other known proteins, analysis of its distribution in tissues, within cells, and over development, and dissection of its role in producing the disease phenotype. Last, the insights gained from studying the molecular biology of the neurofibromatosis type 1 gene have direct impact on other biologic processes, such as neoplasia, cellular differentiation, and growth factor-mediated signal transduction as well as potential application to improved treatments for neurofibromatosis type 1 and cancer. Author(s): Gutmann, D H; Collins, F S Language: eng Source Arch Neurol 1993 Nov;50(11):1185-93.

Neurofibromatosis type 2 and von Hippel-Lindau disease: from gene cloning to function. Abstract: Most of the genes for hereditary tumor syndromes cloned thus far have subsequently been shown to be mutated not only in the germlines and tumors from patients with the relatively rare inherited disease, but also in the much more common sporadic tumor counterparts in the general population. Thus, the isolation and functional characterization of genes associated with hereditary tumor syndromes have emerged as a major strategy to gain insights into some of the most fundamental mechanisms of tumorigenesis. The search for the genes causing two hereditary tumor syndromes of the nervous system, neurofibromatosis type 2 (NF2) and von Hippel-Lindau disease (VHL), has recently culminated in the cloning of both disease genes. This represents another successful application of the so-called positional cloning approach, i.e., the isolation of a hereditary disease gene with unknown function, based on the determination of its chromosomal location in the human genome. The gene for NF2, a syndrome typically associated with vestibular schwannomas and meningiomas, is homologous with a family of genes whose members appear to play an important role in bridging the cell membrane with the intracellular cytoskeleton, including moesin, ezrin, radixin, and protein 4.1. Recent mutation analyses have revealed that the NF2 tumor suppressor gene is frequently mutated not only in vestibular schwannomas and meningiomas from NF2 patients, but also in their sporadic counterparts, which represent approximately one-third of all human brain tumors. Furthermore, malignant human tumors seemingly unrelated to the NF2 syndrome, such as neural crest-derived malignant melanomas, as well as malignant mesotheliomas (a pleural mesoderm-derived tumor), have also been found to be frequently mutated or deleted in the NF2 locus, suggesting a broader role for the NF2 gene in the initiation and progression of human neoplasms. VHL is a rare tumor syndrome characterized by certain types of nervous system tumors (cerebellar and spinal hemangioblastomas as well as retinal angiomas), in conjunction with bilateral renal cell carcinomas and pheochromocytomas. Similar to NF2, recent genetic mutation studies have revealed that the VHL tumor suppressor gene is not only mutated in the hereditary tumors from VHL patients, but also in their sporadic counterparts. Importantly, the VHL gene represents the most frequently mutated cancer-related gene thus far identified in sporadic renal cell carcinoma. In contrast to most other hereditary cancer syndromes, however, VHL mutations are surprisingly specific for tumors typically associated with the VHL syndrome, and have not been detected in any other tumor type unrelated to VHL. The cloning and initial genetic characterization of the NF2 and VHL genes have now provided a rational basis for subsequent functional studies on the elucidation of the normal and tumor-associated cellular signaling pathways of these tumor suppressor genes. Author(s): Kley, N; Whaley, J; Seizinger, B R Language: eng Source Glia 1995 Nov;15(3):297-307.

New advances and future perspectives in mammal cloning Abstract: To the great extent, the study and application of transgenic animal are restricted by the inherent limitation of pronuclear microinjection. Recently, the rapid progresses in gene targeting and cloning of somatic cells have shown that the combination of these two technologies will become a virtual way to producing large transgenic animals. Author(s): Tan, X H; Yang, X; Huang, P T Language: chi. Source Sheng Wu Gong Cheng Xue Bao 2001 Mar;17(2):118-20.

New strategies for isolation of low abundance viral and host cDNAs: application to cloning of the hepatitis E virus and analysis of tissue-specific transcription. Abstract: The ability to clone viruses molecularly has led to dramatic advances in our understanding of this diverse group of agents at the molecular level. These insights are critical to the development of experimental strategies for the containment and control of the various viral pathogens that cause hepatitis in man. Knowledge of the genes and gene products have, for example, assisted in the design of expression systems that have been useful for the cell-free expression and production of viral constituents in the development of a subunit (recombinant) vaccine for HBV. Of particular advantage to the experimenter would be the ability to clone the virus directly from the infectious source without the need to resort to biologic amplification systems that require costly investments of money and manpower. This would obviate the need for the time-intensive development of a tissue culture propagation or the expensive cost of developing an animal model. The methods described here have applications to the discovery and isolation of low abundance transcripts from host (cellular) genes. These host encoded products may have important roles in virus replication and certainly by themselves constitute a growing area of study. The direct selection protocol has already proven its value in the isolation of novel (previously undescribed) gene sequences from complex cellular sources. The application of these methodologies should, at the single cell level, aid in the delineation of those important host-encoded gene products that are critical for the efficient in vitro propagation of hepatotropic viruses. The identification of these low abundance cellular genes will elucidate the biologic interplay between host and obligate intracellular parasite (virus) and potentially lead to the development of new strategies for virus control that take into consideration the role of the cell as host. Author(s): Reyes, G R; Bradley, D W; Lovett, M Language: eng Source Semin Liver Dis 1992 Aug;12(3):289-300.

Nightline. [1993-10-25] / ABC News. Published/Created: United States: ABC-TV, 1993-10-25. Related Names: Vanderbilt Television News Archive Collection (Library of Congress) Description: 1 videocassette of 1 (30 min.): sd., col.; 3/4 in. master. Summary: Re: ethical issues related to scientific advancements in human embryo cloning. Notes: Reference sources used: Television news index and abstracts (Vanderbilt 1993 annual) & container label. LC Classification: VSC 9502 (master)

Nightline. [1997-02-24] / ABC News. Published/Created: United States: ABC-TV, 1997-02-24. Related Names: Vanderbilt Television News Archive Collection (Library of Congress) Description: 1 videocassette of 1 (30 min.): sd., col.; 3/4 in. master. Summary: Issues related to the possibility of human cloning following the announcement of the first successful animal cloning. Notes: Sources used: Vanderbilt Television News Archive website; container label. LC Classification: VSD 4101 (master)

Nightline. [1997-04-04], Fooling with Mother Nature / ABC News. Published: United States: ABC-TV, 1997-04-04. Related Names: Vanderbilt Television News Archive Collection (Library of Congress) Description: 1 videocassette of 1 (30 min.): sd., col.; 3/4 in. master. Summary: Ethical issues related to scientific advances in cloning. Notes: Sources used: Vanderbilt Television News Archive website; container label. LC Classification: VSD 4132 (master)

Nightline. [1998-01-07], Method or madness / ABC News. Published: United States: ABC-TV, 1998-01-07. Related Names: Vanderbilt Television News Archive Collection Description: 1 videocassette of 1 (30 min.): sd., col.; 3/4 in. master. Summary: Issues related to human cloning. Notes: Sources used: Vanderbilt Television News Archive website; container label. LC Classification: VSD 5065 (master)

Nightline. [1998-07-23], Of mice and men / ABC News. Published/Created: United States: ABC-TV, 1998-07-23. Related Names: Vanderbilt Television News Archive Collection (Library of Congress) Description: 1 videocassette of 1 (30 min.): sd., col.; 3/4 in. master. Summary: Issues related to advances in cloning research. Notes: Sources used: Vanderbilt Television News Archive website; container label. LC Classification: VSD 5300 (master)

Nightline. [1998-10-06], Thinking twice / ABC News. Published/Created: United States: ABC-TV, 1998-10-06. Related Names: Vanderbilt Television News Archive Collection (Library of Congress) Description: 1 videocassette of 1 (30 min.): sd., col.; 3/4 in. master. Summary: Taped segments with Robert Krulwich (ABC), Princeton University Professor Lee Silver, Professor Stephen Jay Gould, actors Buck Henry and Jane Curtin, and unidentified others re: issues related to advances in cloning research, with introductory and concluding remarks by Ted Koppel (ABC). Notes: Sources used: Vanderbilt Television News Archive website; container label. LC Classification: VSD 5640 (master)

Nitric oxide synthase isozymes. Characterization, purification, molecular cloning, and functions. Abstract: Three isozymes of nitric oxide (NO) synthase (EC 1.14.13.39) have been identified and the cDNAs for these enzymes isolated. In humans, isozymes I (in neuronal and epithelial cells), II (in cytokine-induced cells), and III (in endothelial cells) are encoded for by three different genes located on chromosomes 12, 17, and 7, respectively. The deduced amino acid sequences of the human isozymes show less than 59% identity. Across species, amino acid sequences for each isoform are well conserved (> 90% for isoforms I and III, > 80% for isoform II). All isoforms use L-arginine and molecular oxygen as substrates and require the cofactors NADPH, 6(R)-5,6,7,8-tetrahydrobiopterin, flavin adenine dinucleotide, and flavin mononucleotide. They all bind calmodulin and contain heme. Isoform I is constitutively present in central and peripheral neuronal cells and certain epithelial cells. Its activity is regulated by Ca2+ and calmodulin. Its functions include long-term regulation of synaptic transmission in the central nervous system, central regulation of blood pressure, smooth muscle relaxation, and vasodilation via peripheral nitrergic nerves. It has also been implicated in neuronal death in cerebrovascular stroke. Expression of isoform II of NO synthase can be induced with lipopolysaccharide and cytokines in a multitude of different cells. Based on sequencing data there is no evidence for more than one inducible isozyme at this time. NO synthase II is not regulated by Ca2+; it produces large amounts of NO that has cytostatic effects on parasitic target cells by inhibiting iron-containing enzymes and causing DNA fragmentation. Induced NO synthase II is involved in the pathophysiology of autoimmune diseases and septic shock. Isoform III of NO synthase has been found mostly in endothelial cells. It is constitutively expressed, but expression can be enhanced, eg, by shear stress. Its activity is regulated by Ca2+ and calmodulin. NO from endothelial cells keeps blood vessels dilated, prevents the adhesion of platelets and white cells, and probably inhibits vascular smooth muscle proliferation. Author(s): Forstermann, U; Closs, E I; Pollock, J S; Nakane, M; Schwarz, P; Gath, I; Kleinert, H Language: eng Source Hypertension 1994 Jun;23(6 Pt 2):1121-31.

Novel approaches and hurdles to somatic cloning in cattle. Abstract: The overall efficiency of somatic cloning in cattle is still low. Many factors are necessary for successful birth of live offspring. Among them, the source of donor cells reveals the importance of the donor genotype but also the influence of the cell line itself. The cell cycle stage has been intensively investigated, and recent results indicate that, in cattle, the G0 stage of the donor nuclei is not a prerequisite for reprogramming, as highly proliferating cultured fibroblasts also result in live offspring after nuclear transfer. A technical approach using direct microinjection of fibroblast nuclei, instead of fusion of the whole cell, has proved to result in high in vitro development rates in cattle. However, full-term development of somatic cloned embryos is still limited by long-lasting effects and a high incidence of losses at periimplantation time (as well as in late gestation and around calving). Author(s): Heyman, Y; Zhou, Qi; Lebourhis, D; Chavatte-Palmer, P; Renard, J P; Vignon, X Language: eng Source Cloning Stem Cells 2002;4(1):47-55.

Novel methods for cloning and engineering genes using the polymerase chain reaction. Abstract: Use of the polymerase chain reaction (PCR) has become increasingly widespread in virtually all aspects of molecular biology. Recently, novel ligation-independent methods have been developed for the cloning of DNA fragments amplified using PCR. Ligation-independent cloning utilizing the enzyme uracil DNA glycosylase (termed UDG cloning) provides an efficient method for gene cloning and recombinant PCR. This technology is now being applied to site-directed mutagenesis, the generation of nested deletions, and the engineering of novel gene constructs. The ease and flexibility of this methodology, combined with PCR amplification, simplify gene cloning and engineering techniques. Author(s): Rashtchian, A Language: eng Source Curr Opin Biotechnol 1995 Feb;6(1):30-6.

Novel strategies of cloning identical and distinct DNA sequences from complex genomes; Zabarovskii, E R Language: rus Source Mol Biol (Mosk) 2000 Jul-Aug;34(4):612-25.

Nuclear cloning and epigenetic reprogramming of the genome. Abstract: Cloning of mammals by nuclear transfer (NT) results in gestational or neonatal failure with at most a few percent of manipulated embryos resulting in live births. Many of those that survive to term succumb to a variety of abnormalities that are likely due to inappropriate epigenetic reprogramming. Cloned embryos derived from donors, such as embryonic stem cells, that may require little or no reprogramming of early developmental genes develop substantially better beyond implantation than NT clones derived from somatic cells. Although recent experiments have demonstrated normal reprogramming of telomere length and X chromosome inactivation, epigenetic information established during gametogenesis, such as gametic imprints, cannot be restored after nuclear transfer. Survival of cloned animals to birth and beyond, despite substantial transcriptional dysregulation, is consistent with mammalian development being rather tolerant to epigenetic abnormalities, with lethality resulting only beyond a threshold of faulty gene reprogramming encompassing multiple loci. Author(s): Rideout, W M 3rd; Eggan, K; Jaenisch, R Language: eng Source Science 2001 Aug 10;293(5532):1093-8.

Nuclear cloning, stem cells, and genomic reprogramming. Abstract: The generation of adult animals by nuclear cloning from adult donor cells is extremely inefficient, with most clones dying soon after implantation. In contrast, cloning from embryonic stem cell donor nuclei is significantly more efficient than from adult donor cells. However, regardless of donor cell type, all clones that survive to birth and beyond suffer serious phenotypic and gene expression abnormalities. All available evidence is consistent with the notion that the anomalous phenotypes of cloned animals are caused by faulty epigenetic reprogramming of the donor nucleus. Faulty reprogramming appears to be caused by the cloning process itself as well as by the epigenetic state of the donor nucleus. In contrast to reproductive cloning, faulty reprogramming of the donor nucleus does not tend to interfere with the application of nuclear transfer technology for therapeutic purposes (therapeutic cloning). Author(s): Jeanisch, Rudolf; Eggan, Kevin; Humpherys, David; Rideout, William; Hochedlinger, Konrad Source Cloning Stem Cells 2002;4(4):389-96.

Nuclear transfer and cell differentiation: contribution of cloning to the study of embryonal stem cells Abstract: The birth of physiologically normal and fertile animal of several mammalian species following the transfer of somatic nuclei into recipient enucleated oocytes has stimulated researches on the functional plasticity of cells available from living organisms. This experimental approach, termed cloning, has up to a low efficiency. It however allows the study of mechanisms compatible with the reacquisition of an undifferentiated nuclear status and contributes to researches aimed at controlling the fate of stem cells for therapeutically applications. Author(s): Renard, J P Language: fre Source Bull Acad Natl Med 2000;184(6):1155-65; discussion 1165-8.

Nuclear transfer and cloning. Abstract: The use of nuclear transfer in human reproductive and therapeutic cloning is reviewed with attention on the origins of this technology from its evolution to the present. The successes and limitations of mammalian reproductive cloning are itemized. A case is made against the use of human reproductive cloning to reproduce an existing person, based on the unacceptable risks to the embryo, fetus, or newborn. However, support is extended for human therapeutic cloning involving the derivation and use of embryonic stem cells to treat human disease. Author(s): Wolf, D P Language: Source Curr Womens Health Rep 2001 Oct;1(2):164-7.

Nuclear transfer technology in mammalian cloning. Abstract: The past several years have witnessed remarkable progress in mammalian cloning using nuclear transfer (NT). Until 1997 and the announcement of the successful cloning of sheep from adult mammary gland or fetal fibroblast cells, our working assumption was that cloning by NT could only be accomplished with relatively undifferentiated embryonic cells. Indeed, live offspring were first produced by NT over 15 years ago from totipotent, embryonic blastomeres derived from early cleavage-stage embryos. However, once begun, the progression to somatic cell cloning or NT employing differentiated cells as the source of donor nuclei was meteoric, initially involving differentiated embryonic cell cultures in sheep in 1996 and quickly thereafter, fetal or adult somatic cells in sheep, cow, mouse, goat, and pig. Several recent reviews provide a background for and discussion of these successes. Here we will focus on the potential uses of reproductive cloning along with recent activities in the field and a discussion concerning current interests in human reproductive and therapeutic cloning. Author(s): Wolf, D P; Mitalipov, S; Norgren, R B Jr Language: eng Source Arch Med Res 2001 Nov-Dec;32(6):609-13.

Nuclear transplantation and cloning in mammals. Author(s):Latham, K E; Westhusin, M E Language: eng Subject(s): Animal; Cell Nucleus; Cloning, Organism/methods; Mammals; Microinjections/methods; Transplantation Source Methods Mol Biol 2000;136:405-25.

Nuclear transplantation as a method for cloning embryos. Author(s): Prather, R S Language: eng Subject(s): Animal; Cell Nucleus/transplantation; Clone Cells; Embryo/cytology; Embryo Transfer; Phenotype; Variation (Genetics) Source Proc Soc Exp Biol Med 1990 Oct;195(1):7-12.

Nucleotide-sugar transporters: cDNA cloning, structures and expression; Ishida, N; Kawakita, M Language: jpn Subject(s): Animal; Carbohydrate Conformation; Carrier Proteins/chemistry/genetics; Cloning, Molecular; Gene Expression Regulation; Golgi Apparatus/metabolism; Membrane Proteins/chemistry/genetics; Monosaccharide Transport Proteins Source Tanpakushitsu Kakusan Koso 1998 Dec;43(16 Suppl):2325-34.

Of needles and haystacks: finding human disease genes by positional cloning. Author(s): Collins, F S Language: eng Subject(s): Amino Acid Sequence;

Chromosome Deletion; Chromosome Mapping; Cloning, Molecular; Cystic Fibrosis/genetics; Genes, Neurofibromatosis 1/genetics; Genetic Markers/genetics; Molecular Sequence Data; Mutation; Translocation (Genetics) Source Clin Res 1991 Dec;39(4):615-23.

On planting alfalfa and growing orchids: the cloning of the gene causing Huntington disease. Author(s): Hayden, M R Language: eng Subject(s): Cells, Cultured; Cloning, Molecular; Genetic Markers; Human; Huntington Disease/genetics; Linkage (Genetics); Polymorphism (Genetics) Source Clin Genet 1993 May;43(5):217-22.

On the application of molecular cloning to developmental neurobiology. Author(s): Schrier, B K; Van Putten, D N Language: eng Subject(s): Animal; Brain/physiology; Brain Chemistry; Cloning, Molecular; DNA/biosynthesis/isolation & purification; Human; Nervous System/growth & development; Nervous System Physiology; Plasmids; Transfection Source Int J Neurol 1982-83;16-17:46-60.

O'Neill, Terry, 1944- Biomedical ethics / Terry O'Neill. Published/Created: San Diego, Calif.: Greenhaven Press, c1999. Description: 144 p.: ill.; 24 cm. ISBN: 1565108744 (pbk.: alk. paper) 1565108752 (lib. bdg.: alk. paper) Summary: Presents opposing viewpoints on various issues of biomedical ethics, including animal testing, human testing, organ transplants, genetic testing, gene therapy, and cloning. Notes: Includes bibliographical references (p. [128]-140) and index. Subjects: Medical ethics--Juvenile literature. Bioethics--Juvenile literature. Medical ethics. Bioethics. Series: Opposing viewpoints digests LC Classification: R724 .O63 1999 Dewey Class No.: 174/.2 21

Ongoing research on mammalian cloning and embryo stem cell technologies: bioethics of their potential medical applications. Abstract: Reproduction by cloning has been achieved by transfer into enucleated oocytes of nuclei from embryonic cells and more recently from cells of adult animals. The efficiency at which embryos produced by such nuclear transfers will develop into healthy newborns is very low but has succeeded in producing some cloned bovines, ovines and mice. Since the first report of sheep cloning from an adult cell in 1997, the potential applications of reproductive cloning in human medicine have been envisaged amidst a flurry of moral debates. Although the technology is still far from being ready for any human use, it has been condemned up front. It has also led to irrational fantasies and fears, based mainly on the misconception that genetic identity means identical twin personalities. Scientific research is ongoing to refine the cloning technology for applications in the production of genetically homogeneous farm animals with useful nutritional or therapeutic genetic traits. A new area of research is non-reproductive therapeutic cloning for the purpose of producing autologous embryonic cells and tissues for transplantation. Author(s): Revel, M Language: eng Subject(s): Animal; Animal Husbandry; Bioethics; Cattle; Cell Nucleus/transplantation; Cloning, Organism; Genetic Engineering; Genome, Human; Human; Mammals; Mice; Oocytes; Personality/genetics; Reproductive Techniques; Sheep; Stem Cells/physiology; Technology, Medical; Tissue Transplantation; Transplantation, Autologous; Twins, Monozygotic/genetics Source Isr Med Assoc J 2000 Jul;2 Suppl:8-14.

Onyeama, Dillibe, 1951- The new man: a perspective in evil / Dillibe Onyeama. Published/Created: Enugu, Nigeria: Delta Publications (Nigeria), 2002. Description: 134 p.; 21 cm. Notes: Includes bibliographical references and indexes. Subjects: Globalization--Religious aspects. Cloning--Religious aspects. International relations.

P2U purinoceptors: cDNA cloning, signal transduction mechanisms and structure-function analysis. Abstract: The cloning of a P2U purinoceptor cDNA has made it possible to use molecular biological approaches to investigate P2U purinoceptor function. Expression of recombinant P2U purinoceptors in mammalian cells lacking endogenous P2U purinoceptors has enabled us to characterize the receptor protein and its downstream effectors, and has allowed a partial analysis of the role of certain amino acid residues in ligand binding. These approaches have placed the pharmacological classification of the P2U purinoceptor on a firm molecular footing and have generated model systems that can be used to investigate receptor-ligand binding, regulation and signal transduction. Author(s): Lustig, K D; Weisman, G A; Turner, J T; Garrad, R; Shiau, A K; Erb, L Language: eng Source Ciba Found Symp 1996;198:193-204; discussion 204-7.

Parathyroid hormone receptors: from cloning to physiological, physiopathological and clinical implications Abstract: It has long been known that parathyroid hormone (PTH) exerts its effects on target tissues via its binding to a membrane receptor. Recently, several types of PTH receptors have been identified. The first receptor which has been cloned and well characterized is "PTH/PTHrP receptor-1". It is activated not only by PTH, but also by PTH-related peptide (PTHrP), via a signal transduction system involving G-proteins, adenylate cyclase and phospholipase C. It is expressed in many tissues, in addition to kidney and bone. The results of recent studies are suggestive of the existence of additional PTH receptors. One or several receptors are probably expressed in the keratinocyte and the glomerular podocyte which are not identical with PTH/ PTHrP receptor-1. A third receptor, which has been cloned recently and called "PTH2 receptor", recognizes solely PTH. It is expressed in brain, pancreas, testis and placenta. Its function is unknown. There is also evidence for a fourth receptor, called "C-PTH receptor", recognizing C-terminal PTH fragments which are generally considered to be biologically inactive. The regulation of these receptors is subject to intensive research. Down-regulation of PTH/PTHrP receptor-1 mRNA expression could explain the well-known resistance to the action of PTH in chronic renal failure. In contrast, the receptor mRNA is up-regulated in vitamin D deficiency, despite a similar tissue resistance to PTH. A mutation of PTH/PTHrP receptor-1 causes Jansen-type metaphyseal chondro-dysplasia. However, no alteration of the PTH/PTHrP receptor-1 gene structure has been found in type 1b pseudo-hypoparathyroidism. Author(s): Mannstadt, M; Drueke, T B Language: fre Source Nephrologie 1997;18(1):5-10.

Parathyroid hormone-related protein: isolation, molecular cloning, and mechanism of action. Abstract: Many factors, such as interleukin 1, TGF alpha, tumor necrosis factor alpha and beta, and PGs, have been implicated in etiological roles in HHM (Martin and Mundy, 1987). Much interest in the past has also centered upon the likelihood of ectopic secretion of PTH in this condition. We have purified a protein (PTHrP) implicated in HHM from a human lung cancer cell line (BEN). Full-length cDNA clones have been isolated and were found to encode a prepropeptide of 36 amino acids and a mature protein of 141 amino acids. Eight of the first 13 amino acids were identical with human PTH, although antisera directed to the NH2 terminus of PTHrP do not recognize PTH; this homology is not maintained in the remainder of the molecule. PTHrP therefore represents a previously unrecognized hormone, possibly related to the PTH gene by a gene duplication mechanism. In support of this notion, the PTHrP gene has been localized to the short arm of chromosome 12; it is believed that chromosome 11, containing the PTH gene, and chromosome 12 are evolutionarily related. In addition, the human PTHrP gene has been isolated, characterized, and shown to have a similar intron--exon organization as the PTH gene. It is possible that the original ancestral gene is indeed the PTHrP gene; resolution of this question awaits studies in lower species. Peptides synthesized to the predicted protein sequence have enabled detailed structure-function studies that have identified NH 2-terminal sequences to be responsible for the biological effects of the molecule. Antibodies raised against the various synthetic peptides have led to the immunohistochemical localization of PTHrP in many human squamous cell carcinomas as well as in a subpopulation of keratinocytes of normal skin. The availability of these antibodies has opened the way for the development of a radioimmunoassay to detect PTHrP in the sera of cancer patients at risk of developing hypercalcemia. The recent characterization of PTHrP-like activity in the ovine fetus suggests some physiological function for PTHrP. It is possible that PTHrP, as the fetal counterpart of PTH, has the role of maintaining the maternal-fetal calcium gradient. The isolation and characterization of PTHrP have added to our understanding of the mechanisms of hypercalcemia and may contribute to the understanding of other metabolic bone diseases, such as osteoporosis and Paget's disease. Finally, and perhaps most importantly, PTHrP may play a hitherto unrecognized role in normal cell physiology. Author(s): Martin, T J; Allan, E H; Caple, I W; Care, A D; Danks, J A; Diefenbach-Jagger, H; Ebeling, P R; Gillespie, M T; Hammonds, G; Heath, J A Language: eng Source Recent Prog Horm Res 1989;45:467-502; discussion 502-6.

Patent, Dorothy Hinshaw. Evolution goes on every day / Dorothy Hinshaw Patent; drawings by Matthew Kalmenoff. Published/Created: New York: Holiday House, c1977. Related Names: Kalmenoff, Matthew. Description: 156 p.: ill.; 24 cm. ISBN: 0823402975 : Summary: Describes some specific evolutionary changes occurring in certain plants and animal species today and generally discusses genetic mechanisms including DNA, mutations, and cloning. Notes: Includes index. Bibliography: p. 147-149. Subjects: Evolution (Biology)--Juvenile literature. Genetics. LC Classification: QH367.1 .P37 Dewey Class No.: 575

PCR and the cloning of receptor subtype genes. Author(s): Watson, S P; James, W Language: eng Source Trends Pharmacol Sci 1989 Sep;10(9):346-8.

PCR cloning of N-terminal RAR isoforms and APL-associated PLZF-RAR alpha fusion proteins. Author(s):Zelent, A Language: eng Subject(s): Animal; Cloning, DNA-Binding Proteins ; Isomerism; Leukemia, Promyelocytic, Acute metabolism; Polymerase Chain Reaction ; Receptors, Retinoic Acid genetics; Recombinant Fusion ; Transcription Factors ; Zinc Fingers Source Methods Mol Biol 1998;89:307-32.

PCR cloning protocols / edited by Bing-Yuan Chen and Harry W. Janes. Edition Information: 2nd ed. Published/Created: Totowa, N.J.: Humana Press, c2002. Related Names: Chen, Bing-Yuan. Janes, Harry W. Description: xiv, 439 p.: ill.; 23 cm. ISBN: 0896039730 (comb.: alk. paper) Notes: Includes bibliographical references and index. Subjects: Molecular cloning--Laboratory manuals. Polymerase chain reaction--Laboratory manuals. Series: Methods in molecular biology (Clifton, N.J.); v. 192. LC Classification: QH442.2 .P37 2002 Dewey Class No.: 572.8/6 21

PCR cloning protocols/ edited by Bruce A. White. Published/Created: Totowa, N.J.: Humana Press, c1997. Related Names: White, Bruce Alan. Description: xiv, 490 p.: ill.; 23 cm. ISBN: 0896033430 (combbound: alk. paper) Notes: Includes bibliographical references and index. Subjects: Molecular cloning--Laboratory manuals. Polymerase chain reaction--Laboratory manuals. Series: Methods in molecular biology (Clifton, N.J.); v. 67. LC Classification: QH442.2 .P37 1997 Dewey Class No.: 574.87/3282 20

PCR cloning using T-vectors. Author(s): Trower, M K; Elgar, G S Language: eng Subject(s): Cloning, Molecular; DNA Primers; DNA-Directed DNA Polymerase ; Escherichia coli; Genetic Vectors genetics; Polymerase Chain Reaction, Genetic Source Methods Mol Biol 1994;31:19-33.

PCR-assisted cDNA cloning: a guided tour of the minefield. Author(s): Hooft van Huijsduijnen, R Language: eng Subject(s): Cloning, Molecular; DNA, Complementary & purification; Disintegrins; Gene Expression Regulation; Human; Metalloendopeptidases; Mutation; Organ Specificity; Placenta; Polymerase Chain Reaction; Taq Polymerase. Source Biotechniques 1998 Mar;24(3):390-2.

PCR-based cloning of cortically localized RNAs from Xenopus oocytes. Author(s):Zhang, J; King, M L Language: eng Source: Methods Mol Biol 2000;136:309-14.

Pence, Gregory E. Who's afraid of human cloning? / Gregory E. Pence. Published/Created: Lanham [Md.]: Rowman & Littlefield, c1998. Description: xv, 181 p.; 24 cm. ISBN: 0847687813 (cloth: alk. paper) 0847687821 (pbk.: alk. paper) Notes: Includes bibliographical references and index. Subjects: Human cloning--Moral and ethical aspects. LC Classification: QH442.2 .P46 1998 Dewey Class No.: 174/.957 21

Perbal, Bernard V. A practical guide to molecular cloning / Bernard Perbal. Edition Information: 2nd ed. Published/Created: New York: Wiley, c1988. Description: xvii, 811 p.: ill.; 29 cm. + 3 folded charts. ISBN: 0471850713 0471850705 (pbk.) Notes: "A Wiley-Interscience publication." Includes bibliographies and index. LC Classification: QH442.2 .P47 1988 Dewey Class No.: 574.87/3282 19

Perbal, Bernard V. A practical guide to molecular cloning / Bernard Perbal. Published/Created: New York: Wiley, c1984. Description: xix, 554 p.: ill.; 24 cm. ISBN: 0471876526 0471876534 (pbk.) Notes: "A Wiley-Interscience publication." Includes bibliographies and index. Subjects: Molecular cloning. DNA, Recombinant. Genetic intervention. LC Classification: QH442.2 .P47 1984 Dewey Class No.: 574.87/3282 19

Perry, Mark L. Frankenstein is alive!: the church's position on cloning / by Mark L. Perry and Gregory V. Jones, Jr. Published: Cleveland Heights, Ohio: Geneva Books, c2001. Related Names: Jones, Gregory V. Description: 59 p.; 20 cm. ISBN: 0970903723 Notes: Includes bibliographical references (p. 55-56). Subjects: Human cloning--Religious aspects--Christianity.

Perspectives on genetic patenting: religion, science, and industry in dialogue / edited by Audrey R. Chapman. Published: Washington, DC: American Association for the Advancement of Science, 1999. Related Names: Chapman, Audrey R. Description: v, 300 p.; 23 cm. ISBN: 0871686287 Notes: Includes bibliographical notes. Subjects: Human genome--Genetic engineering--Human cloning--Patents--Religious aspects--Christianity--Moral and ethical aspects. LC Classification: QH447 .P47 1999 Dewey Class No.: 174/.957 21

Perspectives on methylmalonic acidemia resulting from molecular cloning of methylmalonyl CoA mutase. Abstract: Methylmalonyl CoA mutase deficiency (methylmalonic acidemia) has been a paradigm for biochemical and somatic cell genetic approaches to human disease. Recently, genes encoding this enzyme have been cloned from several species. These studies have provided information about the primary structure & evolution of this enzyme, the mutations which underlie its deficiency state, and the structure-function determinants which are required for its activity. Gene transfer studies now permit restitution of this enzyme to genetically deficient cells and may enable somatic gene therapy to be undertaken. Molecular genetic studies not only provide more detailed information about this enzyme, but introduce new perspectives on the molecular mechanisms and dynamics of its function and raise new questions about the dyshomeostatic consequences of its deficiency. Author(s): Ledley, F D Language: eng Source: Bioessays 1990 Jul;12(7):335-40.

Phage display as a tool for rapid cloning of allergenic proteins. Abstract: Allergic diseases represent an immune disorder associated with the production of immunoglobulin E (IgE) against normally innocuous antigens (allergens). Almost 20% of the population in industrialized countries suffer from type I allergic symptoms such as allergic rhinitis, conjunctivitis, urticaria or asthma. Although the mechanisms responsible for these allergic reactions are quite well understood, knowledge about the repertoire of molecules able to elicit type I symptoms is still limited. To clone and characterize entire allergen repertoires from complex allergenic sources in a fast and efficient way, new technologies are required. The phage surface display of cDNA libraries described here has proven to be a versatile cloning system to selectively isolate allergens physically linked to their genetic information. The screening of cDNA libraries displayed on phage surfaces with immobilised serum IgE from allergic patients reduces the time required for the selection of candidate clones to a few weeks. Robot-assisted high-throughput screening of the enriched library provides a fast and cost-effective way to isolate complete allergen repertoires. The biotechnological production of recombinant allergens derived from these sequences bears a high potential for the improvement of the diagnosis of allergic diseases. Author(s): Appenzeller, U; Blaser, K; Crameri, R Language: eng Source: Arch Immunol Ther Exp (Warsz) 2001;49(1):19-25.

Phenotypic effects of somatic cell cloning in the mouse. Abstract: Although a variety of phenotypes and epigenetic alterations have been reported in animals cloned from somatic cells, the exact nature and consequences of cloning remain unclear. We cloned mice using fresh or short-term cultures of donor cells (cumulus cells, immature Sertoli cells, and fetal or adult fibroblast cells) with defined genetic backgrounds, and then compared the phenotypic and epigenetic characteristics of the cloned mice with those of fertilization-derived control mice. Irrespective of the nucleus-donor cell type, about 50% of the reconstructed embryos developed to the morula/blastocyst stage, but about 90% of these clones showed arrested development between days 5 and 8, shortly after implantation. Most of the clones were alive at term, readily recovered respiration, and did not show any malformations or overgrowths. However, their placentas were two- to threefold larger than those of the controls, due to hyperplasia of the basal (or spongiotrophoblast) layer. Although there was significant suppression of a subset of both imprinted and non-imprinted placental genes, fetal gene suppression was minimal. The seven imprinted genes that we examined were all expressed correctly from the parental alleles. These findings were consistent for every cell type from the midgestation through term stages. Therefore, cloning by nuclear transfer does not perturb the parent-specific imprinting memory that is established during gametogenesis, and the phenotypic and epigenetic effects of cloning are restricted to placental development at the midgestation and term stages. Twelve male mice that were born in a normal manner following nuclear transfer with immature Sertoli cells (B6D2F1 genetic background) were subjected to long-term observation. They died earlier than the genotype-matched controls (50% survival point: 550 days vs. 1028 days, respectively), most probably due to severe pneumonia, which indicates that unexpected phenotypes can appear as a result of the long-term effects of somatic cell cloning. Author(s): Ogura, A; Inoue, K; Ogonuki, N; Lee, J; Kohda, T; Ishino, F Language: eng Source: Cloning Stem Cells 2002;4(4):397-405.

Phytocystatins and their target enzymes-molecular cloning, expression and possible functions. Author(s): Arai, S; Abe, K; Emori, Y Language: eng Source: Adv Exp Med Biol 1996;389:73-8.

Plant cell cloning and culture products. Abstract: Biotechnological developments in the use of plant cells as sources of biochemicals have now brought several applications within commercial range. Plant propagation has developed faster commercially, but now requires similar biotechnology to enable the automated production of large numbers of plants at low cost. Although some of the enabling technology has been developed for pregerminated seeds this has not been coupled to production of plantlets in bioreactors. Our present lack of knowledge of how to control gene expression in plant cells, which stems from our ignorance of the organization and regulation of the plant genome, is a critical factor limiting all applications of plant cell cultures. Author(s): Jones, L H Language: eng Source: Biochem Soc Symp 1983;48:221-32.

Planting alfalfa and cloning the Huntington's disease gene. Author(s): Goodfellow, P N Language: eng Source: Cell 1993 Mar 26;72(6):817-8.

Plasmid instability and molecular cloning in Bacillus subtilis. Author(s): Bron, S; Meijer, W; Holsappel, S; Haima, P Language: eng Source: Res Microbiol 1991 Sep-Oct;142(7-8):875-83.

Plasmids and DNA cloning in streptomycetes; Dekhtiarenko, T D Language: rus; Antibiotics/biosynthesis; Bacillus subtilis/genetics; Cloning, Molecular/drug effects; DNA Restriction Enzymes/pharmacology; DNA, Bacterial/genetics; Escherichia coli/genetics; Gene Expression Regulation; Genetic Markers; Genetic Vectors; Microscopy, Electron; Nucleic Acid Hybridization. Source: Mikrobiol Zh 1983 Mar-Apr;45(2):96-107.

Plasmids as cloning vehicles (author's transl) Abstract: Plasmids are regulated autonomously replicating DNA elements, which have been used successfully for the in vitro insertion and the transfer and stable maintenance of defined foreign DNA fragments into bacteria and lower eukaryontes. For Escherichia coli a large set of ColE1 derivatives have been developed and have been used for cloning of DNA fragments from many biological systems. More recently plasmids have been also isolated and modified for the use as cloning vehicle for other bacteria like Bacillus subtilis, Pseudomonas and Streptomyces and for the yeast Saccharomyces cerevisiae. Author(s): Goebel, W Language: Source: Arzneimittelforschung 1980;30(3a):533-40.

Plasmids as vectors for gene cloning. Author(s): Helinski, D R Language: eng Subject(s): Chromosomes, Bacterial/metabolism; Colicins/biosynthesis; Conjugation, Genetic; DNA Replication; DNA Restriction Enzymes.Source Basic Life Sci 1977;9:19-49.

Polidoro, J. P. Project Samuel / J.P. Polidoro. Published/Created: Laconia, NH: Longtail Pub., c2001. Description: 299 p.; 23 cm. ISBN: 0967761913 Subjects: Human cloning--Fiction. Baseball players--Fiction. Biotechnologists--Fiction. Boston (Mass.)--Fiction. San Diego (Calif.)--Fiction. LC Classification: PS3616.O567 P65 2001 Dewey Class No.: 813/.6 21

Population biology and evolution of clonal organisms / edited by Jeremy B.C. Jackson, Leo W. Buss, Robert E. Cook; contributors, Jeffrey W. Ashmun ... [et al.]. Published/Created: New Haven: Yale University Press, c1985. Related Names: Jackson, Jeremy B. C., 1942- Buss, Leo W. Cook, Robert E. (Robert Edward), 1946- Ashmun, Jeffrey W. Description: xi, 530 p.: ill.; 25 cm. ISBN: 0300033796 (alk. paper) 0300036507 (pbk.) Notes: Includes bibliographies and indexes. Subjects: Evolution (Biology) Cloning. Population biology. LC Classification: QH371 .P66 1985 Dewey Class No.: 575 19

Positional candidate gene cloning of CLN1. Abstract: Mutations in the CLN1 gene encoding palmitoyl-protein thioesterase (PPT) underlie the recessive neurodegenerative disorder, infantile Batten disease, or infantile neuronal ceroid lipofuscinosis (INCL). The CLN1 gene was mapped to chromosome 1p32 in the vicinity of a microsatellite marker HY-TM1 in a cohort of Finnish INCL families, and mapping of the PPT gene to the CLN1 critical region (and the discovery of mutations in PPT in several unrelated families) led to conclusive identification of PPT as the disease gene. PPT is a lysosomal thioesterase that removes fatty acids from fatty-acylated cysteine residues in proteins. The accumulation of fatty acyl cysteine thioesters can be reversed in INCL cells by the exogenous administration of recombinant PPT, which enters the cells through the mannose 6-phosphate receptor pathway. Over two dozen PPT mutations have been found in PPT-deficient patients worldwide. In the United States, all PPT-deficient patients show "GROD" histology but the age of onset of symptoms is later in some children due to the presence of missense mutations that result in enzymes with residual PPT activity. Now that INCL is known to be caused by a defect in a soluble lysosomal enzyme, appropriate therapies may be forthcoming. Prospects for therapy include enzyme replacement, stem cell transplantation, gene therapy, and metabolic therapy aimed at depleting the abnormal substrate accumulation in the disease. Author(s): Hofmann, S L; Das, A K; Lu, J Y; Soyombo, A A Language: eng Source Adv Genet 2001;45:69-92.

Positional cloning and functional analysis of the gene responsible for Nijmegen breakage syndrome, NBS1. Abstract: Nijmegen breakage syndrome (NBS) is a rare autosomal recessive disorder characterized by microcephaly, combined immunodeficiency, and a high incidence of lymphoid tumor. Cells from NBS patients show chromosomal instability, hypersensitivity to ionizing radiation and abnormal p53-mediated cell cycle regulation. We cloned the underlying gene for NBS, designated NBS1, by complementation-assisted positional cloning from the candidate region 8q21. Large genomic sequencing, as well as a search using computer programs, provides a powerful approach for identifying the underlying gene for a disease. The NBS1 gene encodes a protein of 754 amino acids that has FHA and BRCT domains which often are conserved in cell-cycle checkpoint proteins. The gene has weak homology to the yeast (Saccharomyces cerevisiae) Xrs2 protein in the N-terminus region. Like yeast Xrs2, the NBS1 protein forms a complex with hRAD50/hMRE11, and the complex is condensed as foci in the nucleus after irradiation, indicative that this triple-complex is a crucial factor in DNA repair. Functional analysis of the NBS1 protein is in progress and it should provide further clues to understanding the repair mechanism of radiation-induced DNA double-strand breaks. Author(s): Tauchi, H Language: eng Source: J Radiat Res (Tokyo) 2000 Mar;41(1):9-17.

Positional cloning and multicolor in situ hybridization. Principles and protocols. Author(s): Lebo, R V; Su, Y Language: eng Source: Methods Mol Biol 1994;33:409-38.

Positional cloning and the molecular basis of immunodeficiency. Author(s): Smith, C I; Hammarstrom, L Language: eng Subject(s): Agammaglobulinemia/genetics; Cloning, Molecular; Genes; Granulomatous Disease, Chronic/genetics; Human; Immunologic Deficiency Syndromes/genetics Source: Ann Med 1995 Aug;27(4):477-9.

Positional cloning approach to the dominant polycystic kidney disease gene, PKD1. Abstract: Positional cloning is a powerful strategy for identifying the site of disease-producing mutations when the underlying biochemical defect is unknown. The approach also offers new methods for the presymptomatic diagnosis of genetic disease. Using these methods we have localized the PKD1 gene, mutated in the majority of PKD1 families, to a small (500 kb) segment of chromosome 16, band p13.3. Virtually all of this interval has been cloned in cosmids and lambda bacteriophage. Over 20 sets of non-overlapping cDNA clones have been isolated from the region. Sequence and mutational analyses are currently underway. In addition, a set of polymorphic clones has been identified for presymptomatic diagnosis. Included in this set are several highly variable [CA]n microsatellite repeats. These highly informative markers can be rapidly assayed from a small amount of genomic DNA using the polymerase chain reaction. Despite these advances, presymptomatic diagnosis cannot be established with certainty in many families. However, identification of the PKD1 gene itself will eventually allow diagnosis by direct detection of mutations. Author(s): Germino, G G; Somlo, S; Weinstat-Saslow, D; Reeders, S T Language: eng Subject(s): Base Sequence; Chromosome Mapping; Chromosomes, Human, Pair 16; Cloning, Molecular; DNA/genetics; DNA Mutational Analysis; Human; Linkage (Genetics); Molecular Sequence Data; Pedigree; Polycystic Kidney, Autosomal Dominant/genetics Source: Kidney Int Suppl 1993 Jan;39:S20-5.

Positional cloning by Exon trapping and cDNA selection / [edited by] Bernhard Korn, ... [et al.]. Published/Created: New York: Wiley; Heidelberg: Spektrum, c1999. Related Names: Korn, Bernhard. Description: x, 76 p.: ill.; 26 cm. ISBN: 0471297976 (pbk.: alk. paper) Notes: "An EMBO practical course"--Cover. Includes bibliographical references (p. [75]-76) and index. Subjects: Gene mapping. Exons (Gentetics) Molecular cloning. LC Classification: QH445.2 .P67 1999 Dewey Class No.: 572.8/633 21

Positional cloning in Arabidopsis. Why it feels good to have a genome initiative working for you. Abstract: Positional (or map-based) cloning techniques are widely used to identify the protein products of genes defined by mutation. In Arabidopsis the information generated by the Genome Initiative is giving this approach a decisive boost. A wealth of sequence polymorphisms and molecular markers is now available and can be exploited for fine mapping with technically simple and robust polymerase chain reaction-based methods. As a result it has become possible to complete positional cloning projects in a short time and with relatively little effort. Author(s): Lukowitz, W; Gillmor, C S; Scheible, W R Language: eng Subject(s): Arabidopsis/genetics; Chromosome Mapping; Cloning, Molecular; Genome, Plant; Physical Chromosome Mapping; Source: Plant Physiol 2000 Jul;123(3):795-805.

Positional cloning of anomaly-syndrome-causing genes; Niikawa, N Language: jpn Subject(s): Abnormalities, Chromosome Aberrations; Chromosomes, Human, Pair 5/genetics; Cloning, Molecular methods; Craniofacial Abnormalities genetics; Diaphyseal Dysplasia, Progressive; Human; Mutation; Peptide Fragments; Protein Precursors. Source Tanpakushitsu Kakusan Koso 2001 Dec;46(16 Suppl):2313-8.

Positional cloning of genes involved in the Beckwith-Wiedemann syndrome, hemihypertrophy, and associated childhood tumors. Abstract: The Beckwith-Wiedemann syndrome (BWS) is an overgrowth malformation syndrome that occurs with an incidence of 1:13,700 births. There is a striking incidence of childhood tumors found in BWS patients. Various lines of investigation have localized "imprinted" genes involved in BWS and associated childhood tumors to 11p15. High resolution mapping of 8 rare balanced chromosomal BWS rearrangements enabled us to identify three distinct regions on chromosome 11p15 that might harbor genes involved in the above-mentioned disorders. These results suggest genetic heterogeneity that correlates with the clinical heterogeneity seen in the patients studied. Expressed candidate gene sequences from these regions have been cloned and partly sequenced. These transcripts are either disrupted by or are at least within a few kb of these BWS chromosome breakpoints. So far, zinc-finger sequences and one Kruppel-associated box (KRAB) domain were found in independent candidate genes which are compatible with a regulating function of growth promoting genes. The abundance of expression of these genes varies from low abundant in all adult and fetal tissues tested to detectable on Northern blots of adult tissues. In addition to our 11p15 studies we have analyzed additional chromosome regions, in particular 1p. Cytogenetic, loss of heterozygosity (LOH) and comparative genomic hybridization (CGH) studies have identified 1p35 as a region of interest. A positional cloning effort to identify a balanced 1p35 translocation found in a Wilms tumor has led to the isolation of a YAC, crossing this breakpoint. Author(s): Mannens, M; Alders, M; Redeker, B; Bliek, J; Steenman, M; Wiesmeyer, C; de Meulemeester, M; Ryan, A; Kalikin, L; Voute, T; De Kraker, J; Hoovers, J; Slater, R; Feinberg, A; Little, P; Westerveld, A Language: eng Source: Med Pediatr Oncol 1996 Nov;27(5):490-4.

Positional cloning of genes responsible for epilepsy and autism using a patient with chromosomal translocation; Isomura, M; Nakamura, Y Language: jpn Subject(s): Autistic Disorder/genetics; Chromosome Mapping; Cloning, Molecular/methods; Cosmids/genetics; Epilepsy/genetics; Genome, Human; Human; Sequence Analysis, DNA; Translocation (Genetics) Source Tanpakushitsu Kakusan Koso 1997 Dec;42(17 Suppl):2786-90.

Positional cloning of genes responsible for hereditary tumors Abstract: Recently, remarkable progress in molecular biology has enabled isolation of genes responsible for hereditary tumors such as retinoblastoma (RB), Wilms' tumor (WT), von Recklinghausen neurofibromatosis (NF 1), and familial adenomatous polyposis (FAP). Since patients with FAP develop multiple adenomatous polyps in the colon, some of which progress to colon cancer, isolation of the FAP gene allows us a rare opportunity to study genetic events underlying the well defined morphological changes during progression of colorectal tumors. In this report, we presented an approach called "positional cloning" which has become a powerful tool for identifying genes responsible for hereditary tumors, as well as characteristics of some of such genes. Author(s): Yamakawa, K; Nakamura, Y Language: jpn Source: Gan To Kagaku Ryoho 1992 Jan;19(1):1-7.

Positional cloning of Lps, and the general role of toll-like receptors in the innate immune response. Abstract: In mice (and by inference, in all mammals), a single pathway exists to serve lipopolysaccharide (LPS) signal transduction, and as such, allelic mutations at a single locus entirely abolish responses to LPS in C3H/HeJ and C57BL/10ScCr mice. Positional cloning of this locus, known as Lps, revealed that mutations of the Toll-like receptor 4 gene (Tlr4) are responsible for endotoxin resistance. A quick succession of studies have shown Tlr4 to be the critical transmembrane component of the LPS signal transduction complex. As LPS sensing by Tlr4 depends on physical contact between the two molecules, Tlr4 is a direct interface with the microbial world. Eight other molecules with strong similarity to Tlr4 are presently known in mammals, and taking Tlr4 as a model, all may be guessed to participate in the early detection of invasive pathogens. Acting together, the Toll-like receptors may be assumed to present macrophages with a comprehensive "picture" of the micobial world, and thus comprise the principal sensing molecules utilized by cells of the innate immune system. Author(s): Beutler, B; Poltorak, A Language: eng Source Eur Cytokine Netw 2000 Jun;11(2):143-52.

Positional cloning of mutated zebrafish genes. Author(s):Talbot, W S; Schier, A F Language: eng Subject(s): Animal; Cloning, Molecular/methods; Human; Mutagenesis; Zebrafish/genetics Source Methods Cell Biol 1999;60:259-86.

Positional cloning of the causative gene for the Medaka mutation, double anal fin (Da), defective in dorsoventral patterning; Ohtsuka, M; Kimura, M Language: jpn Source: Tanpakushitsu Kakusan Koso 2000 Dec;45(17 Suppl):2864-71.

Positional cloning of the genes for progeroid syndromes; Miki, T; Nakura, J; Kamino, K; Ogihara, T Language: jpn Subject(s): Alzheimer Disease/genetics; Chromosome Mapping; Cloning, Molecular/methods; Genome, Human; Human; Muscular Dystrophies/genetics; Trinucleotide Repeats; Werner Syndrome/genetics Source Tanpakushitsu Kakusan Koso 1996 Mar;41(3):245-54.

Positional cloning of the HYP gene: a review. Author(s): Econs, M J Language: eng Subject(s): Animal; Chromosome Mapping; Cloning, Molecular; Hypophosphatemia/genetics; Rickets/genetics Source Kidney Int 1996 Apr;49(4):1033-7.

Positional cloning of the JNCL gene, CLN3. Author(s): Lerner, T J Language: eng Subject(s): Animal; Chromosomes, Human, Pair 16; Cloning, Molecular; DNA Mutational Analysis; Exons; Gene Deletion; Human; Mutation; Neuronal Ceroid-ipofuscinosis/genetics/metabolism; Physical Chromosome Mapping; Proteins/biosynthesis/genetics; Tissue Distribution Source: Adv Genet 2001;45:107-21.

Positional cloning of the PEX gene: new insights into the pathophysiology of X-linked hypophosphatemic rickets. Abstract: X-linked hypophosphatemic rickets (HYP) is the most common form of hereditary renal phosphate wasting. The hallmarks of this disease are isolated renal phosphate wasting with inappropriately normal calcitriol concentrations and a mineralization defect in bone. Studies in the Hyp mouse, one of the murine models of the human disease, suggest that there is an approximately 50% decrease in both message and protein of NPT-2, the predominant sodium-phosphate cotransporter in the proximal tubule. However, human NPT-2 maps to chromosome 5q35, indicating that it is not the disease gene. Positional cloning studies have led to the identification of a gene, PEX, which is responsible for the disorder. Further studies have led to identification of the murine Pex gene, which is mutated in the murine models of the disorder. These studies, in concert with other studies, have led to improved understanding of the pathophysiology of HYP and a new appreciation for the complexity of normal phosphate homeostasis. Author(s): Econs, M J; Francis, F Language: eng Source: Am J Physiol 1997 Oct;273(4 Pt 2):F489-98.

Positional cloning of the putative gene responsible for transient abnormal myelopoiesis and that for multiple cartilaginous exostoses Abstract: Positional cloning of the putative gene responsible for transient abnormal myelopoiesis (TAM) and that for multiple cartilaginous exostoses (MEX) is described. TAM is a leukemoid reaction occurring frequently in Down syndrome (DS) newborn infants and they often develop true leukemia several years later. The previous findings of "disomic homozygosity in trisomic cells" and tentative mapping of the TAM gene to 21q11.1, and an encounter of a unique DS-associated TAM patient with inv(21) (q11.1q22.13) let us start positional cloning of the TAM gene. One type of MEX is an autosomal dominant disorder and patients with MEX sometimes develop chondrosarcoma. The MEX gene has been mapped to 8q24. We encountered a sporadic case of MEX with de novo t(8q; 13q). Thus, we hypothesized that in both patients, the TAM and the MEX genes are disrupted by the structural chromosome abnormalities. For TAM, we first mapped the proximal breakpoint of inv(21) between 2 STSs using 7 cosmid clones as FISH probes that were isolated on the basis of STS markers at the 21q11.1 region, isolated their corresponding YACs, and then analyzed them. However, since YACs corresponding to 2 other STSs between the two markers could not be isolated, we carried out a chromosome walking to construct a cosmid contig between the 2 STSs. Southern analysis with a cosmid clone within the contig detected EcoRI-/HindIII extra bands on the patient's DNA. The cDNA screening and exon trapping to isolate a gene from the region are underway. Similarly, in the MEX patient we mapped the 8q breakpoint between 2 cosmid markers, then isolated YACs and cosmid subclones. By exon trapping after detection of a cosmid covering the breakpoint, an exon-like sequence was isolated. The 3'-RACE/5'-RACE revealed a novel transcript from this cosmid. Whether the transcript is the MEX gene remains to be determined. Author(s): Niikawa, N Language: jpn Source: Rinsho Byori 1996 Jan;44(1):13-8.

Positional cloning reaches maturity. Abstract: The identification of genes involved in human genetic disease is no longer the province of those who would make a career of 'not finding' a gene. Developments from the human genome initiative have vastly facilitated the process of localizing genetic intervals segregating mutations, as well as that of obtaining the physical regents necessary for characterizing the region. In a few years' time, efforts aimed at the assignment of genes to the physical map, coupled with increasing quantities of sequence data from both cDNA and genomic sources, will provide numerous candidate genes for analysis, with consequences for the approaches used to define the gene and mutations(s) involved in the disease of interest. Author(s): Nelson, D L Language: eng Source Curr Opin Genet Dev 1995 Jun;5(3):298-303.

Positional cloning uncovers a new old oncogene. Author(s): Frischauf, A M Language: eng Subject(s): Cloning, Molecular; Human; Multiple Endocrine Neoplasia/genetics; Oncogenes; Proto-Oncogenes; Thyroid Neoplasms/genetics Source: Hum Mol Genet 1993 Jul;2(7):847-8.

Positional cloning. Author(s):Xu, L; Li, Y Language: eng Subject(s): Animal; Cloning, Molecular; Human Source: Methods Mol Biol 2000;136:285-96.

Positional cloning; Aburatani, H Language: jpn Subject(s): Animal; BRCA2 Protein; Breast Neoplasms, Male genetics; Chromosome Mapping; Cloning, Molecular methods; Human; Human Genome Project; Huntington Disease genetics; Male; Mutation; Neoplasm Proteins genetics; Sequence Analysis, DNA; Transcription Factors genetics Source: Tanpakushitsu Kakusan Koso 1996 Nov;41(15 Suppl):2424-32.

Positional cloning--current status and future directions Abstract: Several strategies for isolating disease genes have been developed. In functional cloning, the gene is isolated based on information on the function of gene products (proteins). In contrast, positional cloning is based on the isolation of the disease gene, starting from the knowledge of its location on the genome. The flow chart of positional cloning was described as follows; 1) identification of the location of the disease gene by linkage analysis. 2) construction of the physical map spanning the region. 3) isolation of the candidate genes from the region. 4) identification of mutations in the candidate genes, which is specific to the disease. The current concept and strategies on the positional cloning is discussed. Author(s): Onodera, O; Tsuji, S Language: jpn Source Nippon Rinsho 1993 Sep;51(9):2225-33.

Positional-candidate cloning of genes from mouse mutants. Author(s): Avraham, K B Language: eng Source Methods Mol Biol 2001;158:369-79.

Positional-cloning efforts to identify the tumor suppressor gene at 3p Abstract: Lung cancer is known to carry multiple cytogenetic alterations. Among such chromosomal abnormalities, deletions at the short arm of chromosome 3 has been shown to be the most frequent and can be detected even in hyperplasia of the bronchial epithelium, suggesting its important roles in the pathogenesis of lung cancer. In this chapter, we describe current status of the positional-cloning attempts to identify the target tumor suppressor gene at 3p in our laboratories as well as in other laboratories. Author(s): Hibi, K; Takahashi, T Language: jpn Source Nippon Rinsho 1996 Feb;54(2):478-81.

Potential benefits of cell cloning for human medicine. Abstract: The successful cloning of a mammal from an adult somatic cell nucleus opens new avenues for major advances in reproductive medicine, biotechnology and cellular-based transplantation therapies for degenerative diseases. At the same time, this breakthrough has generated much heated discussion concerning the ethics of cloning. Twinning is a form of cloning, and there are instances in clinical assisted reproduction in which the deliberate formation of twins by embryo dissection would seem ethically acceptable. Nuclear transfer technology might facilitate the derivation of human embryonic stem cells, capable of differentiation into a wide variety of somatic cell lineages. Directed differentiation of human embryonic stem cells into specific cell types in vitro could provide a universal source of cells for transplantation therapy. The potential benefits of therapeutics based on cloning technologies are considerable, and hasty legislation to ban all such procedures could block progress in critical arenas of biomedical research. Author(s): Trounson, A; Pera, M Language: eng Source Reprod Fertil Dev 1998;10(1):121-5.

Pox viruses as eukaryotic cloning and expression vectors: future medical and veterinary vaccines. Abstract: It was proposed at the beginning of this article that the use of poxviruses as expression vectors provided a means of optimizing the conditions and prerequisites of vaccination and of elucidating a number of the associated problems. These advantages will now be summarized. Vaccinia virus recombinants carrying one or more multiple foreign genes have been shown to induce a sufficient immune response to that antigen in laboratory animals to protect against subsequent challenge by the infectious agent. In those animals tested the duration of immunity has been long-lasting with no apparent detrimental side-effects. Vaccinia virus is very stable in normal environmental situations and no cold-chain is required in the distribution network. There is no evidence for genetic instability of the vaccinia 'carrier' strain nor of loss of the inserted genetic elements. The vaccine is administered intradermally and it is proposed that one inoculation will be sufficient to induce immunity to multiple disease antigens; however, re-inoculation at a later time with the same or new antigenic determinants is not precluded. The efficiency of vaccinia virus as the vaccine strain used to eliminate the disease of smallpox with minimal side-effects is well documented. The recent studies described here on the development of vaccinia as a viral vector confer other advantages. Since the expression of the inserted genetic element is under vaccinia virus control, different levels of expression can be engineered by substituting the promoter sequence placed proximal to the insert. This leads to the possibility of high levels of expression when required for a single antigen or more modulated levels when multiple foreign antigens are expressed and 'genetic overload' could become a problem. Much work remains to be done on the body's immune response to this proposed genetic intrusion. The ability to insert large amounts of genetic information without disrupting virus replication and assembly indicates a great degree of flexibility in the genome strategy of the virus. This may indicate that the replication system can be further disrupted to eliminate or modify those sequences coding for virulence and wide host range to make the poxvirus vector as safe as possible for vaccine use. One of the reasons traditional vaccine strategy has failed for some virus groups is the degree of variability found in virus strains in nature. Influenza viruses are a good example, as are the bluetongue virus group, rotaviruses and bunyaviruses.(ABSTRACT TRUNCATED AT 400 WORDS); Taylor, J; Paoletti, E Language: eng Source Prog Vet Microbiol Immunol 1988;4:197-217.

Practical aspects of donor cell selection for nuclear cloning. Abstract: Choosing the right nuclear donor is the most critical decision in cloning by nuclear transfer (NT), or nuclear cloning, because the cloned animal will be a genetic copy of the donor cell genome used for NT. Both donor cell type and cell cycle stage are important methodological parameters and influence nuclear cloning efficiency. Cloning, however, is a multi-step procedure and the exact contribution of the nuclear donor to overall cloning success must be determined in comparative studies. This requires strict standardization of isolation, purification, and culture protocols, and application of stringent identification criteria in order to obtain a homogenous donor cell population. In all these respects, the standards in the cloning field are currently poor. The aim of this review is to provide a brief guideline for the major practical aspects of donor cell selection, cell cycle synchronization and preparation for NT. Author(s): Oback, Bjorn; Wells, David Language: eng Source Cloning Stem Cells 2002;4(2):169-74.

Predictive drug testing on human tumor cells / edited by V. Hofmann, M.E. Berens, G. Martz. Published/Created: Berlin; New York: Springer-Verlag, 1984. Related Names: Hofmann, V. (Victor), 1948- Berens, M. E. (Michael E.), 1953- Martz, G. (Georg) Description: xii, 285 p.: ill.; 25 cm. ISBN: 0387134972 (U.S.) Notes: Includes bibliographical references and index. Subjects: Antineoplastic agents--Testing. Cancer cells--Growth--Regulation. Cancer--Chemotherapy--Evaluation. Antineoplastic Agents--pharmacodynamics. Colony-Forming Units Assay. Neoplasms--drug therapy. Series: Recent results in cancer research; 94 LC Classification: RC261 .R35 vol. 94 RC271.C5 Dewey Class No.: 616.99/4 s 616.99/2061 19

Preservation of endangered species and populations: a role for genome banking, somatic cell cloning, and androgenesis?; Corley-Smith, G E; Brandhorst, B P Language: eng Subject(s): Biological Specimen Banks; Cell Nucleus; Cloning, Organism; Conservation of Natural Resources; Cryopreservation; Human; Male; Sperm Banks; Spermatozoa. Source Mol Reprod Dev 1999 Jul;53(3):363-7.

Principles of cloning / edited by Jose Cibelli ... [et al.]. Published/Created: Amsterdam; Boston: Academic Press, c2002. Related Names: Cibelli, Jose B., 1963- Description: xxii, 531 p.: ill. (some col.); 29 cm. ISBN: 012174597X (alk. paper) Notes: Includes bibliographical references and index. LC Classification: QH442.2 .P75 2002 Dewey Class No.: 660.6/5 22

Principles of DNA cloning. Author(s): Miles, J S; Wolf, C R Language: eng Subject(s): Amino Acid Sequence; Base Sequence; Cloning, Molecular/methods; DNA/genetics; DNA Restriction Enzymes; DNA, Recombinant; Gene Library; Human; Molecular Sequence Data; Nucleic Acid Hybridization Source BMJ 1989 Oct 21;299(6706):1019-22.

Procreation by cloning: crafting anticipatory guidelines. Abstract: Author argues in favor of a nonlegislative approach to cloning that distinguishes among types of cloning and recognizes the conditions under which cloning would likely be sought. Author(s): Bonnicksen, A L Language: eng Subject(s): Advisory Committees; Cloning, Organism/legislation & jurisprudence/methods/standards; Commodification; Dehumanization; Embryo Research; Federal Government; Fetal Research; Government Regulation; Guidelines; Health Policy; Human; Internationality; Oocyte Donation; Risk Assessment; Spermatozoa; Twinning; United States Source J Law Med Ethics 1997 Winter;25(4):273-82, 231.

Production of genetically identical sets of mammals: cloning? Abstract: In the past few years, there has been phenomenal progress in producing genetically identical mammalian multiplets by asexual means. The simplest method is to divide embryos into two, three, or four parts, and transfer the parts to surrogate mothers for gestation to term. Another approach is to inject nuclei from totipotent embryonic cells into fertilized one-cell embryos and to remove both male and female pronuclei. When these embryos reach the blastocyst stage, nuclei from their totipotent cells can be removed and injected into one-cell embryos, thus amplifying the process ad infinitum. Subsets of these embryos can be stored in liquid nitrogen to make genetic material for additional copies available indefinitely. Some of these techniques are fairly efficacious, others are not. No reliable techniques are available for making genetic copies of nonhomozygous adult mammals (or other vertebrates) unless samples of embryonic cells were cryopreserved when the individuals were embryos. However, injection of nuclei of spermatogonia into enucleated one-cell embryos is a promising strategy. The genetic totipotency of these nuclei is obvious since they become gamete nuclei. Moreover, the birth of mice from the diploidized genetic material of a single X-bearing spermatozoan after removal of the female pronucleus from a one-cell fertilized embryo suggests that we should eventually be able to make genetic copies of adult male mammals. This approach would not work for adult females because their oogonia have all become oocytes, and the DNA configuration in oocytes is not identical to somatic cells due to meiotic crossing over. Author(s): Seidel, G E Jr Language: eng Source J Exp Zool 1983 Nov;228(2):347-54.

Progress in cloning embryos from domesticated livestock. Author(s): Prather, R S Language: eng Subject(s): Animal; Blastomeres/ultrastructure; Cell Cycle; Cell Nucleus/transplantation; Cloning, Molecular; Embryo/ultrastructure Source Proc Soc Exp Biol Med 1996 May;212(1):38-43.

Progress toward cloning of the gene responsible for childhood spinal muscular atrophy. Author(s): Kleyn, P W; Gilliam, T C Language: eng Subject(s): Child; Chromosome Mapping; Cloning, Molecular; Human; Spinal Muscular Atrophies of Childhood/genetics Source Semin Neurol 1993 Sep;13(3):276-82.

Progress towards cloning the cystic fibrosis gene. Abstract: Genetic linkage analysis with polymorphic DNA markers (restriction fragment length polymorphisms: RFLPS) has allowed the assignment of the cystic fibrosis (CF) locus to the long arm of chromosome 7, within the region of band q31. Two of these markers, MET and D7S8, are tightly linked to the disease locus. Although recent data suggest that they are located on opposite sides of CF, the two can be separated by as much as 5 centimorgans. To obtain a better description of the CF locus and, eventually, to identify the affected gene, additional DNA markers are required to connect MET and D7S8, physically. We have screened the flow-sorted chromosome-7-specific library and thus far isolated 28 new probes from the 7q31 region by DNA hybridization analysis that uses a series of somatic cell hybrids containing various portions of human chromosome 7. Together with the previously identified markers, MET, D7S8, D7S13 and D7S16, these new markers should provide a fine genetic and physical map for the chromosomal region surrounding CF. DNA segments can then be sequentially cloned by chromosome walking from points closest to the CF locus and examined for genes that are preferentially expressed in tissues known to be affected in the disease. Author(s): Tsui, L C; Rommens, J M; Burns, J; Zengerling, S; Riordan, J R; Carlock, L R; Grzeschik, K H; Buchwald, M Language: eng Source Philos Trans R Soc Lond B Biol Sci 1988 Jun 15;319(1194):263-73.

Properties of the beta 1- and beta 2-adrenergic receptor subtypes revealed by molecular cloning. Abstract: The beta 1- and beta 2-adrenergic receptor subtypes are biochemically and functionally similar, because both receptors mediate the catecholamine-dependent activation of adenylate cyclase through the GTP-binding protein, Gs. Pharmacologically, the two receptors can be distinguished on the basis of their relative affinities for the agonists epinephrine and norepinephrine as well as their affinities for several selective antagonists. The primary structures of the human beta 1- and beta 2-adrenergic receptors have recently been deduced from the cloning of their genes and (or) cDNAs, revealing high sequence homology and a membrane topography of seven putative transmembrane regions similar to that of rhodopsin. Chimeric beta 1/beta 2-adrenergic receptor cDNAs have been constructed by site-directed mutagenesis and the chimeric RNA transcripts expressed in Xenopus laevis oocytes. The pharmacological properties of the expressed chimeric receptor proteins were assessed by radioligand binding utilizing subtype-selective agonists and antagonists. Apparently, several of the putative transmembrane regions contribute significantly to the determination of subtype selectivity, presumably by formation of a ligand-binding pocket, with determinants for agonist and antagonist binding being distinguishable. Author(s): Frielle, T; Caron, M G; Lefkowitz, R J Language: eng Source Clin Chem 1989 May;35(5):721-5.

Protein traps: using intracellular localization for cloning. Abstract: A much-sought goal - the rapid cloning of genes whose protein products have specific intracellular localizations - has now been made possible. A visual screen of cells expressing fusions between coding DNA sequences and a reporter, such as green-fluorescent protein (GFP) or beta-galactosidase, identifies cells with the pattern of interest. The DNA sequences encoding these targeted fusions can then be cloned either directly from these cells or by repeated rounds of screening and subdivision of library pools. It is expected that systematic screenings based on these methods will identify additional components for every compartment and define new domains, thus facilitating a more comprehensive understanding of the cellular architecture. Author(s): Gonzalez, C; Bejarano, L A Language: eng Subject(s): Cell Compartmentation; Cloning, Molecular/methods; Eukaryotic Cells; Intracellular Fluid/metabolism; Proteins/genetics; Recombinant Fusion Proteins/genetics/metabolism; Saccharomyces cerevisiae; Schizosaccharomyces Source Trends Cell Biol 2000 Apr;10(4):162-5.

Pseudomonas lipases: biochemical properties and molecular cloning. Abstract: Lipases (glycerol ester hydrolases; EC 3.1.1.3) are important enzymes which, due to their ability to catalyze a number of reactions, are receiving considerable interest from both academia and industry. The bacterial genus Pseudomonas is a prolific producer of a number of extracellular enzymes including lipase. This review summarizes the biochemical properties and recent advances in the molecular genetic analysis of a wide variety of Pseudomonas lipases. In particular, a comparison is made between the amino acid sequences of the various lipases as well as their secondary gene products, which are thought to be essential for secretion of the enzyme. Author(s): Gilbert, E J Language: eng Subject(s): Amino Acid Source Enzyme Microb Technol 1993 Aug;15(8):634-45.

Purification and cloning of brain proteases capable of degrading the beta-amyloid precursor protein. Author(s): Abraham, C R; Razzaboni, B L; Papastoitsis, G; Picard, E; Kanemaru, K; Meckelein, B; Mucke, L Language: eng Source Ann N Y Acad Sci 1992 Dec 31;674:174-9.

Purification and cloning of opioid receptors. Author(s): Li, L Y; Su, Y F; Zhang, Z M; Wong, C S; Chang, K J Language: eng Subject(s): Cloning, Molecular; Human; Receptors, Opioid chemistry isolation & purification Source NIDA Res Monogr 1993;134:146-64.

Purification, reconstitution, and cloning of an NMDA receptor-ion channel complex from rat brain synaptic membranes: implications for neurobiological changes in alcoholism. Author(s): Michaelis, E K; Michaelis, M L; Kumar, K N; Tilakaratne, N; Joseph, D B; Johnson, P S; Babcock, K K; Aistrup, G L; Schowen, R L; Minami, H Language: eng Source Ann N Y Acad Sci 1992 Jun 28;654:7-18.

Purinergic receptors in human blood platelets: chemical modification and cloning investigations. Abstract: Platelet aggregation is important for maintaining normal hemostasis. However, aberrant platelet aggregation plays a major role in acute coronary artery diseases, myocardial infarction, unstable angina, and stroke. ADP is one of the earliest and most important platelet agonists. ADP induces platelet aggregation, shape change, secretion, influx and intracellular mobilization of Ca2+, and inhibition of the adenylyl cyclase stimulated by prostaglandins. Binding of ADP to purinergic receptor(s) is required for elicitation of the ADP-induced platelet responses. But the platelet ADP receptor(s) has not been purified, largely due to the unavailability of the reagents that can be used to selectively label the platelet ADP receptor. The ADP receptor responsible for the ADP-induced platelet aggregation and inhibition of stimulated adenylyl cyclase activity has not been cloned due to difficulties in screening responsive clones generated from a cDNA library. Since the purified ADP-receptor protein is not available, antibodies that can be used as alternative tools to purify the ADP receptor or screen the clones expressing the receptor could not be made. In addition, the problem may be compounded by the low copy number and the susceptibility of the receptor to proteolysis. Therefore, signal transduction mechanisms underlying biochemical transformations in ADP-induced platelet responses remain less well defined and/less well understood. In the past decade efforts have been made to identify a platelet ADP receptor(s) by photoaffinity as well as affinity labeling by the ADP-affinity analogs. More recently efforts have been directed to clone the platelet ADP receptors. These investigations, however, have not produced definite results. The purpose of this review is to examine the results obtained by the photoaffinity- and affinity-labeling investigations and cloning experiments to identify a platelet ADP receptor(s). Author(s): Puri, R N; Colman, R W Language: eng Source J Protein Chem 1998 Jul;17(5):429-51.

QTL mapping to QTL cloning: mice to the rescue. Author(s): Georges, M Language: eng Subject(s): Animal; Chromosome Mapping; Genome; Mice Source Genome Res 1997 Jul;7(7):663-5.

Quantitation and cloning of cytolytic T lymphocytes and their precursors. Author(s): MacDonald, H R; Cerottini, J C; Ryser, J E; Maryanski, J L; Taswell, C; Widmer, M B; Brunner, K T Language: eng Source Immunol Rev 1980;51:93-123.

Rapid in silico cloning of genes using expressed sequence tags (ESTs). Abstract: Expressed sequence tags (ESTs) are short single-pass DNA sequences obtained from either end of cDNA clones. These ESTs are derived from a vast number of cDNA libraries obtained from different species. Human ESTs are the bulk of the data and have been widely used to identify new members of gene families, as markers on the human chromosomes, to discover polymorphism sites and to compare expression patterns in different tissues or pathologies states. Information strategies have been devised to query EST databases. Since most of the analysis is performed with a computer, the term "in silico" strategy has been coined. In this chapter we will review the current status of EST databases, the pros and cons of EST-type data and describe possible strategies to retrieve meaningful information. Author(s): Gill, R W; Sanseau, P Language: eng Source Biotechnol Annu Rev 2000;5:25-44.

Recent developments in the molecular cloning and characterization of hepatitis C and E viruses. Abstract: The molecular properties of the genomes of both hepatitis C virus (HCV) and hepatitis E virus (HEV), the major etiologic agents of non-A, non-B hepatitis, are briefly described. The organization of the genome of each of these viruses is discussed and compared to those of other related or distantly related viruses that contain single-stranded, positive-sense RNA genomes. In the case of HEV, the reactivity of expressed proteins and genetic divergence of geographically distinct isolates are also described within the context of sequence variation, type-common epitopes and type-specific epitopes. Author(s): Bradley, D W; Beach, M J; Purdy, M A Language: eng Source Microb Pathog 1992 Jun;12(6):391-8.

Recent progress and problems in animal cloning. Abstract: It is remarkable that mammalian somatic cell nuclei can form whole individuals if they are transferred to enucleated oocytes. Advancements in nuclear transfer technology can now be applied for genetic improvement and increase of farm animals, rescue of endangered species, and assisted reproduction and tissue engineering in humans. Since July 1998, more than 200 calves have been produced by nuclear transfer of somatic cell nuclei in Japan, but half of them were stillborn or died within several months of parturition. Morphologic abnormalities have also been observed in cloned calves and embryonic stem cell-derived mice. In this review, we discuss the present situation and problems with animal cloning and the possibility for its application to human medicine. Author(s): Tsunoda, Y; Kato, Y Language: eng Source Differentiation 2002 Jan;69(4-5):158-61.

Recent progress in mammalian cloning. Abstract: PURPOSE: Our purpose was to review recent progress in the use of nuclear transfer technology to produce genetically identical mammals. METHODS: A literature review was conducted. RESULTS: The reasons for cloning nonhuman mammals are manyfold including commercial, biomedical, and basic research applications. Individual steps in the nuclear transfer process are itemized, along with a detailed description of the specific approaches used in the production of Dolly, NETI, and DITTO. The potential application of nuclear transfer in the treatment of human infertility is also considered, along with bioethical concerns. Finally, insights are provided concerning the future application of cloning technology in rhesus macaques. CONCLUSIONS: The cloning of a lamb (Dolly) from an adult, mammary gland cell coupled with the successful production of rhesus monkeys (NETI and DITTO) by nuclear transfer of embryonic cells marks the beginning of a "Golden Age" in the development and application of somatic cell cloning technology in mammals. Author(s): Wolf, D P; Meng, L; Ely, J J; Stouffer, R L Language: eng Source J Assist Reprod Genet 1998 May;15(5):235-9.

Recent progress in molecular cloning of glycosyltransferase genes of eukaryotes. Author(s): Narimatsu, H Language: eng Source Microbiol Immunol 1994;38(7):489-504.

Receptor cloning and heterologous expression--towards a new tool for drug discovery. Abstract: The explosion in the number of cloned receptors presents the pharmaceutical industry with challenges to discover new drugs targeting those receptors; to find more-selective drugs for all novel receptor subtypes; and to learn more about the function of the receptors in order to discern the conditions where such drugs may be applied usefully as therapeutics. At the same time, receptor cloning affords an unprecedented opportunity to address these challenges: heterologously expressed recombinant human receptors can be used for drug screening and - through an improved understanding of structure-function relationship - possibly for drug design, while the receptor clones permit mobilization of the full power of molecular biology to elucidate the function of the receptors in health and disease. Author(s): Luyten, W H; Leysen, J E Language: eng Source Trends Biotechnol 1993 Jun;11(6):247-54.

Receptors for parathyroid hormone and parathyroid hormone-related peptide: from molecular cloning to definition of diseases. Abstract: The parathyroid hormone/parathyroid hormone-related peptide receptor belongs to a distinct family of G protein-coupled receptors, the members of which usually signal through at least two second messenger systems, adenylate cyclase and phospholipase C. The parathyroid hormone/ parathyroid hormone-related peptide receptor is most abundantly expressed in bone, kidney and growth-plate chondrocytes, and, at lower levels, in a variety of fetal and adult tissues. To search for human diseases that are caused by parathyroid hormone/parathyroid hormone-related peptide receptor defects, genomic DNA of patients with pseudohypoparathyroidism type Ib and of patients with Jansen's metaphyseal chondrodysplasia was screened for mutations in all coding exons of the receptor gene. Inactivating parathyroid hormone/parathyroid hormone-related peptide receptor mutations were excluded in patients with pseudohypoparathyroidism type Ib. However, a receptor mutation that causes agonist-independent, constitutive cAMP accumulation was identified in a patient with Jansen's metaphyseal chondrodysplasia, a rare form of short-limbed dwarfism associated with hypercalcemia despite normal or low concentrations of parathyroid hormone and parathyroid hormone-related peptide. These findings allow the conclusion to be drawn that parathyroid hormone/parathyroid hormone-related peptide receptors mediate the endocrine actions of parathyroid hormone, which are required for the control of calcium homeostasis and the autocrine-paracrine actions of parathyroid hormone-related peptide, which are required for normal growth-plate development. Author(s): Juppner, H; Schipani, E Language: eng Source Curr Opin Nephrol Hypertens 1996 Jul;5(4):300-6.

Recombinant DNA markets.
Published/Created: Norwalk, Conn. (30 High St., Norwalk 06851): International Resource Development, 1981. Related Names: International Resource Development, inc. Description: vi, 189 leaves; 28 cm. Subjects: Genetic engineering industry. Pharmaceutical industry. Molecular cloning--Industrial applications. Series: Report (International Resource Development, inc.); #163. LC Classification: HD9999.G452 R42 Dewey Class No.: 338.4/7574873282 19

Recombinant DNA principles and methodologies / edited by James J. Greene, Venigalla B. Rao. Published/Created: New York: Marcel Dekker, c1998. Related Names: Greene, James J. (James Joseph) Rao, Venigalla B. Description: ix, 750 p.: ill.; 24 cm. ISBN: 0824799895 (alk. paper) Notes: Includes bibliographical references and index. Subjects: Molecular cloning. Recombinant DNA. LC Classification: QH442.2 .R43 1998 Dewey Class No.: 660.6/5 21

Recombinant DNA research and viruses: cloning and expression of viral genes / edited by Yechiel Becker. Published/Created: Boston: M. Nijhoff, c1985. Related Names: Becker, Yechiel. Description: x, 373 p.: ill.; 24 cm. ISBN: 0898386837 Notes: Includes bibliographies and index. Subjects: Recombinant viruses. Recombinant DNA. Cloning. Gene expression. Molecular virology. Cloning, Molecular. DNA, Recombinant. Genes, Viral. Series: Developments in molecular virology; 5 LC Classification: QR392 .R44 1985 Dewey Class No.: 576/.64 19

Recombinant hematopoietic growth factors. Results of molecular cloning and potential therapeutic applications; Steinheider, G Language: ger; Cloning, Molecular; Hematopoietic Cell Growth Factors/genetics/therapeutic use; Human; Recombinant Proteins/therapeutic use Source Med Klin 1991 Apr 15;86(4):209-16.

Recombinant hemopoietic growth factors. Molecular cloning and potential therapeutic applications. Abstract: In analogy with data obtained from other mammalian species it appears that the majority of physiologically relevant human hemopoietic growth factors (e.g.

erythropoietin and colony stimulating factors) has been characterized in biological systems, defined by biochemical methods and manufactured by molecular cloning of the corresponding genetic elements and expression in suitable biological systems. A review of the present situation of production of recombinant hemopoietic growth factors is presented and potential therapeutic applications of these factors will be discussed. In addition some preliminary clinical results, due to the medical application of recombinant hemopoietic growth factors, will be described. Author(s): Steinheider, G Source Arzneimittelforschung 1988 Jun;38(6):851-4.

Regulation and cloning of microbial chitinase genes. Abstract: A range of chitinase genes from microorganisms have been cloned and the potential uses of these genetically manipulated organisms are being investigated by various researchers. Fungi and yeast are better producers of chitinase than bacteria. Since fungi grow at a slower rate, there have been efforts to clone the fungal chitinase genes into fast-growing bacteria. This review gives a brief survey of recent progress in the regulation and cloning of microbial chitinase genes. Emphasis is placed on the post-translational modification and localization of the recombinant protein in the host. Various amino acid domains are present in this protein. The mode of catalytic activity of the recombinant protein in comparison to the wild-type protein is discussed in the available literature. The different mechanisms involved in the regulation of chitinase genes from various microorganisms is discussed by the researchers. The scope of future research and conclusions yet to be obtained in this particular area are also outlined in this review. Author(s): Felse, P A; Panda, T Language: eng Source Appl Microbiol Biotechnol 1999 Feb;51(2):141-51.

Replicable RNA vectors: prospects for cell-free gene amplification, expression, and cloning. Author(s): Chetverin, A B; Spirin, A S Language: eng Source Prog Nucleic Acid Res Mol Biol 1995;51:225-70.

Report on human cloning through embryo splitting: an amber light. Language: eng Subject(s): Advisory Committees; Beginning of Human Life; Child; Cloning, Organism; Cryopreservation; Dehumanization; Dissent and Disputes; Embryo; Embryo Research; Embryo Transfer; Embryo and Fetal Development; Ethics; Family Relations; Fees and Charges; Fertilization in Vitro; Fetal Tissue Transplantation; Government Regulation; Group Processes; Guidelines; Individuality; Internationality; Life; Methods; Motivation; Parent-Child Relations; Personhood; Politics; Preimplantation Diagnosis; Public Policy; Reproductive Techniques, Assisted; Research; Risk; Risk Assessment; Self Concept; Social Change; Social Control, Formal; Twins; United States; Value of Life Source Kennedy Inst Ethics J 1994 Sep;4(3):251-82.

Representational difference analysis in cloning of aberrant genes. Author(s):Yang, L; Yuan, A L; Li, Y Language: chi; Cloning, Molecular/methods; DNA, Neoplasm/genetics, Human; Oncogene Proteins/genetics; Oncogenes; Sequence Analysis, DNA Source Sheng Li Ke Xue Jin Zhan 1999 Jul;30(3):271-4.

Reproductive semi-cloning respecting biparental embryo origin: embryos from syngamy between a gamete and a haploidized somatic cell. Abstract: Embryos formed by somatic cell nuclear transfer to enucleated oocytes (cloning) have given rise to viable offspring in several mammalian species. The possibility of future application of this technique to human assisted reproduction (reproductive cloning) has been widely debated. On this background there is current discussion of the potential for a cloning-derived technique, which aims at syngamy between a gamete nucleus from one parent and a somatic cell nucleus from the other. Critical analysis of the clinical indications, the current state of the art, biological concerns and ethical considerations relative to this technique, called here reproductive semi-cloning, are presented. Such a technique requires validation by further research before it can be considered as a treatment option. This debate explores issues raised by the technique. Author(s): Tesarik, J Language: eng Source Hum Reprod 2002 Aug;17(8):1933-7.

Reprogramming nuclei: insights from cloning, nuclear transfer and heterokaryons. Abstract: Mammals and amphibians can be cloned following the transfer of embryonic nuclei into enucleated eggs or oocytes. As nuclear functions become more specialized in the differentiated cells of an adult, successful cloning using these nuclei as donors becomes more difficult. Differentiation involves the assembly of specialized forms of repressive chromatin including linker histones, Polycomb group proteins and methyl-CpG-binding proteins. These structures compartmentalize chromatin into functional domains and maintain the stability of the differentiated state through successive cell divisions. Efficient cloning requires the erasure of these structures. The erasure can be accomplished through use of molecular chaperones and enzymatic activities present in the oocyte, egg or zygote. We discuss the mechanisms involved in reprogramming nuclei after nuclear transfer and compare them with those that occur during remodeling of somatic nuclei after heterokaryon formation. Finally we discuss how one might alter the properties of adult nuclei to improve the efficiency of cloning. Author(s): Kikyo, N; Wolffe, A P Language: eng Source J Cell Sci 2000 Jan;113 (Pt 1):11-20.

Reverse genetics of positional cloning; Douar, A M; Aubourg, P Language: fre Subject(s): Animal; Chromosome Mapping/methods; Cloning, Molecular; DNA/genetics; Gene Library; Human Source Diabete Metab 1992 Jul-Aug;18(4):322-8.

Review of the President's Commission's recommendations on cloning: hearing before the Committee on Science, Subcommittee on Technology, U.S. House of Representatives, One Hundred Fifth Congress, first session, June 12, 1997. Published/Created: Washington: U.S. G.P.O.: For sale by the U.S. G.P.O., Supt. of Docs., Congressional Sales Office, 1997. Description: iii, 227 p.: ill.; 24 cm. ISBN: 0160558816 Notes: Distributed to some depository libraries in microfiche. Shipping list no.: 98-0117-P. "No. 19." Includes bibliographical references. Subjects: Cloning--Government policy--Moral and ethical aspects--United States. Genetic engineering. Bioethics. LC Classification: KF27 .S39957 1997k Govt. Doc. No.: Y 4.SCI 2:105/19 GPO Item No.: 1025-A-01 1025-A-02 (MF)

Review of the research of glia maturation factor and cloning of human and rat glia maturation factor-gamma (GMFG) cDNA Abstract: Glia maturation factor-beta (GMFB) is a 17-kDa protein that was initially identified as a growth and differentiation factor acting on neurons as well as glia in the vertebrate brain. We isolated human and rat glia maturation factor-gamma (GMFG) cDNA and examined the tissue distribution of GMFG in human and rat by Northern blots and Western blots. Sequence analysis of the entire cDNA revealed an open reading frame of 426 nucleotides with a deduced protein of 142 amino acid residues. The deduced amino acid sequence of its putative product is highly homologous to GMFB. Northern blot analysis indicated that a 0.9 kb mRNA is predominantly expressed in rat thymus, testis, and spleen. In comparison with GMFB, the current study demonstrated that the tissue distribution of GMFG is not the same as that of GMFB, and GMFG is predominantly in proliferative and differentiative organs. Author(s): Asai, K Language: jpn Source Nihon Shinkei Seishin Yakurigaku Zasshi 2001 Feb;21(1):15-20.

Review of the Streptomyces lividans/vector pIJ702 system for gene cloning. Abstract: Interest in the biology of the Streptomyces and application of these soil bacteria to production of commercial antibiotics and enzymes has stimulated the development of efficient cloning techniques and a variety of streptomycete plasmid and phage vectors. Streptomyces lividans is routinely employed as a host for gene cloning, largely because this species recognizes a large number of promoters and appears to lack a restriction system. Vector pIJ702 was constructed from a variant of a larger autonomous plasmid and is often used as a cloning vehicle in conjunction with S. lividans. The host range of vector pIJ702 extends beyond Streptomyces spp., and its high copy number has been exploited for the overproduction of cloned gene products. This combination of host and vector has been used successfully to investigate antibiotic biosynthesis, gene structure and expression, and to map various Streptomyces mutants. Author(s): Gusek, T W; Kinsella, J E Language: eng Source Crit Rev Microbiol 1992;18(4):247-60.

Revolutions in rapid amplification of cDNA ends: new strategies for polymerase chain reaction cloning of full-length cDNA ends. Abstract: Rapid amplification of cDNA ends (RACE) is a polymerase chain reaction (PCR)-based technique which was developed to facilitate the cloning of full-length cDNA 5'- and 3'-ends after a partial cDNA sequence has been obtained by other methods. While RACE can yield complete sequences of cDNA ends in only a few days, the RACE procedure frequently results in the exclusive amplification of truncated cDNA ends, undermining efforts to generate full-length clones. Many investigators have suggested modifications to the RACE protocol to improve the effectiveness of the technique. Based on first-hand experience with RACE, a critical review of numerous published variations of the key steps in the RACE method is presented. Also included is a detailed, effective protocol based on RNA ligase-mediated RACE/reverse ligation-mediated PCR, as well as a demonstration of its utility. Author(s): Schaefer, B C Language: eng Source Anal Biochem 1995 May 20;227(2):255-73.

rhDNase: scientific background, cloning and production Abstract: Despite the hopes raised by the first attempts in gene therapy, direct correction of the defect in CFTR protein associated with cystic fibrosis is still beyond clinical reach. Therefore we have to set upon the consequences of the defect. Respiratory distress and progressive lung destruction in cystic fibrosis can be accounted for by infectious exacerabations and the accumulation of viscous purulent secretions in the airways. For a long time we have known that purulent secretions that accumulate in the airways of patients with cystic fibrosis contain large amounts of DNA, a complex macromolecule that contributes mostly to the viscosity and hinders the mucociliary function. Hence we hypothesized that enzymatic cleaving of DNA molecules by desoxyribonuclease (DNase) should reduce the viscosity of sputum, and slow or prevent the deterioration of pulmonary function. Using the techniques of molecular biology and genetic engineering, we identified the gene of human DNase I, which was cloned in mammalian cells to produce large amounts of a glycosylated protein for therapeutic use. Catalytic amounts of rhDNase greatly reduce the viscosity of purulent cystic fibrosis sputum, transforming it within minutes from a nonflowing viscous gel to a flowing liquid. This effect was associated with a decrease in size of DNA fragments in the sputum. Our studies suggested that inhalation of a rhDNase aerosol might be a simple direct approach to reduce the viscosity of purulent secretions... Author(s): Shak, S Language: fre Source Arch Pediatr 1995 Jul;2(7):666-9.

Richardson, Hazel. How to clone a sheep / by Hazel Richardson; illustrated by Andy Cooke. Edition Information: 1st American ed. Published: New York: Franklin Watts, 2001. Related Names: Cooke, Andy, ill. Description: 92 p.: ill.; 21 cm. ISBN: 0531146456 (lib. bdg.) 0531162001 (pbk.) Notes: Originally published: Oxford: Oxford University Press, 1999. Subjects: Sheep--Cloning--Juvenile literature. Series: How to LC Classification: QH442.2 .R534 2001 Dewey Class No.: 660.6/5 21

Roberts, Melinda A., 1954- Child versus childmaker: future persons and present duties in ethics and the law / Melinda A. Roberts. Published/Created: Lanham: Rowman & Littlefield, c1998. Description: xv, 235 p.; 24 cm. ISBN: 084768900X (cloth: alk. paper) 0847689018 (pbk.: alk. paper) Notes: Includes bibliographical references (p. 221-226) and index. Subjects: Human reproductive technology--Moral and ethical aspects. Cloning--Moral and ethical aspects. Series: Studies in social, political, and legal philosophy LC Classification: RG133.5 .R62 1998 Dewey Class No.: 176 21

Robinson, Freddie, 1934- Christianity and cloning / [Freddie Robinson, biblical researcher]. Published/Created: [Highland Park, MI]: Transfiguration Production Co., c1998. Description: 174 p.; 21 cm. ISBN: 0964048736 Subjects: Cloning--Human cloning--Religious aspects--Christianity. LC Classification: QH442.2 .R63 1998 Dewey Class No.: 241/.674957 21

Rorvik, David M. In his image: the cloning of a man / David M. Rorvik. Edition Information: 1st ed. Published/Created: Philadelphia: Lippincott, c1978. Description: 239 p.; 22 cm. ISBN: 0397012551 : Notes: Includes index. Bibliograhy: p. 223-234. Subjects: Human cloning. LC Classification: QH442.2 .R67 Dewey Class No.: 612.6

Rorvik, David M. In his image: the cloning of a man / David M. Rorvik. Published: London: H. Hamilton, 1978. Description: 207 p.; 23 cm. ISBN: 0241100100 : Notes: Bibliography: p. 197-207. LC Classification: QH442.2 .R67 1978b Dewey Class No.: 612.6

Roy, Ron, 1940- Kidnapped at the Capital / by Ron Roy; illustrated by Liza Woodruff. Edition Information: 1st Random House ed. Published/Created: New York: Random House, c2003. Related Names: Woodruff, Liza, ill. Description: p. cm. ISBN: 0307265145 (pbk.) 0307465144 (lib. bdg.) Summary: KC's mother and the clone of the President of the United States are kidnapped by disgruntled astronauts who want to take over the International Space Station. Notes: "A Stepping Stone Book." Subjects: Kidnapping--Fiction. Cloning--Fiction. Series: Capital mysteries; #2 LC Classification: PZ7.R8139 Ki 2003 Dewey Class No.: [Fic] 21

Roy, Ron, 1940- Kidnapped at the Capital / by Ron Roy; illustrated by Liza Woodruff. Published/Created: New York: Golden Books, c2002. Related Names: Woodruff, Liza, ill. Description: 73 p.: ill.; 21 cm. ISBN: 0307465144 (GB) 0307265145 (pbk.) Summary: KC's mother and the clone of the President of the United States are kidnapped by disgruntled astronauts who want to take over the International Space Station. Subjects: Kidnapping--Cloning--Fiction. Series: Capital mysteries; 2 Road to reading. Mile 5 LC Classification: PZ7.R8139 Ki 2002 Dewey Class No.: [Fic] 21

Roy, Ron, 1940- Who cloned the President? / by Ron Roy; illustrated by Liza Woodruff. Edition Information: 1st Random House ed. Published/Created: New York: Random House, 2003. Related Names: Woodruff, Liza, ill. Description: p. cm. ISBN: 0307265102 (pbk.) 0307465101 (lib. bdg.) Summary: KC discovers that the President of the United States has been replaced by a clone and sets out with her friend Marshall on a dangerous mission to set things right. Notes: "A Stepping Stone Book." Subjects: Cloning--Fiction. Series: Capital mysteries; #1 LC Classification: PZ7.R8139 Wl 2003 Dewey Class No.: [Fic] 21

Roy, Ron, 1940- Who cloned the President? / by Ron Roy; illustrated by Liza Woodruff. Published/Created: New York: Golden Books, 2001. Related Names: Woodruff, Liza, ill. Description: 75 p.: ill.; 20 cm. ISBN: 0307265102 (pbk.) 0307465101 (gb) Summary: KC discovers that the President of the United States has been replaced by a clone and sets out with her friend Marshall on a dangerous mission to set things right. Series: Road to reading. Mile 5 LC Classification: PZ7.R8139 Wl 2001 Dewey Class No.: [Fic] 21

Rubik, the amazing cube / [production company unknown]; produced by Steve Werner, Mark Jones. Published/Created: United States: [s.n.], 1983; United States: RCA/Columbia Pictures Home Video, [1985?] Related Names: Coca-Cola Company Collection (Library of Congress) Description: 1 videocassette of 1 (Beta): sd., col.; 1/2 in. viewing copy. Summary: Based on the internationally popular puzzle, comes to life on this videocassette as the friend and protector of Carlos, Lisa and Reynaldo Rodriguez. The Rodriguez kids stumble upon a sinister scientist testing a cloning machine on stolen diamonds who captures Rubik in order to tap his powers. Notes: Copyright: Ruby/Spears Enterprises, Inc. DCR 1983; PUB 10 Sep83; REG 1Mar84; PA219-446. Volume I appears on cassette box. On cassette with: Rubik and the mysterious man. Summary taken from cassette box. Animation. Reissue date is on copyright package design. Sources used: cassette box and data base. LC Classification: VAC 9472 (viewing copy)

S-adenosylmethionine decarboxylase from Escherichia coli and from Saccharomyces cerevisiae: cloning and overexpression of the genes. Author(s): Kashiwagi, K; Taneja, S K; Xie, Q W; Tabor, C W; Tabor, H Language: eng Source Adv Exp Med Biol 1988;250:73-9.

SAG/ROC/Rbx/Hrt, a zinc RING finger gene family: molecular cloning, biochemical properties, and biological functions. Abstract: The RING (really interesting new gene) finger proteins containing a characteristic C3HC4 or C3H2C3 motif appear to act as E3 ubiquitin ligase and play important roles in many processes, including cell-cycle progression, oncogenesis, signal transduction, and development. This review is focused on SAG/ROC/Rbx/Hrt (sensitive to apoptosis gene/regulator of cullins/RING box protein), an evolutionarily conserved RING finger family of proteins that were cloned recently by several independent laboratories through differential display, yeast two-hybrid screening, or biochemical purification. SAG/ROC2/Rbx2/Hrt2 is expressed in multiple mouse adult tissues, as well as early embryos. In humans, both SAG and ROC1 are ubiquitously expressed at a very high level in heart, skeletal muscle, and testis. Expression of both SAG and ROC1 is induced by mitogenic stimulation. SAG is also induced by a redox agent in cultured cells, as well as in in vivo mouse brain upon ischemia/reperfusion. Structurally, SAG consists of four exons and three introns with at least one splicing variant and two pseudogenes. The SAG gene promoter is enriched with multiple transcription factor binding sites. Biochemically, SAG binds to RNA, has metal-ion binding/free radical scavenging activity, and is redox-sensitive. Most importantly, like ROC1, SAG/ROC2 binds to cullins and acts as an essential component of E3 ubiquitin ligase. Biologically, SAG is a growth-essential gene in yeast. In mammalian cells, SAG protects apoptosis mainly through inhibition of cytochrome c release/caspase activation, and promotes growth under serum deprivation at least in part by inhibiting p27 accumulation. Blocking SAG expression via antisense transfection inhibits tumor cell growth. Thus, SAG appears to be a valid drug target for anticancer therapy. Author(s): Sun, Y; Tan, M; Duan, H; Swaroop, M Language: eng Source Antioxid Redox Signal 2001 Aug;3(4):635-50.

Sambrook, Joseph. Molecular cloning: a laboratory manual / J. Sambrook, E.F. Fritsch, T. Maniatis. Edition Information: 2nd ed. Published/Created: Cold Spring Harbor, N.Y.: Cold Spring Harbor Laboratory, 1989. Related Names: Fritsch, E. F. Maniatis, Tom. Description: 3 v.: ill; 28 cm. ISBN: 0879693096 Notes: Includes bibliographical references and indexes. Subjects: Molecular cloning--Laboratory manuals. Eukaryotic cells--Laboratory manuals. LC Classification: QH442.2 .S26 1989 Dewey Class No.: 574.87/3224 19

Sambrook, Joseph. Molecular cloning: a laboratory manual / Joseph Sambrook, David W. Russell. Edition Information: 3rd ed. Published/Created: Cold Spring Harbor, N.Y.: Cold Spring Harbor Laboratory Press, c2001. Related Names: Russell, David W. (David William), 1957- Description: 3 v.: ill.; 28 cm. ISBN: 0879695773 (paper) 0879695765 (cloth) Notes: Includes bibliographical references and index. Subjects: Molecular cloning--Cloning, Molecular--Laboratory Manuals. LC Classification: QH442.2 .S26 2001 Dewey Class No.: 572.8 21

Saunders, Robin. Sons of Anubis / Robin Saunders. Edition Information: 1st ed. Published/Created: Cape Town: Tafelberg, 1998. Description: 129 p.; 22 cm. ISBN: 0624037118 Summary: A father and daughter are in danger when an experiment in genetic engineering escapes from the Perlemoen, South Africa institute where he and others like him are being studied. Notes: 1998 Sanlam Prize for Young Literature Subjects: Genetic engineering--Cloning--Blacks--South Africa--Fiction. LC Classification: PZ7.S2573 So 1998 Dewey Class No.: [Fic] 21

Scientific and medical aspects of human reproductive cloning / Committee on Science, Engineering, and Public Policy, National Academy of Sciences, National Academy of Engineering, Institute of Medicine. Published/Created: Washington, D.C.: National Academy Press, c2002. Related Names: Committee on Science, Engineering, and Public Policy (U.S.) Description: xxi, 272 p.; 23 cm. ISBN: 0309076374 Notes: Includes bibliographical references (p. 154-257). Subjects: Human cloning. Human reproductive technology. Cloning. LC Classification: RG133.5 .S385 2002 Dewey Class No.: 612.6 21

Scientific discoveries in cloning: challenges for public policy: hearing before the Subcommittee on Public Health and Safety of the Committee on Labor and Human Resources, United States Senate, One Hundred Fifth Congress, first session ... March 12, 1997. Published/Created: Washington: U.S. G.P.O.: For sale by the U.S. G.P.O., Supt. of Docs., Congressional Sales Office, 1997. Description: iii, 87 p.; 23 cm. ISBN: 0160549388 Notes: Distributed to some depository libraries in microfiche. Shipping list no.: 97-0227-P. Includes bibliographical references. Subjects: Cloning--Government policy--United States. Series: United States. Congress. Senate. S. hrg.; 105-22. LC Classification: KF26 .L278 1997 Govt. Doc. No.: Y 4.L 11/4:S.HRG.105-22 GPO Item No.: 1043-A 1043-B (MF)

Scope of DNA cloning and chemical methods in development of chemotherapeutic agents. Author(s): Narang, S A; Brousseau, R; Georges, F Language: eng Source Pharmacol Ther 1984;26(2):163-89.

Scott, Jefferson. Fatal defect: a genetic thriller / Jefferson Scott. Published/Created: Sisters, Or.: Multnomah Publishers, c1998. Description: 339 p.; 21 cm. ISBN: 1576734528 (alk. paper) Subjects: Human cloning--Fiction. Chemical weapons--Fiction. Biotechnologists--Fiction. LC Classification: PS3569.C6356 F38 1998 Dewey Class No.: 813/.54 21

Screening and cloning of susceptible genes related to acute lung injury. Author(s):Feng, Gang; Wang, Zheng-guoSource Zhongguo Wei Zhong Bing Ji Jiu Yi Xue 2003 Jun;15(6):372-5.

Sensing of extracellular Ca2+ by parathyroid and kidney cells: cloning and characterization of an extracellular Ca(2+)-sensing receptor. Abstract: The ability of the parathyroid cell to sense minute fluctuations in the extracellular ionized calcium concentration (Ca2+ o) is essential for maintaining mineral ion homeostasis. However, the mechanism(s) through which the parathyroid cell and other cells recognize and respond to changes in Ca2+ o has remained unclear. We recently isolated a cDNA encoding a Ca2+ o-sensing receptor from bovine parathyroid using expression cloning in Xenopus laevis oocytes. The receptor shows pharmacologic properties that are almost identical to those of the receptor on the parathyroid cell and, like the latter, stimulates phospholipase C in a G-protein-dependent manner. The amino acid sequence of the cloned receptor deduced from this cDNA predicts a protein with a molecular mass of 121 kd, which has three principal structural domains. The first is a 613 amino acid, putatively extracellular amino terminus which has several regions rich in acidic amino acids that may potentially be involved in binding Ca2+ and other polycationic agonists. The second comprises seven membrane-spanning segments that are characteristic of the superfamily of G-protein-coupled receptors, and the third is a 222 amino acid cytoplasmic tail. Transcripts for this Ca2+ o-sensing receptor are present in the parathyroid as well as in the kidney, thyroid, and brain. We next investigated the hypercalcemic disorders, familial hypocalciuric hypercalcemia and neonatal severe hyperparathyroidism, as possible examples of inherited abnormalities in this Ca2+ o-sensing receptor, since both disorders show abnormal Ca2+ o-sensing and/or handling in the kidney and parathyroid. (ABSTRACT TRUNCATED AT 250 WORDS); Brown, E M; Pollak, M; Hebert, S C Language: eng Source Am J Kidney Dis 1995 Mar;25(3):506-13.

Shostak, Stanley. Becoming immortal: combining cloning and stem-cell therapy / Stanley Shostak. Published/Created: Albany: State University of New York Press, c2002. Description: xii, 309 p.: ill.; 23 cm. ISBN: 0791454010 0791454029 (pbk.) Notes: Includes bibliographical references (p. [263]-292) and index. Subjects: Human cloning. Stem cells--Research. Cloning, Organism. Evolution. Stem Cells. LC Classification: QH442.2 .S564 2002 Dewey Class No.: 612.6 21

Shotgun phage display cloning. Abstract: Shotgun phage display cloning is a useful tool for studying interactions between bacterial and host proteins. Libraries are constructed by cloning randomly fragmented prokaryotic DNA into phage mid-vectors. Theoretically, these libraries will consist of phages that together display all proteins encoded by the bacterial genome. Selecting a gene III-based library, made from Staphylococcus aureus DNA, against IgG and fibronectin resulted in 20-40% positive clones after two pannings. Increasing the number of fusion proteins per phage particle by using gene VIII-based display, increased the frequency of correct clones to 75-100%. Author(s): Jacobsson, K; Frykberg, L Language: eng Source Comb Chem High Throughput Screen 2001 Apr;4(2):135-43.

Should surgeons clone genes? The strategy behind the cloning of the human inducible nitric oxide synthase gene. Abstract: Nitric oxide (NO) is a recently recognized biological mediator that has triggered an explosion of scientific research over the last 5 years, resulting in NO being named "Molecule of the Year" by Science in 1992. Much of the excitement is due to the physiological and pathophysiological functions exerted by NO as a vasodilator, neurotransmitter, and antimicrobial effector molecule. The purpose of this review is to describe the cloning of the inducible NO synthase gene from human hepatocytes, with particular emphasis on the importance to surgeons of cloning genes and on the utility of applying modern molecular biology to the study of diseases relevant to surgical practice. Overall cloning techniques & the specific strategy used to clone the human inducible NO synthase gene are reviewed. Applications of gene cloning to clinical surgery are discussed. Isolating the inducible human NO synthase gene has provided important information regarding regulation of induced NO synthesis in human cells provided a critical tool for further studies to define the role of NO in normal & disease processes. Author(s): Geller, D A; Billiar, T R Language: eng Source Arch Surg 1993 Nov;128(11):1212-20.

Significance of anthraquinone formation resulting from the cloning of actinorhodin genes in heterologous streptomycetes. Abstract: This review explores the underlying biochemical and genetic principles leading to the formation of hybrid anthraquinones by recombinant anthracycline-producing streptomycetes transformed with genes encoding the early steps in actinorhodin biosynthesis. Experiments indicate that simple aromatic polyketides are probably synthesized using very similar mechanisms, allowing for the interspecies cloning of polyketide synthase genes for the potential production of novel aromatic polyketide structures. Author(s): Strohl, W R; Connors, N C Language: eng Source Mol Microbiol 1992 Jan;6(2):147-52.

Silver, Lee M. Remaking Eden: cloning and beyond in a brave new world / Lee M. Silver. Edition Information: 1st ed. Published: New York: Avon Books, c1997. Description: viii, 317 p.; 24 cm. ISBN: 0380974940 (hardcover) Notes: Includes bibliographical references (p. 255-306) and index LC Classification: RG133.5 .S56 1997 Dewey Class No.: 176 21

Silverstein, Alvin. Cells / Alvin Silverstein, Virginia Silverstein, Laura Silverstein Nunn. Published: Brookfield, Conn.: Twenty-First Century Books c2002. Related Names: Silverstein, Virginia B. Nunn, Laura Silverstein. Description: 64 p.: col. ill.; 26 cm. ISBN: 076132254X (lib. bdg.) Summary: A detailed presentation of the structure and function of different types of cells and of research into developing technologies such as cloning and the use of stem cells. Notes: Includes bibliographical references (p. 60) and index. Series: Silverstein, Alvin. Science concepts. LC Classification: QH582.5 .S538 2002 Dewey Class No.: 571.6 21

Simian virus 40 as a eukaryotic cloning vehicle. Author(s): Elder, J T; Spritz, R A; Weissman, S M Language: eng Source Annu Rev Genet 1981;15:295-340.

Similarities and differences between animal and human cloning; Huber, G Language: fre Source Gynecol Obstet Fertil 2000 Sep;28(9):612-9.

Singer, Peter, 1946- Making babies: the new science and ethics of conception / Peter Singer and Deane Wells. Published/Created: New York: C. Scribner's Sons, 1985. Related Names: Wells, Deane. Singer, Peter, 1946- Reproduction revolution. Description: ix, 245 p.; 22 cm. ISBN: 0684183714 : Notes: Rev. ed. of: The reproduction revolution. 1st ed. 1984. Includes index. Bibliography: p. 231-238. Subjects: Fertilization in vitro, Human. Surrogate mothers. Sex preselection. Cloning. Genetic engineering. Fertilization in vitro, Human--Moral and ethical aspects. Surrogate motherhood--Moral and ethical aspects. Sex preselection--Moral and ethical aspects. Cloning--Moral and ethical aspects. Genetic engineering--Moral and ethical aspects. LC Classification: RG135 .S557 1985 Dewey Class No.: 613.9/4 19

Singer, Randy (Randy D.) Irreparable harm: a novel / Randy Singer. Edition Information: 1st ed. Published/Created: Colorado Springs, Colo.: WaterBrook Press, c2003. Description: x, 421 p.; 21 cm. ISBN: 1578566347 (pbk.) Subjects: Cloning--Fiction. Stem cells--Fiction. Attorney and client--Fiction. LC Classification: PS3619.I5725 I77 2003 Dewey Class No.: 813/.6 21

Single cell sorting and cloning. Abstract: Cell sorters now allow the selection of cells and other bodies according to a range of quite diverse criteria. The additional refinement that allows the sorting of individual cells based on these criteria has seen application in many fields of research. Single cells may be sorted for microscopy, for culture and for genetic analysis by way of single cell PCR (polymerase chain reaction). In practical terms, in the setting up of an instrument for single cell sorting, there are additional requirements to ensure that each detected event is indeed a single cell or body, that this cell can be reliably sorted via saline droplet, separate from its fellow travelers, that the aiming of the droplet deflection is sufficiently precise to find the target vessel and that the cell will be undamaged on arrival. Among the diverse reported applications of the technique, two fields which have benefited greatly are lymphocyte development and haemopoiesis. In the former case, the analysis of gene rearrangements in lymphocytes, both in the pre- and post-antigenic phases of development, has been enabled by the combined technologies of single cell sorting and PCR. It is argued that such experiments could not have been done without that partnership. In a similar way, the single cell sorting technique has been found to be the perfect way to demonstrate precursor/progeny relationships between haemopoietic cells and, further, to demonstrate rigorously the effects of particular cytokines on the haemopoietic system. Author(s): Battye, F L; Light, A; Tarlinton, D M Language: eng Source J Immunol Methods 2000 Sep 21;243(1-2):25-32.

Single-stranded DNA plasmid, vector construction and cloning of Bacillus stearothermophilus alpha-amylase in Lactobacillus. Abstract: Vector plasmids were constructed by ligating chloramphenicol and erythromycin resistance genes to TaqI-digested DNA of a cryptic plasmid from Lactobacillus plantarum. The minimal region of Lactobacillus plasmid DNA that was required for DNA replication was defined and a single-stranded DNA intermediate replication system was observed. Homologies with other origins of replication of plasmids from Gram-positive bacteria, replicating via rolling circle mechanism, were found. It was shown that the constructed vectors, named pPSC20 and pPSC22, were transformable into L. plantarum, Lactobacillus acidophilus, Lactobacillus reuteri, Lactobacillus fermentum, Lactobacillus helveticus, Lactococcus lactis subsp. lactis, Bacillus subtilis, and Escherichia coli. Using plasmid pPSC22, the alpha-amylase gene of Bacillus stearothermophilus was cloned and expressed in several Lactobacillus species. Author(s): Cocconcelli, P S; Gasson, M J; Morelli, L; Bottazzi, V Language: eng Source Res Microbiol 1991 Jul-Aug;142(6):643-52.

Singleton, Linda Joy. The killer / L.J. Singleton. Published/Created: New York: Berkley Jam Books, c2001. Related Names: Copyright Paperback Collection (Library of Congress) Description: 180 p.; 18 cm. ISBN: 0425178641 (pbk.) Subjects: Human cloning--Fiction. Cloning--Fiction. Science fiction. Mystery and detective stories. Series: Singleton, Linda Joy. Regeneration series. LC Classification: CPB Box no. 1943 vol. 7

Sjogren's syndrome nuclear antigen B (La): cDNA cloning, structural domains, and autoepitopes. Abstract: SS-B/La is a major antigenic target for autoantibodies in patients with Sjogren's syndrome. Its transient association with nascent RNA polymerase III transcripts in the cell nucleus suggest a functional role of SS-B/La in RNA processing and maturation. Human SS-B/La autoantibodies recognize at least two distinct epitopes on two separate structural domains of the SS-B/La protein and these epitopes are conserved among mammalian species. In contrast, murine monoclonal antibodies produced though immunization with purified bovine SS-B/La recognize different epitopes. To elucidate these differences, cDNA sequences of SS-B/La were cloned from several mammalian species including human, bovine, and rabbit. Complete human SS-B/La cDNAs including the coding sequence (1227bp) and untranslated sequences were isolated. The complete bovine sequence was determined from two overlapping partial cDNA clones. Comparison of the complete protein sequences encoded by the human and bovine cDNAs revealed a high degree of conservation of amino acid sequence showing only 26 substitutions/deletions out of 408 residues. RNA blot analysis indicated the presence of two size species of transcripts in bovine and rabbit cells, 1.8 kb and greater than 2.5 kb, in contrast to a single 1.8 kb mRNA species in human cells. The results of cDNA cloning support our previous finding of SS-B/La as a two-domain protein, and the RNA-binding site is confirmed to be located within N-terminal domain designated X. Epitope mapping using recombinant SS-B/La fusion proteins confirmed the findings of at least two autoepitopes located on different domains. Author(s): Chan, E K; Sullivan, K F; Fox, R I; Tan, E M Language: eng Source J Autoimmun 1989 Aug;2(4):321-7.

Skurzynski, Gloria. The clones / Gloria Skurzynski. Edition Information: 1st ed. Published/Created: New York: Atheneum Books for Young Readers, 2002. Description: 181 p.; 22 cm. ISBN: 0689842635 Summary: Having won the Virtual War for the Western Hemisphere Federation, fifteen-year-old Corgan finds himself raising a clone of the young mutant genius who helped him win before dying. Notes: "The virtual war chronologs book 2." Sequel to: Virtual War. Subjects: Cloning--Fiction. Science fiction. LC Classification: PZ7.S6287 Co 2002 Dewey Class No.: [Fic] 21

Slote, Alfred. Clone catcher / by Alfred Slote; illustrated by Elizabeth Slote. Edition Information: 1st ed. Published/Created: New York: Lippincott, c1982. Related Names: Slote, Elizabeth, ill. Description: 154 p.: ill.; 22 cm. ISBN: 0397320175: 0397320183 (lib. bdg.) : Summary: In the twenty-first century Arthur Dunn is summoned to Australia to track down a runaway clone urgently needed to provide vital organ transplants for her parent, the superb actress Lady Kate. Subjects: Cloning--Fiction. Mystery and detective stories. Science fiction. LC Classification: PZ7.S635 Cl 1982 Dewey Class No.: [Fic] 19

Smith, Michael A. (Michael Anthony), 1942 Dec. 15- The inheritors / Michael A. Smith. Edition Information: 1st ed. Published/Created: New York: Forge, 2000. Description: 349 p.; 25 cm. ISBN: 0312866399 (acid-free paper) Notes: "A Tom Doherty Associates book." Subjects: Terrorists--United States--Fiction. Cloning--Fiction. LC Classification: PS3569.M53783 I54 2000 Dewey Class No.: 813/.54 21

Sodium- and chloride-dependent transporters in brain, kidney, and gut: lessons from complementary DNA cloning and structure-function studies. Abstract: The family of Na(+)- and Cl(-)-dependent, 12 transmembrane domain transporter proteins now includes transporters for neurotransmitter molecules in the brain and for substances important in extraneuronal tissues, including adrenal, kidney, and gut. Transported substrates include monoamine and amino acid neurotransmitters and nonperturbing osmolytes. A common protein topology is predicted and features intracellular N- and C-termini possessing phosphorylation sites and at least one large extramembranous loop with N-linked glycosylation. Using the rat dopamine transporter as a template, molecular modeling of putative transmembrane domains coupled with amino acid sequence conservation analysis indicates amino acid residues potentially involved in substrate and/or ion recognition. Targeting such residues with site-directed mutagenesis will help clarify substrate and ion binding sites and should facilitate rational design of therapeutics to combat depression, locomotor disorders, and substance abuse. Author(s): Surratt, C K; Wang, J B; Yuhasz, S; Amzel, M; Kwon, H M; Handler, J S; Uhl, G R Language: eng Source Curr Opin Nephrol Hypertens 1993 Sep;2(5):744-60.

Somatic cell nuclear transfer (cloning): implications for the medical practitioner. Abstract: The current century will bring tremendous changes to the science and the practice of medicine. This century will be acknowledged as the century of Biology as the fusion of molecular genetics and experimental embryology pushes the barriers of science beyond perimeters that we have thought existed, as much as the past century was the century of Physics, with all the exact scientific calculations and predictions, resulting in electricity, nuclear power and quantum physics. The first major breakthrough has been the pioneering work of Wilmut and Campbell, first with the birth of Megan and Moran in 1995 (1), followed by the birth of Dolly the sheep, the first reported mammalian clone from a fully differentiated adult cell, reported in July 1996 (2). However, current cloning techniques are an extension of over 40 years of research using nuclei derived from non-human embryonic and fetal cells. However, following the birth of Dolly, the prospects of cloning technology have extended to ethically hazier areas of human cloning and embryonic stem cell research. This review hopes to bring the reader closer to the science and the ethics of this new technology, and what the implications are for the medical practitioner. Author(s): Tong, W F; Ng, Y F; Ng, S C Language: eng Source Singapore Med J 2002 Jul;43(7):369-76.

Somatic cloning in mammals Abstract: Somatic cloning in mammals became possible due to refinement of the nuclear transfer technique consisting of using nuclear donor cells in GO or activating reconstituted oocytes few hours after nuclear transfer (post-activation). Sheep, goats, pigs, cattle and mice have been cloned this way; the latter two species - of both sexes. Cloning of transgenic mammals producing human therapeutic proteins in milk (or urine) can find application in pharmacology and medicine. Somatic cloning of a goat producing human antithrombine III in its milk has already been achieved. Author(s): Modlinski, J A; Karasiewicz, J Language: pol Source Med Wieku Rozwoj 2001;5(1 Suppl 1):9-25.

Spangler, Brenda D. Methods in molecular biology and protein chemistry: cloning and characterization of an enterotoxin subunit / Brenda D. Spangler. Published: Chichester, UK; New York: Wiley, c2002. Description: x, 209 p.: ill.; 28 cm. ISBN: 0470843608 (pbk.) Notes: Includes bibliographical references and index. Subjects: Molecular biology--Proteins--Experiments. Enterotoxins--chemistry--genetics--Cloning, Molecular--Molecular Biology--Protein Subunits--Proteins--chemistry--Laboratory Manuals. LC Classification: QP519.7 .S67 2002 Dewey Class No.: 572.8 21

Special uses of lambda phage for molecular cloning. Author(s): Murray, N E Language: eng Subject(s): Bacteriophage lambda/genetics; Cloning, Molecular; Escherichia coli/genetics; Genes, Bacterial; Genes, Viral; Genetic Vectors; Recombination, Genetic Source Methods Enzymol 1991;204:280-301.

Specialized plasmid vectors for cloning cDNA. Author(s): Denhardt, D T; Edwards, D R; Kowalski, J; Parfett, C L; Waterhouse, P Language: eng Subject(s): Cloning, Molecular; DNA/genetics; Escherichia coli/genetics; Genetic Vectors; Oligodeoxyribonucleotides/genetics; Plasmids; RNA, Messenger/genetics Source Biotechnology 1988;10:237-51.

Stanley, Debbie. Genetic engineering: the cloning debate / Debbie Stanley. Edition Information: 1st ed. Published/Created: New York: Rosen, 2000. Description: 64 p.: col. ill.; 25 cm. ISBN: 0823932117 Summary: Examines the nature, history, and ethical aspects of cloning, discussing both humans and other animals. Notes: Includes bibliographical references (p. 57) and index. Subjects: Cloning--Juvenile literature--Moral and ethical aspects--Juvenile literature. Series: Focus on science and society LC Classification: QH442.2 .S73 2000 Dewey Class No.: 660.6/5 21

Staphylococcus carnosus: a new host organism for gene cloning and protein production. Author(s): Gotz, F Language: eng Source Soc Appl Bacteriol Symp Ser 1990;19:49S-53S.

Steel, Danielle. The Klone and I: a high-tech love story / Danielle Steel. Published/Created: New York, N.Y.: Delacorte Press, c1998. Description: 232 p.; 20 cm. ISBN: 0385323921 0385332955 (lg. print ed.) Subjects: Triangles (Interpersonal relations)--Cloning--Fiction. LC Classification: PS3569.T33828 K57 1998 Dewey Class No.: 813/.54 21

Stem cells. Published: Dayton, Ohio: AlphaMed Press, c1993- Description: v.: ill.; 26 cm. Vol. 11, no. 1 (Jan. 1993)- ISSN: 1066-5099 Notes: Title from cover. Supplements with their own numbering, e.g., v. 11, supplement 1, v. 11, supplement 2, etc. accompany some vols. SERBIB/SERLOC merged record Subjects: Cloning--Clone cells--Stem Cells. LC Classification: QH442.2 .I57 Dewey Class No.: 574.87/6/05 19

Strategies and methods of cloning disease genes. Author(s):San, H Y; Zhou, Y G Language: chi Source Sheng Li Ke Xue Jin Zhan 2001 Apr;32(2):180-4.

Strategies for cloning mammalian DNA repair genes. Author(s):Thompson, L H Language: eng Subject(s): Animal; CHO Cells; Cloning, Molecular/methods; DNA Repair/genetics; Databases, Factual; Gene Library; Genetic Diseases, Inborn/genetics; Hamsters; Human; Mammals; Open Reading Frames; Transfection/methods Source Methods Mol Biol 1999;113:57-85.

Strategies for cloning new MMPs and TIMPs. Author(s):Velasco, G; Lopez-Otin, C Language: eng Source Methods Mol Biol 2001;151:25-44.

Strategies for cloning unknown cellular flanking DNA sequences from foreign integrants. Abstract: Many virus and transposon DNAs can integrate into the host genome. In this review, techniques, including inverse polymerase chain reaction (IPCR), novel Alu-PCR and vectorette- or splinkerette-PCR are introduced as possible strategies for cloning flanking DNA regions of the integrants. Targeted gene-walking PCR, restriction-site PCR, capture PCR, and panhandle PCR and boomerang DNA amplification are also described. The principles, advantages and limitations of each approach are discussed. Author(s): Hui, E K; Wang, P C; Lo, S J Language: eng Source Cell Mol Life Sci 1998 Dec;54(12):1403-11.

Strategies for gene cloning. Abstract: Cancer, including genitourinary malignancy, is a consequence of accumulated genetic aberrations in genes involved in crucial regulatory pathways. The result is a deregulation of cellular behaviour, leading to neoplastic transformation, uncontrolled cell proliferation and acquisition of metastatic ability. The development and perfection of techniques in the field of recombinant DNA technology, gene cloning, (differential) analysis of gene expression, and sequencing of genes and proteins have provided a wealth of information about the genetic aberrations associated with cancer development. This "recombinant DNA and gene cloning" technology and the recently developed DNA chip technology may provide new molecular diagnostic tools. Furthermore, the technology of gene cloning in combination with the progress in in vivo gene delivery techniques offers new treatment modalities, like gene therapy, additional to conventional therapies. This review is intended to provide a general introduction to the fundamentals or strategies of recombinant DNA and gene cloning techniques as a basis for understanding the rapidly expanding range of new diagnostic and therapeutic opportunities. Some illustrative examples are provided, addressing basic and biomedical research and possible clinical applications in genitourinary oncology. Author(s): Schamhart, D H; Westerhof, A C Language: eng Source Urol Res 1999 Apr;27(2):83-96.

Strategies for the analysis of accessory-cell function: the in vitro cloning and characterization of the P cell. Author(s): Schrader, J W; Nossal, G J Language: eng Source Immunol Rev 1980;53:61-85.

Streptomyces cloning: possible construction of novel compounds and regulation of antibiotic biosynthetic genes. Author(s): Tomich, P K Language: eng Source Antimicrob Agents Chemother 1988 Oct;32(10):1472-6.

Streptomyces cloning: useful recombinant DNA systems and a summation of cloned genes. Author(s): Tomich, P K Language: eng Source Antimicrob Agents Chemother 1988 Oct;32(10):1465-71.

Structural features of the core proteins of human airway mucins ascertained by cDNA cloning. Abstract: Tracheobronchial secretions are one of the most important elements of the mucociliary system that protects the respiratory mucosa. They contain bronchial mucus, which is composed of a group of macromolecules secreted by the goblet cells of the epithelium and the submucosal glands. Bronchial mucins are the most characteristic molecules of this mucus. They form a group of complex, polydispersed O-linked glycoproteins containing sugars, which make up 80% of their weight. The protein core of human airway mucin has been difficult to sequence by traditional technologies because of its high content of serine and threonine residues linked to numerous oligosaccharide chains. We therefore prepared a lambda gt11 cDNA library from one sample of human tracheobronchial mucosa and screened this library with a polyclonal antibody directed against the apopeptides of human bronchial mucins. We obtained 20 positive clones that were sequenced. These sequences were classified into three different types. The use of the nucleotide probes from these clones in Northern blot analysis showed that the RNA messages were extremely polydispersed. At the current time, four of these probes allow us to map human tracheobronchial mucins genes to at least three different chromosomes. These results suggest that the peptide moiety of the human airway mucin is very heterogeneous. Author(s): Porchet, N; Dufosse, J; Audie, J P; Duperat, V G; Perini, J M; Nguyen, V C; Degand, P; Aubert, J P Language: eng Source Am Rev Respir Dis 1991 Sep;144(3 Pt 2):S15-8.

Structural features of two kininase I-type enzymes revealed by molecular cloning. Abstract: Kininase I-type carboxypeptidases remove a single C-terminal Arg residue from kinins. The circulating kininase I (carboxypeptidase N) contains two types of subunits: a 50 kDa catalytic subunit and an 83 kDa carrier subunit which protects the active subunit in blood. The 83 kDa subunit contains 12 leucine-rich tandem repeats, similar in sequence to other proteins with binding functions. Human carboxypeptidase M is a widely distributed "tissue kininase I" bound to plasma membranes. It has 41% sequence identity with the 50 kDa subunit of carboxypeptidase N and may regulate the activity of kinins and other peptides at the cell surface. Author(s): Skidgel, R A; Tan, F Language: eng Source Agents Actions Suppl 1992;38 (Pt 1):359-67.

Structure of aminoacyl-tRNA-synthetase of higher eukaryotes from molecular cloning data Abstract: This review summarizes the data on the primary structure of eukaryotic aminoacyl-tRNA synthetases (EC 6.1.1) obtained during last decade by means of molecular cloning of respective cDNAs. This structural information is compared with that available for bacterial species. The alignment of E. coli, yeast and mammalian sequences are presented. The N- and C-terminal extensions of higher-eukaryotic aminoacyl-tRNA synthetases are discussed in view of their chimeric origin and potential functions. For some enzymes, the exon/intron organization of their genes are compared. A possible relationship between the domain structure of the protein and the exon/intron organization of the gene is considered. Author(s): Grigor'eva, A Iu Language: rus; Source Mol Biol (Mosk) 1994 Sep-Oct;28(5):978-90.

Study of Bacillus subtilis as a cloning system; Zhang, Z Y Language: chi Source Sheng Li Ke Xue Jin Zhan 1983 Jul;14(3):209-15.

Subtractive cDNA cloning and characterization of genes induced by all-trans retinoic acid. Author(s):Chen, Y; Talmage, D A Language: eng Subject(s): Molecular; DNA, Complementary. Source Ann N Y Acad Sci 1999;886:225-8.

Subtractive cloning: past, present, and future. Abstract: Subtractive cloning is a powerful technique for isolating genes expressed or present in one cell population but not in another. This method and a related one termed positive selection have their origins in nucleic acid reassociation techniques. We discuss the history of subtractive techniques, and fundamental information about the nucleic acid composition of cells that came out of reassociation analyses. We then explore current techniques for subtractive cloning and positive selection, discussing the merits of each. These techniques include cDNA library-based techniques and PCR-based techniques. Finally, we briefly discuss the future of subtractive cloning and new approaches that may augment or supersede current methods. Author(s): Sagerstrom, C G; Sun, B I; Sive, H L Language: eng Source Annu Rev Biochem 1997;66:751-83.

Syngeneic monoclonal antimelanoma antibodies and their application for analysis of tumor antigens, gene cloning, and in vitro/in vivo diagnosis. Abstract: In this article, we summarized syngeneic monoclonal antimelanoma antibodies and their application for chemical characterization of mouse melanoma antigens, cloning of genomic DNA controlling antigen expression, and in vivo/in vitro tumor diagnosis. The melanoma antigen is composed of a protein complex in association with GM3(NeuAc)-like sugar moiety. The GM3 structure expresses the cross-species epitopes shared in various mammalian species, whereas the mouse specific melanoma epitope is present on protein molecules. By using the monoclonal antimelanoma reactive with GM3 epitope, we developed a very sensitive sandwich radioimmunoassay system detecting soluble melanoma antigens equivalent to 10(2)-10(3) cells/ml. The antibody was also useful in imaging tumor in vivo. These results indicate that the antibody with cross-species reactivity has a potential for tumor targeting. The monoclonal antibody M562 recognizing protein molecule with species specific epitope but not other antimelanoma antibodies, however, effectively inhibited experimental lung metastasis of melanoma cells, indicating that the M562 epitope seems to possess important biological functions. Recently, the genomic DNA controlling the antigen expression was successfully isolated by DNA transfection and expression technique with monoclonal anti-melanoma M562 and the fluorescence-activated cell sorter. We also found that genomic DNA possesses transformation-related activity in NIH3T3 cells. Author(s): Taniguchi, M; Wakabayashi, S; Tagawa, M Language: eng Source Pigment Cell Res 1989 Jul-Aug;2(4):254-8.

Synthesis and cloning of operator DNA of the lac operon of E. coli. Author(s): Narang, S A; Bahl, C P; Wu, R Language: eng Source Can J Biochem 1977 Nov;55(11):1125-33.

Synthesis, cloning, and expression of hormone genes in Escherichia coli. Author(s): Riggs, A D; Itakura, K; Crea, R; Hirose, T; Kraszewski, A; Goeddel, D; Kleid, D; Yansura, D G; Bolivar, F; Heyneker, H L Language: eng Subject(s): Animal; Cloning, Molecular; DNA/biosynthesis; Drug Industry; Escherichia coli/genetics; Genes; Hormones, Synthetic; Human; Insulin/biosynthesis/genetics/isolation & purification Source Recent Prog Horm Res 1980;36:261-76.

TA cloning: a rapid and efficient method for cloning of the PCR products; Yasuda, J Language: jpn Subject(s): Cloning, Molecular/methods; DNA-Directed DNA Polymerase; Polymerase Chain Reaction; Taq Polymerase Source Tanpakushitsu Kakusan Koso 1996 Apr;41(5):518-21.

T-cell cloning in human type I diabetes. Author(s): Giordano, C Language: eng Source Diabetes Metab Rev 1992 Apr;8(1):39-51.

Techniques for cloning cDNAs encoding interactive transcriptional regulatory proteins. Abstract: Several approaches aimed at detecting and cloning interactive transcriptional regulatory proteins have been presented. All of the techniques can effectively identify specific interactions between two transcription proteins. However, interaction cloning and the two hybrid system have the added advantage of yielding a cDNA expression clone directly. The other methods, EMSA-mediated cloning, co-immunoprecipitation, oligonucleotide/PCR-facilitated cloning, Southwestern, and Farwestern, require additional manipulations to obtain a cDNA clone. Clearly, the interactive cloning system of choice will depend on the proteins under investigation. Author(s): Silver, S C; Hunt, S W 3rd Language: eng Source Mol Biol Rep 1993 Apr;17(3):155-65.

Techniques: Recombinogenic engineering--new options for cloning and manipulating DNA. Abstract: Driven by the needs of functional genomics, DNA engineering by homologous recombination in Escherichia coli has emerged as a major addition to existing technologies. Two alternative approaches, RecA-dependent engineering and ET recombination, allow a wide variety of DNA modifications, including some which are virtually impossible by conventional methods. These approaches do not rely on the presence of suitable restriction sites and can be used to insert, delete or substitute DNA sequences at any desired position on a target molecule. Furthermore, ET recombination can be used for direct subcloning and cloning of DNA sequences from complex mixtures, including bacterial artificial chromosomes and genomic DNA preparations. The strategies reviewed in this article are applicable to modification of DNA molecules of any size, including very large ones, and present powerful new avenues for DNA manipulation in general. Author(s): Muyrers, J P; Zhang, Y; Stewart, A F Language: eng Source Trends Biochem Sci 2001 May;26(5):325-31.

Telomere cloning and mammalian chromosome analysis. Abstract: Although eucaryotic chromosomes vary in size over five orders of magnitude and are constituents of diverse genetic systems the fundamental features of their telomeres appear to be almost completely conserved. This can be exploited to enable molecular cloning of human telomeres in yeast and suggests that many of the ideas that will arise from studies of telomeres in the experimentally tractable ciliates and yeasts will hold true of mammalian telomeres. The particular value of cloned mammalian telomeres is that they contribute reagents for mapping mammalian chromosomes and that they provide one set of elements for the construction of artificial mammalian chromosomes. Author(s): Brown, W R; Dobson, M J; MacKinnon, P Source J Cell Sci 1990 Apr;95 (Pt 4):521-6.

The 6th day / Phoenix Pictures presents a Jon Davison production; directed by Roger Spottiswoode; producer, Mike Medavoy; screenplay, Cormac Wibberley and Marianne Wibberley. Published/Created: United States: Sony Pictures, 2000. Related Names: Copyright Collection (Library of Congress) Description: 13 film reels of 13 on 7 (ca. 124 min., 11,160 ft.): sd., col.; 35 mm. viewing print. Summary: In the near future when the cloning of all species except for humans is legal, family man Adam Gibson comes home from work one day to discover that a clone has replaced him. He must next save himself from the assassins who need to destroy him in order to protect their secret. Then he must uncover who and what is behind the horrible things happening to him. Notes: Copyright: Phoenix Pictures, Inc. NM: entire motion picture except as described above. DCR 2000; PUB 17Nov00; REG 18Dec00; PA1-008-689. Summary taken from Baseline document; 08/28/01. Sources used: copyright datasheet; copyright database; Baseline database. Cast: Arnold Schwarzenegger, Michael Rapaport, Tony Goldwyn, Michael Rooker, Sarah Wynter, Wendy Crewson. Credits: Editors, Mark Conte, Dominique Fortin, Michel Arcand; music, Trevor Rabin; director of photography, Pierre Mignot. Source of Acquisition: Received: 01/26/2001; viewing print; copyright deposit--RNR; Copyright Collection. LC Classification: CGD 0826-0832 (viewing print)

The cannabinoid receptor-pharmacologic identification, anatomical localization and cloning. Author(s): Johnson, M R; Rice, K C; Howlett, A; Melvin, L S; Herkenham, M Language: eng ource NIDA Res Monogr 1992;119:86-90.

The cloning and biological characterization of recombinant human interleukin 11. Abstract: Interleukin 11 (IL-11) is a pleiotropic hematopoietic growth factor with stimulatory effects on multiple hematopoietic progenitor cells. An immortalized primate bone marrow stromal cell line was established to facilitate the analysis of interactions between hematopoietic progenitors and the microenvironment. This line was found to produce a novel growth factor that directed the proliferation of a murine plasmacytoma cell line. A cDNA encoding this activity was isolated through functional expression cloning in mammalian cells. Preliminary characterization of the cytokine revealed that IL-11 stimulated T cell-dependent B cell immunoglobulin secretion and synergized with IL-3 in murine megakaryocyte formation. Subsequent analysis of the clone has demonstrated that it also synergizes with IL-3 and steel factor in the stimulation of early progenitors and affects early megakaryocyte formation and maturation. In addition, IL-11 induces secretion of acute phase proteins in the liver and may function in the hematopoietic microenvironment as an adipogenic antagonist in a paracrine manner. Further in vitro and in vivo analysis of IL-11 will be necessary to determine the biologic function and potential therapeutic use of the cytokine. Author(s): Paul, S R; Schendel, P Language: eng Source Int J Cell Cloning 1992 May;10(3):135-43.

The cloning and characterization of a novel cytochrome P450 family, CYP26, with specificity toward retinoic acid. Abstract: A new family of cytochrome P450 enzymes, CYP26, has been cloned and characterized in zebra fish, human, and mouse tissues. CYP26 displays specificity toward retinoic acid and it may function as an important regulator or differentiation and a possible modulator of disease states by controlling retinoid concentration and homeostasis. Author(s): Haque, M; Anreola, F Language: eng Source Nutr Rev 1998 Mar;56(3):84-5.

The cloning and characterization of interleukin-7. Author(s): Goodwin, R G; Namen, A E Language: eng Subject(s): Amino Acid Sequence; Animal; Cloning, Molecular; Human; Interleukin-7/genetics; Molecular Sequence Data Source Year Immunol 1989-90;6:127-39.

The cloning issue / editor, Craig Donnellan. Edition Information: New ed. Published/Created: Cambridge: Independence, 2002. Related Names: Donnellan, Craig. Description: 44 p.: ill.; 30 cm. ISBN: 1861681909 (pbk.) Notes: Previous ed.: Cloning. Subjects: Cloning--Human cloning--Moral and ethical aspects. Series: Issues (Cambridge, England); v. 12. LC Classification: QH442.2 .C5673 2002 Dewey Class No.: 174/.957 21

The cloning of a mechano-gated membrane ion channel. Author(s): Hamill, O P; McBride, D W Jr Language: eng Subject(s): Animal; Cloning, Molecular; Comparative Study; Escherichia coli/metabolism; Human; Ion Channel Gating; Ion Channels/metabolism; Mechanoreceptors/physiology Source Trends Neurosci 1994 Nov;17(11):439-43.

The cloning of antigen genes from malarial parasites and Leishmania species. Author(s): Scaife, J G Language: eng Source Genet Eng 1988;(7):57-90.

The cloning of eukaryotic DNA segments as a means of studying the organization and function of chromosomes. Author(s): Thomas, C A Jr Language: eng Source Adv Pathobiol 1976;(3):42-54.

The cloning of extracellular Ca(2+)-sensing receptors from parathyroid and kidney: molecular mechanisms of extracellular Ca(2+)-sensing. Abstract: The parathyroid cell detects changes in the extracellular ionized calcium concentration (Ca2 + o) with exquisite sensitivity, but the mechanisms through which it senses Ca2 + o have remained obscure. Recently, we isolated a cDNA encoding a Ca2 + o-sensing receptor from bovine parathyroid using expression cloning in Xenopus laevis oocytes. The expressed receptor stimulates phospholipase C and has a pharmacological profile almost identical to that of the native receptor. Furthermore, its deduced amino acid sequence confirms that it belongs to the superfamily of G-protein-coupled receptors. Receptor transcripts are present in parathyroid and other tissues sensing Ca2 + o (e.g., kidney and thyroidal C-cells) as well as those not known to be involved in Ca2+ homeostasis (viz., in the brain). We have also shown that mutations in the receptor cause three inherited disorders of calcium metabolism: Familial hypocalciuric hypercalcemia (FHH) and neonatal severe hyperparathyroidism (NSHPT) result from inactivating mutations, when present in the heterozygous and homozygous states, respectively, whereas an autosomal dominant form of hypocalcemia is due to an activating mutation. Thus this Ca2 + o-sensing receptor permits Ca2+o to act as an extracellular, first messenger in addition to its better known role as an intracellular second messenger. Author(s): Brown, E M; Pollak, M; Chou, Y H; Seidman, C E; Seidman, J G; Hebert, S C Language: eng Source J Nutr 1995 Jul;125(7 Suppl):1965S-1970S.

The cloning of schistosome antigens. Author(s): Simpson, A J; Knight, M; Kelly, C; Hackett, F; Omer-Ali, P; Smithers, S R Language: eng Source Acta Trop Suppl 1987 Jun;12:83-9.

The cloning of schistosome genes encoding antigenic and maturation-linked polypeptides. Author(s): Knight, M; Simpson, A J; Kelly, C; Smithers, S R Language: eng Source Parasitology 1986;92 Suppl:S39-51.

The cloning of the steroid hormone receptors: basic and clinical implications. Author(s): Fuller, P J Language: eng Source Aust N Z J Med 1988 Dec;18(7):890-6.

The cloning sourcebook / edited by Arlene Judith Klotzko. Published/Created: New York: Oxford University Press, 2003. Related Names: Klotzko, Arlene Judith. Description: p. cm. ISBN: 0195128834 (pbk.: alk. paper) Notes: Rev. and updated for the paperback ed. Includes bibliographical references and index. Subjects: Human cloning. Human cloning--Moral and ethical aspects. Human reproductive technology--Moral and ethical aspects. LC Classification: QH442.2 .C568 2003 Dewey Class No.: 176 21

The cloning sourcebook / edited by Arlene Judith Klotzko. Published/Created: Oxford; New York: Oxford University Press, 2001. Related Names: Klotzko, Arlene Judith. Description: xix, 328 p.; 24 cm. ISBN: 0195128826 Notes: Includes bibliographical references and index. Subjects: Human cloning. Human cloning--Moral and ethical aspects. Human reproductive technology--Moral and ethical aspects. LC Classification: QH442.2 .C568 2000 Dewey Class No.: 176 21

The dangers of cloning and the promise of regenerative medicine: hearing before the Committee on Health, Education, Labor, and Pensions, United States Senate, One Hundred Seventh Congress, second session on examining cloning research, focusing on the clarification of how stem cell research, or therapeutic cloning, differs from human reproductive cloning, and the ethical and public-policy issues related to both, and related issues of S. 1853 to ban human cloning while protecting stem cell research, March 5, 2002. Published/Created: Washington: U.S. G.P.O.: For sale by the Supt. of Docs., U.S. G.P.O., [Congressional Sales Office], 2002. Description: iii, 63 p.; 24 cm. Notes: "Printed for the use of the Committee on Health, Education, Labor, and Pensions." Subjects: Human cloning--Research--United States. Human reproductive technology--Moral and ethical aspects

United States. Stem cells--Research--Moral and ethical aspects--United States. Stem cells--Research--Government policy--United States. Series: United States. Congress. Senate. S. hrg.; 107-497. LC Classification: KF26 .L27 2002r Govt. Doc. No.: Y 4.L 11/4:S.HRG.107-497

The development and applications of the bacterial artificial chromosome cloning system. Abstract: The development of the Bacterial Artificial Chromosome (BAC) system was driven in part by the Human Genome Project as a means to construct genomic DNA libraries and physical maps for genomic sequencing. The BAC system is based on the well-characterized Escherichia coli F-factor, a low copy plasmid that exists in a supercoiled circular form in host cells. The structural features of the F-factor allow stable maintenance of individual human DNA clones as well as easy manipulation of the cloned DNA. BACs are currently used in a wide array of applications from genome sequencing to gene discovery. This paper describes the key elements in the development of the BAC system and its current notable applications. Author(s): Shizuya, H; Kouros-Mehr, H Language: eng Source Keio J Med 2001 Mar;50(1):26-30.

The difficult challenge of cloning the angiotensin II receptor. Abstract: The vasopressor peptide angiotensin II exerts its cellular effects through a membrane-bound receptor coupled to a G protein. Biochemical and pharmacological analyses of this receptor already identify two different membrane-bound receptors and one cytosoluble angiotensin-II-binding protein. Nevertheless, the purification of the membrane-bound form(s) appears to be difficult. In the absence of purified protein, two cloning strategies of the gene have been explored: (1) expression cloning, identifying the functions of the protein expressed from a cDNA library in COS cells or Xenopus oocytes, has been unsuccessful until now; (2) analogical cloning, trying to identify related members of the seven transmembrane segment receptor family, which could be related to angiotensin receptors, identifies the mas oncogene and two related genes. However, there are accumulating data to exclude their involvement in angiotensin binding. Author(s): Weber, V; Monnot, C; Bihoreau, C; Corvol, P; Clauser, E Language: eng Source Horm Res 1990;34(3-4):101-4.

The ethics of human cloning / William Dudley, editor. Published/Created: San Diego, CA: Greenhaven Press, c2001. Related Names: Dudley, William, 1964- Description: 77 p.; 24 cm. ISBN: 0737704713 (pbk.: alk. paper) 0737704721 (lib. bdg.: alk. paper) Notes: Includes bibliographical references (p. 72-74) and index. Subjects: Human cloning--Moral and ethical aspects. Series: At issue (San Diego, Calif.) LC Classification: QH442.2 .E847 2001 Dewey Class No.: 174/.25 21

The future is now: America confronts the new genetics / edited by William Kristol and Eric Cohen. Published/Created: Lanham [Md.]: Rowman & Littlefield: Distributed by National Book Network, c2002. Related Names: Kristol, William. Cohen, Eric, 1977- Description: xix, 357 p.; 24 cm. ISBN: 0742521958 (alk. paper) 0742521966 (pbk.: alk. paper) Notes: Includes index. Subjects: Human cloning. Stem cells--Research. Bioethics. Genetics--Moral and ethical aspects. LC Classification: QH442.2 .F88 2002

The future of cloning. Author(s):Gurdon, J B; Colman, A Language: eng Subject(s): Animal; Bioethics; Cell Nucleus/transplantation; Cloning, Organism/legislation & jurisprudence/trends; Embryo Research; Forecasting; Gene Expression Regulation; Genome; Human; Internationality; Risk Assessment; Social Change; Therapeutic Human Experimentation; Tissue Culture; Tissue Transplantation Source Nature 1999 Dec 16;402(6763):743-6.

The genetic revolution and human rights / edited by Justine Burley. Published/Created: Oxford; New York: Oxford University Press, 1999. Related

Names: Burley, Justine. Description: xxviii, 220 p.; 20 cm. ISBN: 0019286201 Notes: Includes bibliographical references (p. 194-215) and index. Subjects: Medical genetics--Moral and ethical aspects. Human cloning--Moral and ethical aspects. Eugenics--Moral and ethical aspects. Series: The Oxford amnesty lectures; 1998 LC Classification: RB155 .G3869 1999 Dewey Class No.: 174/.25 21

The gonadotrophin receptors: insights from the cloning of their cDNAs. Abstract: The cloning and expression of the cDNA's for the gonadotrophin hormone receptors has confirmed that these receptors are each composed of a single polypeptide which can both bind hormone and activate adenylyl cyclase when occupied with agonist. Although some studies by others have suggested that the LH/CG receptor and the FSH receptor are composed of multiple subunits (Ascoli and Segaloff 1989; Reichert and Dattatreyamurty 1989; Shin and Ji 1985a, b, and c; Smith et al. 1985; Smith et al. 1986), biochemical studies on the LH/CG receptor have shown that it is composed of a single polypeptide with a molecular weight of 93,000 when analyzed on sodium dodecyl sulfate gels in the presence or absence of disulfide reducing agents (Ascoli and Segaloff 1989). As the LH/CG receptor has been shown to be readily proteolyzed into smaller-sized fragments (Ascoli and Segaloff 1986; Ascoli and Segaloff 1989), it is reasonable to postulate that the FSH receptor may be similarly susceptible to proteolysis, and that this may account for the discrepant reports on its structure. Clearly, the molecular cloning and functional expression of the cDNA's for the rat ovarian LH/CG receptor (McFarland et al. 1989) and the rat testicular receptor (Sprengel et al. 1990) demonstrate conclusively that the gonadotrophin receptors are indeed single polypeptides. As shown schematically in Fig. 5.12, the gonadotrophin hormone receptors (together with the thyrotrophin receptor, see Frazier et al. 1990; Libert et al. 1989; Nagayami et al. 1989; Parmentier et al. 1989) define a unique subclass of G protein-coupled receptors. Thus, in addition to the seven membrane-spanning domains which appear to be the hallmark of G protein-coupled receptors, the gonadotrophin receptors (and thyrotrophin receptor) also contain large extracellular domains. These extracellular domains are composed of a repeating leucine-rich motif which is found in a number of widely-diverse proteins, collectively termed leucine-rich glycoproteins. We have shown that the extracellular domain of the LH/CG receptor is entirely responsible for binding hormone with high affinity. This is in contrast to the rhodopsin-like G protein-coupled receptors which have relatively small amino-terminal extracellular domains and in which the ligands interact directly with amino acids within the transmembrane helices. As with other G protein-coupled receptors, however, it is thought that the gonadotrophin receptors interact with and activate Gs through residues located within the cytoplasmic loops and the amino-terminal region of the cytoplasmic tail.(ABSTRACT TRUNCATED AT 400 WORDS); Segaloff, D L; Ascoli, M Language: eng Source Oxf Rev Reprod Biol 1992;14:141-68.

The Growth and form of modular organisms: proceedings of a Royal Society discussion meeting, held on 27 and 28 June 1985 / organized and edited by J.L. Harper, B.R. Rosen, and J. White. Published/Created: London: Royal Society; Great Neck, N.Y.: Distributed by Scholium International, 1986. Related Names: Harper, John L. Rosen, Brian R. (Brian Roy) White, J. Royal Society (Great Britain) Description: 250 p., [4] p. of plates: ill.; 31 cm. ISBN: 0854032819 Notes: Includes bibliographies. Subjects: Morphology--Congresses. Growth--Congresses. Cloning--Congresses. LC Classification: QH351 .G76 1986 Dewey Class No.: 574.4 20

The human cloning debate / edited by Glenn McGee. Edition Information: 3rd ed. Published/Created: Berkeley, Calif.: Berkeley Hills Books, c2002. Related Names: McGee, Glenn, 1967- Description:

305 p.: ill.; 23 cm. ISBN: 1893163415 (alk. paper) Contents: The science of cloning / Potter Wickware -- The question of human cloning / John Robertson -- If ethics won't work here, where? / Arthur Caplan -- The wisdom of repugnance / Leon Kass -- What exactly is wrong with cloning people? / Ronald Bailey -- Ban cloning? / Susan M. Wolf -- The uncertain art / Sherwin B. Nuland -- Human cloning / Philip Kitcher -- The confusion over cloning / Richard Lewontin -- Would cloned humans really be like sheep? / Leon Eisenberg -- Cloning and infertility / Carson Strong -- Entitlement to cloning? / Timothy F. Murphy -- A model for regulating cloning / Glenn McGee and Ian Wilmut -- Human cloning and child welfare / Ronald Lindsay -- Creation and cloning in Jewish thought / Jonathan R. Cohen -- A Protestant perspective on cloning / Gilbert Meilaender -- A Catholic perspective on cloning / John Haas -- Buddhist perspectives on cloning / Ravi Ravindra ... [et al.] -- The Islamic view / Munawar Ahmad Anees. Notes: Includes bibliographical references. Subjects: Human cloning--Moral and ethical aspects. Cloning, Organism. Bioethical Issues. Religion. LC Classification: QH442.2 .H85 2002 Dewey Class No.: 174/.25 21

The human cloning debate / edited by Glenn McGee. Edition Information: 2nd ed. Published/Created: Berkeley, Calif.: Berkeley Hills Books, c2000. Related Names: McGee, Glenn, 1967- Description: 305 p.: ill.; 23 cm. ISBN: 1893163121 Notes: Includes bibliographical references. Subjects: Human cloning--Moral and ethical aspects. Human reproductive technology--Moral and ethical aspects. LC Classification: QH442.2 .H85 2000 Dewey Class No.: 174/.25 21

The human cloning debate / edited by Glenn McGee. Published/Created: Berkeley, Calif.: Berkeley Hills Books, c1998. Related Names: McGee, Glenn, 1967- Description: vii, 270 p.: ill.; 23 cm. ISBN: 0965377482 (pbk.: alk. paper) 0965377474 (hbk.: alk. paper) Notes: Includes index. Subjects: Human cloning--Moral and ethical aspects. Human reproductive technology--Moral and ethical aspects. LC Classification: QH442.2 .H85 1998 Dewey Class No.: 174/.25 21

The Human Cloning Prohibition Act of 2001 and the Cloning Prohibition Act of 2001: hearing before the Subcommittee on Health of the Committee on Energy and Commerce, One Hundred Seventh Congress, first session on H.R. 1644 and H.R. 2172,, June 20, 2001. Published/Created: Washington: U.S. G.P.O.: For sale by the Supt. of Docs., U.S. G.P.O., [Congressional Sales Office], 2001. Description: iii, 110 p.; 23 cm. ISBN: 0160661668 Notes: Distributed to some depository libraries in microfiche. Shipping list no.: 2002-0018-P. "Serial no. 107-41." Includes bibliographical references. Subjects: Human cloning--Research--Law and legislation--United States. LC Classification: KF27 .E553 2001a Govt. Doc. No.: Y 4.C 73/8:107-41

The human immune response to red blood cell antigens as revealed by repertoire cloning. Abstract: A major goal of current immunologic research is to develop specific therapeutic strategies by which the enormous diversity in immune response can be enhanced, attenuated, or eliminated, depending on the particular disease process. For nearly a century, the human immune response to red blood cell antigens has served as a paradigm for understanding the pathophysiology of autoimmune disorders and alloimmune reactions to foreign cells and tissues. Recent developments in molecular biology have facilitated the expression of immune repertoires in the form of immunoglobulin Fab fragments on the surface of filamentous bacteriophage. Such approaches have provided powerful means for producing monoclonal antibodies for research, clinical, and therapeutic applications. Our laboratory has combined these techniques with novel cell-surface selection methods to isolate extraordinarily large arrays of human antibodies to the clinically relevant red blood cell Rh(D) antigen. Our results have provided a comprehensive genetic and serologic analysis of anit-Rh(D) antibodies within single alloimmunized individuals thereby offering new insights into the development of human immune repertoires. Author(s): Siegel, D L Language: eng Source Immunol Res 1998;17(1-2):239-51.

The human interferons--from protein purification and sequence to cloning and expression in bacteria: before, between, and beyond. Author(s): Pestka, S Language: eng Source Arch Biochem Biophys 1983 Feb 15;221(1):1-37.

The human tumor cloning assay in cancer drug development. A review. Abstract: Animal studies with transplantable tumor cell lines suggested that the sensitivities of the in vitro tumor cultures to certain anticancer drugs agreed with the drug sensitivities of the same tumors in vivo. Soft agar cloning techniques have been established for human myeloma and ovarian cancer cells. Refinement of the techniques now permits cloning of most human malignancies. Drug sensitivity studies have been conducted measuring the reduction in the number of tumor colonies formed. Clinical trials showed that approximately one half of patients whose cells demonstrate in vitro sensitivity to a drug will obtain a clinical response to the same drug. Patients whose cells appear resistant to a certain drug in vitro rarely respond to the drug clinically. In vitro drug sensitivity testing may eventually allow some individualization of chemotherapy. In addition it already appears that these techniques may be excellent for screening new drugs with specific anticancer potentials. Author(s): Agre, P; Williams, T E Language: eng Source Invest New Drugs 1983;1(1):33-45.

The human tumor cloning assay in cancer research and therapy: a review with clinical correlations. Author(s): Hanauske, A R; Hanauske, U; Von Hoff, D D Language: eng Source Curr Probl Cancer 1985 Dec;9(12):1-66.

The identification and cloning of genes encoding haloaromatic catabolic enzymes and the construction of hybrid pathways for substrate mineralization. Abstract: This paper reviews the genetic basis of haloaromatic biodegradation by bacteria, with a focus on the genetic analysis of Alcaligenes eutrophus JMP134, an organism which can utilize 3-chlorobenzoate, 2,4-dichlorophenoxyacetate (2,4-D) and related compounds as sole carbon and energy sources, and Pseudomonas sp. B13, a chlorobenzoate degrader. The involvement of transmissible plasmids pJP4 and pWR1, isolated from strains JMP134 and B13, respectively, in chloroaromatic mineralization has been examined, and restriction fragments of both plasmids have been cloned on the broad host range plasmid vector pKT231. Transposon Tn5 mutagenesis of these and other soil isolates enriched in and purified from mixed cultures utilizing 2,4,5-trichlorophenoxyacetate (2,4,5-T) as sole carbon and energy source, has been carried out using a "suicide" transposon donor, pLG221 (Col1bdrd-1::Tn5). Mapping of Tn5 insertions in mutants which accumulate pathway intermediates has facilitated the identification and cloning of genes encoding chlorocatechol 1,2-dioxygenase, and other key enzymes in haloaromatic catabolism. There are good prospects for the genetic construction of hybrid haloaromatic catabolic pathways by combining genes encoding broad specificity enzymes, capable of transforming halogenated analogues of their natural substrates, with genes for halocatechol degradation. Author(s): Weightman, A J; Don, R H; Lehrbach, P R; Timmis, K N Language: eng Source Basic Life Sci 1984;28:47-80.

The immunotherapy of solid cancers based on cloning the genes encoding tumor-rejection antigens. Abstract: Cellular immune reactions play a major role in the host reaction to growing cancers. Tumor infiltrating lymphocytes (TIL) can be isolated from melanomas and can specifically recognize unique tumor antigens. The adoptive transfer of TIL plus interleukin-2 can mediate tumor regression in patients with metastatic melanoma. TIL capable of mediating tumor regression have been used to clone and sequence a variety of the genes that encode the tumor-regression antigens recognized by these TIL. This information has opened new opportunities for the development of cancer immunotherapies. These gene products can be used to generate lymphocytes, in vitro, with improved antitumor activity for use in adoptive transfer. Active immunization can be performed using either the immunodominant peptides present in these proteins or by incorporating the tumor antigen genes into recombinant viruses. Cancer vaccine trials using many of these approaches have recently begun. Attempts to apply a similar strategy to epithelial tumors such as breast and ovarian cancer are underway. Author(s): Rosenberg, S A Language: eng Source Annu Rev Med 1996;47:481-91.

The impact of new cloning techniques on the diagnosis and treatment of infectious diseases. Author(s): Engleberg, N C; Eisenstein, B I Language: eng Source N Engl J Med 1984 Oct 4;311(14):892-901.

The implications of the cloning of the XLA gene. Author(s): Flinter, F A; Vetrie, D; Bobrow, M Language: eng Subject(s): Agammaglobulinemia/diagnosis/genetics/therapy; Child, Preschool; Chromosome Mapping; Cloning, Molecular; Female; Gene Therapy; Human; Infant; Linkage (Genetics); Male; Prenatal Diagnosis; X Chromosome Source J R Coll Physicians Lond 1993 Jul;27(3):233-5.

The male-specific histocompatibility antigen, H-Y: a history of transplantation, immune response genes, sex determination and expression cloning. Abstract: H-Y was originally discovered as a transplantation antigen. In vivo primary skin graft responses to H-Y are controlled by immune response (Ir) genes mapping to the MHC. In vitro T cell responses to H-Y are controlled by MHC class I and II Ir genes, which-respectively, restrict CD8 and CD4 T cells: These can be isolated as T cell clones in vitro. T cell receptor (TCR) transgenic mice have been made from the rearranged TCR genes of several of these, of which that specific for H-Y/Db is the best studied. Non-MHC Ir genes also contribute to the control of in vitro CTL responses to H-Y. The Hya/HYA gene(s) encoding H-Y antigen have been mapped using translocations, mutations, and deletions to Yq in humans and to the short arm of the Y chromosome in mice, where they lie in the deletion defined by the Sxrb mutation between Zfy-1 and Zfy-2. Hya/HYA has been separated from the testis-determining gene, Sry/SRY, in both humans and mice and in humans the azoospermia factor AZF has been separated from HYA. In mice transfection of cosmids and cDNAs mapping to the Sxrb deletion has identified two genes encoding H-Y peptide epitopes. Two such epitopes, H-Y/K(k) and H-Y/D(k), are encoded within different exons of Smcy and a third, H-Y/D(b), by a novel gene, Uty. Peptide elution approaches have isolated a human H-Y epitope, H-Y/HLA-B7, and identified it as a product of SMCY. Each of the Hya genes in mice is ubiquitously expressed but of unknown function. Their X chromosome homologues do not undergo X inactivation. Author(s): Simpson, E; Scott, D; Chandler, P Language: eng Source Annu Rev Immunol 1997;15:39-61.

The molecular biology of thyroid peroxidase: cloning, expression and role as autoantigen in autoimmune thyroid disease. Author(s): McLachlan, S M; Rapoport, B Language: eng Source Endocr Rev 1992 May;13(2):192-206.

The nociceptin (ORL1) receptor: molecular cloning and functional architecture. Abstract: Nociceptin and the ORL1 receptor share high sequence similarity with opioid peptides, particularly dynorphin A, and their receptors. However, nociceptin and dynorphin A may use distinct molecular pathways to bind and activate their cognate receptors. Activation of the kappa-opioid receptor by dynorphin A is thought to require interactions of its

N-terminal hydrophobic domain (Y(1)GGF) with the receptor opioid binding pocket, located within the transmembrane helix bundle, while activation of the ORL1 receptor appears to require interactions of the positively charged core (R(8)KSARK) of nociceptin with the negatively charged second extracellular receptor loop. Author(s): Meunier, J; Mouledous, L; Topham, C M Language: eng Source Peptides 2000 Jul;21(7):893-900.

The P1 cloning system: past and future. Author(s): Sternberg, N Language: eng Subject(s): Animal; Chimera; Cloning, Molecular; DNA Transposable Elements; Genomic Library; Human; Plasmids Source Mamm Genome 1994 Jul;5(7):397-404.

The prohibition of federal government funding of human cloning research: hearing before the Committee on Science, Subcommittee on Technology, U.S. House of Representatives, One Hundred Fifth Congress, first session, July 22, 1997. Published/Created: Washington: U.S. G.P.O.: For sale by the U.S. G.P.O., Supt. of Docs., Congressional Sales Office, 1997 [i.e. 1998] Description: iii, 156 p.; 24 cm. ISBN: 0160561124 Notes: Distributed to some depository libraries in microfiche. Shipping list no.: 98-0155-P. "No. 32." Includes bibliographical references. Subjects: Cloning--Moral and ethical aspects--Government policy United States. Cloning--Research--Government policy--United States. Genetic engineering--Research--Government policy--United States. Federal aid to medical research--United States. LC Classification: KF27 .S39957 1997f Govt. Doc. No.: Y 4.SCI 2:105/32

The proof of the cloning is in the eating. Author(s): Geisow, M Language: eng Source Trends Biotechnol 1991 Jan;9(1):5-7.

The pros and cons of human cloning. Author(s): LaBar, Martin Language: eng Subject(s): Cloning, Organism; Embryo; Ethics; Eugenics; Gene Pool; Health Care Rationing; Parent-Child Relations; Resource Allocation; Risk; Risk Assessment; Sex Preselection; Social Change Source Thought 1984 Sep;59(234):319-33.

The Pvu II restriction-modification system: cloning, characterization and use in revealing an E. coli barrier to certain methylases or methylated DNAs. Author(s): Blumenthal, R M Language: eng Source Gene Amplif Anal 1987;5:227-45.

The question of human cloning. Author(s): Robertson, J A Language: ource Hastings Cent Rep 1994 Mar-Apr;24(2):6-14.

The recent history of somatic cloning in mammals. Abstract: The history of somatic cell nuclear transfer (NT) in mammals is full of exciting experiments and findings regarding the technique and outcome of NT, despite only covering a period of 6 years. The production of Dolly, for the first time demonstrating cloning from an adult somatic cell, had a great impact on subsequent studies. However, the more progress we make, the more obvious it becomes how little we know about the processes during NT, specifically how reprogramming events occur. Therefore, it is certainly challenging to continue investigating every step of somatic cell NT more intensively, starting from the donor cell, (type, cell cycle, synchronization, population doublings) and continuing until the cloned offspring are born and even further, to see how and if NT has an influence on health, viability, quantitative traits, and reproduction of cloned individuals. Author(s): Brem, Gottfried; Kuhholzer, Birgit Language: eng Source Cloning Stem Cells 2002;4(1):57-63.

The role of the preimplantation geneticist in human cloning. Abstract: If human cloning is to become a reality, the preimplantation geneticist must be responsible for determining the indications for undertaking cloning and for establishing the risks and benefits of human cloning. The unresolved issue is whether a compelling argument can be made for cloning a human for therapeutic reasons while outweighing legal, moral and ethical objections. At present, 'whole person' cloning does not seem justified under any circumstance, whereas cloning for the replacement of diseased cells, tissues or organ systems, i.e. 'spare parts', seems to be a likely, acceptable application of cloning strategies for humans. Author(s): Fiddler, M; Pergament, D; Pergament, E Language: eng Source Prenat Diagn 1999 Dec;19(13):1200-4.

The search for truth and freedom: ethical issues surrounding human cloning and stem cell research. Abstract: The reality of cloning and stem cell research has provoked wonder, fear and anger. These developments have the potential fundamentally to alter humanity. But how well informed is the range of views being expressed? Is progress being threatened by understandable but uninformed fears? Or are scientists rushing toward an ethical abyss, so concerned with what they can do that they never stop to ask what they should do? This article identifies some of the fears and hopes surrounding cloning and stem cell research. It aims to provoke ethical debate in evaluating such research. Author(s): Bruce, Alex Language: eng Source J Law Med 2002 Feb;9(3):323-35.

The thyrotropin receptor 25 years after its discovery: new insight after its molecular cloning. Abstract: The molecular cloning and functional expression of the TSH receptor has led to rapid advances in understanding the structure and function of the molecule. Knowledge of its genomic structure provides information on the evolutionary origin of the TSH receptor as well as on the functional organization of its extracellular domain, which is responsible for ligand binding. A beginning has been made in defining the discontinuous contact points for TSH in this extracellular region, but determination of all of the amino acids involved will be difficult. The binding sites of TSH receptor autoantibodies do not appear to be identical to the TSH binding site. Two of the six potential glycosylation sites in the extracellular domain are important in the expression of a functional receptor. Disulfide bonding contributes toward maintenance of the three-dimensional structure of the receptor. Recent evidence suggests that the TSH receptor exists as a single polypeptide chain without subunits. Significant progress has been made in understanding the intracellular regions of the TSH receptor that are involved in signal transduction. Although still in the distant future, we are closer to the goal of understanding precisely how TSH interacts with and activates its receptor. More importantly from the clinical perspective, we are closer to defining the B cell, and ultimately T cell, epitopes on the TSH receptor that are recognized by the immune system. This information may ultimately facilitate the development of immunological approaches to treating Graves' disease, which will be an improvement over thyroid gland destruction and consequent hypothyroidism, the most common form of therapy at the present time. Author(s): Nagayama, Y; Rapoport, B Language: eng Source Mol Endocrinol 1992 Feb;6(2):145-56.

The thyrotropin-releasing hormone gene 1998: cloning, characterization, and transcriptional regulation in the central nervous system, heart, and testis. Abstract: Mechanisms of triiodothyronine (T3) negative regulation of the human thyrotropin-releasing hormone (TRH) gene were investigated with a chimeric construct of the 5' flanking region fused to a luciferase reporter gene, transfected into human neuroblastoma cells (HTB-11). Maximum negative regulation was achieved with constructs containing bases -242 to +54. Four sequences in this region exhibited homology with half sites of thyroid hormone response elements (TRE) (AGGTCA). The most important site was a sequence with an overlapping TRE/CRE, involving bases -53 to -60 (TGACCTCA). Potential combinatorial interactions of thyroid hormone receptors and CREB at this site were explored. Modest promoter stimulation was achieved with dibutyryl cyclic adenosine monophosphate (cAMP) (10(-3) M) plus IBMX (0.5 mM). Stimulation was greatly enhanced (+820%) by cotransfection of a constitutively activated protein kinase A (pPKA) construct. Cotransfection with pCREB increased stimulation further to 1350% above control. Stimulation of pPKA and pCREB interfered with stimulation by unliganded TRbeta1, and co-transfected pPKA and pCREB blocked T3 negative inhibition by TRbeta1-T3 complexes. When this site was mutated by polymerase chain reaction (PCR) mutagenesis, the mutant construct failed to respond to unliganded TRbeta1, and stimulation by pPKA and/or pCREB was inhibited markedly, from 12.5- to 2.1-fold, p < 0.001. Moreover, TRbeta1-T3 complexes failed to show any inhibition of the mutated promoter. These results suggest that negative regulation is achieved by inhibition of CREB stimulation of the TRH promoter at this overlapping TRE/CRE site. The two cosuppressors, NCoR and SMRT, were able to augment stimulation of the TRH promoter by unliganded TRbeta1 and enhance the magnitude of T3 inhibition. The potential role of the TRH gene and the pathophysiology of thyroid hormone resistance was investigated with three mutant TRbeta1 constructs. Thyroid hormone resistance was found to be expressed at the level of TRH gene regulation, due to lowered inhibition by mutant TRbeta1-T3 complexes and by their dominant negative effects on wild-type TRbeta1-T3 inhibition. TRH gene expression has been identified in the heart. Cardiac TRH mRNA was not regulated by T3, in contrast to HTB-11 cells, but cardiac TRH mRNA density could be augmented by glucocorticoids and by testosterone. TRH receptors were identified using Scatchard blots that showed a kilodalton of 1.4 nM and a bmax of 10 pmol/mg protein. TRH-R mRNA was identified also by reverse transcription polymerase chain reaction (RT-PCR). Enhanced ventricular contractility by TRH was demonstrated in both an open-chested dog preparation and in ex vivo ventricular myocytes, using video edge cinematography. Under controlled conditions, myocyte shortening was 13.3%, and TRH (10(-6) M) caused muscle shortening to increase 140%, (p < 0.005). TRH gene expression was demonstrated exclusively in Leydig cells of the testis. High affinity binding sites were identified in testicular membranes with a kilodalton of 1.6 x 10(-6) M. TRH was able to inhibit LH and HCG-activated testosterone secretion significantly. Thus, one paracrine role of TRH in the testis may be to serve as inhibitory modulator of gonadotropin-stimulated testosterone secretion. Author(s): Wilber, J F; Xu, A H Language: eng Source Thyroid 1998 Oct;8(10):897-901.

The use of M13 phages in the cloning and sequencing of DNA Author(s): Plaza, G Language: pol Source Postepy Biochem 1987;33(4):489-500.

The use of terminal transferase for molecular cloning and nucleic acids analysis. Author(s): Roychoudhury, R Language: eng Subject(s): Animal; Cattle; Cloning, Molecular; DNA; DNA Nucleotidylexotransferase; Nucleic Acids/analysis; Oligoribonucleotides Source Gene Amplif Anal 1981;2:41-83.

The utility of streptomycetes as hosts for gene cloning. Author(s): Tomich, P K; Yagi, Y Language: eng Source Genet Eng (N Y) 1990;12:53-72.

The X-linked lymphoproliferative disease: from autopsy toward cloning the gene 1975-1990. Abstract: Although X-linked lymphoproliferative disease (XLP) is rare (1-2 males per 1 x 10(6)), it serves as a model for discerning diverse diseases caused by Epstein-Barr virus (EBV) ranging from agammaglobulinemia to fatal infectious mononucleosis following infection with the virus. The study of patients with XLP has also paved the way to understanding how EBV induce diseases in children with primary immunodeficiency diseases, organ transplant recipients, and those with acquired immunodeficiency syndrome. This review is dedicated to the memory of Gordon Vawter, M.D., who generously provided insights into the causes of pathogenesis of immune deficiency and lymphoproliferative disorders. Author(s): Purtilo, D T; Grierson, H L; Davis, J R; Okano, M Language: eng Source Pediatr Pathol 1991 Sep-Oct;11(5):685-710.

The yeast cell--prospective recipient system for cloning eukaryotic cellular and viral genes; Zhdanov, B M; Granovskii, N N Language: rus; Cloning, Molecular; Genes; Genes, Viral; Saccharomyces cerevisiae/genetics Source Vopr Virusol 1982 Sep-Oct;27(5):9-13.

Therapeutic cloning in debate Abstract: Human embryos can be conceived by cell nuclear transfer in order to isolate human embryonic stem cells (hES cells) for research into autologous cell therapy (therapeutic cloning). However, this technique broaches the major ethical problem concerning the instrumental use of human preimplantation embryos. From the viewpoint of subsidiarity, it is argued that various potential alternatives for therapeutic cloning should first be investigated further. The question as to whether therapeutic cloning should be allowed only becomes apparent when research with surplus embryos obtained in the course of in-vitro fertilization suggests that usable transplants can be obtained in vitro from hES cells, and when the potential alternatives for therapeutic cloning are either less promising or need more time for development than is currently expected. Author(s): de Wert, G Language: dut Source Ned Tijdschr Geneeskd 2001 Nov 3;145(44):2109-11.

Therapeutic cloning: concepts and practicalities. Abstract: The concept of using embryonic stem (ES) cells as a source of multiple cell types for use in tissue repair has existed for approximately 20 years. Recent breakthroughs in somatic nuclear transfer and human ES cell derivation have produced a flurry of new activity in this area, with the recognition that ES cell lines that are customized and genetically identical to those of the patient are a distinct possibility. This article examines the background of and prospects for these exciting new developments. Author(s): Colman, A; Kind, A Language: eng Source Trends Biotechnol 2000 May;18(5):192-6.

Therapeutic cloning: far from application at this stage Abstract: Therapeutic cloning has become possible since the discovery that nuclei from somatic cells of adult animal tissue can successfully be used for cloning and the fact that human embryonic stem cell lines have been established from preimplantation embryos. When nuclei from healthy tissue of a patient are transplanted into enucleated oocytes, these oocytes can be artificially activated so that embryos develop from which embryonic stem cells of the donor can be derived. These embryonic stem cells can be cultured as permanent lines in unlimited numbers and remain pluripotent, i.e. they can be induced to differentiate into the required cell type by adding one or more specific factors. These cells can then be transplanted back into the patient suffering from either a lack or dysfunction of these cells. This approach prevents the rejection of the transplanted cells by the patient's immunological system. As this type of cloning has a very low efficiency, a large number of unfertilized donor oocytes is required. It is questionable whether enough donors are or will be available for this purpose. The cultured cells must satisfy certain conditions before they can be used for transplantation. They must be checked for chromosomal abnormalities, and a complete differentiation of the embryonic stem cells into the cells types needed by the patient is necessary as after the transplantation, undifferentiated stem cells will form teratomas. Furthermore, it is difficult to ensure that the cells end up in the right place and to ensure that they fully integrate into the existing tissue to form functional connections. Due to this array of technical problems the question remains as to whether therapeutic cloning will become feasible in the near future. Author(s): De

Both, N J Language: dut. Source Ned Tijdschr Geneeskd 2001 Nov 3;145(44):2111-5.

Therapeutic cloning: needs and prospects. Abstract: There has been widespread speculation regarding the possibility of reprogramming human somatic nuclei by transfer into oocytes. Stem cells derived from such reconstructed embryos could theoretically provide a source of autologous tissue for therapeutic transplantation. However the biological, technical and ethical hurdles which must be overcome before this can be realised are considerable. Here we review the background and prospects for such work. Author(s): Kind, A; Colman, A Language: eng Source Semin Cell Dev Biol 1999 Jun;10(3):279-86.

Thiel, Teresa. Biotechnology DNA, to protein: a laboratory project in molecular biology / Teresa Thiel, Shirley Bissen, Eilene M. Lyons. Edition Information: 1st ed. Published/Created: Boston: McGraw-Hill, c2001. Related Names: Bissen, Shirley. Lyons, Eilene M. Description: x, 182 p.: ill.; 28 cm. ISBN: 0072416645 Subjects: Molecular biology--Laboratory manuals. Amylases. Biology projects. Molecular cloning--Laboratory manuals. LC Classification: QH506 .T48 2002 Dewey Class No.: 572.8 21

Thomas, Lewis, 1913- There's hope in future cancer research [sound recording] / Lewis Thomas. Published/Created: [New York]: Encyclopedia Americana/CBS News Audio Resource Library, p1979. Related Names: Reeves, Dick. Description: on side A of 1 sound cassette (29 min., 48 sec.): analog. Summary: Dr. Lewis Thomas, President of the Sloan-Kettering Cancer Center, discusses the present state of cancer research and prospects for the future. He also discusses such matters as cloning of human beings ("nonsense"), the danger of nuclear destruction, and the importance of government support for medical research. Notes: Distributed by Grolier Education Corp, New York. With: Turmoil in Nicaragua: a special report by Dan Rather -- On signing the SALT II treaty / Pres. Breshnev & Pres. Carter. Recorded June 14, 1979, in New York. Preservation master. Washington, D.C. Library of Congress Magnetic Recording Laboratory, 1979. On 1 sound reel: analog, 7 1/2 ips, 2 track, mono.; 10 in. Subjects: Cancer--Research--United States. Series: Vital history cassettes; June 1979, no. 2. LC Classification: RZA 0780 RWA 0377 A1 (preservation master)

Three technical approaches for cloning opioid receptors. Author(s): Machida, C A; Salon, J; Grandy, D; Bunzow, J; Albert, P; Hanneman, E; Civelli, O Language: eng Subject(s): Amino Acid Sequence; Cloning, Molecular/methods; DNA/genetics; Eukaryotic Cells/metabolism; Human; Receptors, Opioid/genetics; Receptors, Sensory/metabolism; Sequence Homology, Nucleic Acid; Transfection Source NIDA Res Monogr 1988;87:93-110.

Thyrotropin-releasing hormone receptor: cloning and regulation of its expression. Author(s): Gershengorn, M C Language: eng Source Recent Prog Horm Res 1993;48:341-63.

Toward cloning and mapping the genome of Drosophila. Abstract: An ultimate goal of Drosophila genetics is to identify and define the functions of all the genes in the organism. Traditional approaches based on the isolation of mutant genes have been extraordinary fruitful. Recent advances in the manipulation and analysis of large DNA fragments have made it possible to develop detailed molecular maps of the Drosophila genome as the initial steps in determining the complete DNA sequence. Author(s): Merriam, J; Ashburner, M; Hartl, D L; Kafatos, F C Language: eng Source Science 1991 Oct 11;254(5029):221-5.

Toward positional cloning with SNPs. Abstract: Efficient methods are available for positional cloning of major genes by linkage and linkage disequilibrium (LD). The application of these methods to complex diseases is more demanding, but advances in sequencing the human genome, SNP discovery, high-throughput genotyping techniques, LD maps, and haplotype annotation are revolutionizing positional cloning of oligogenes for common multifactorial disease. Author(s): Morton, Newton E; Collins, Andrew Language: Source Curr Opin Mol Ther 2002 Jun;4(3):259-64.

Towards cloning the familial breast-ovarian cancer gene on chromosome 17. Abstract: The past year has seen a great deal of excitement in the field of breast cancer genetics. Since linkage of the familial breast-ovarian cancer gene (BRCA1) to chromosome 17, the critical region has been narrowed to 1.0-1.5 Mb by recombination studies, a detailed physical map has been constructed and much of the region has been cloned in yeast artificial chromosome, bacteriophage P1 and cosmid vectors. The focus now lies on identifying the genes housed within the BRCA1 region and scanning them for oncogenic mutations. Author(s): Brown, M A; Solomon, E Language: eng Source Curr Opin Genet Dev 1994 Jun;4(3):439-45.

Towards the cloning of genes underlying murine QTLs. Abstract: Many quantitative trait loci (QTLs), including those for ethanol-related traits, have been mapped in the mouse. In light of rapidly developing tools and resources, we briefly review the strategy for identifying the genes underlying these QTLs. We note that positional cloning will soon be a matter of testing candidate genes rather than discovering genes; therefore, we describe a "congenic test" to support that a candidate gene is indeed a QTL. Considering the rapid development of congenics and mutants, we also identify four areas of investigation-phenotypes, ethanol specificity, environment, and gene interactions-that might be exploited during the course of positional cloning to gain insights into QTL pathways. In particular, we note that multiple mutants of nearly every major neurotransmitter pathway have now been made. These mutants are not only useful for phenotypic tests, but also could be used to conduct "gene dependence" tests of QTLs. We also consider potential applications for the very recently developed ability to clone mice. Author(s): Rikke, B A; Johnson, T E Language: eng Source Mamm Genome 1998 Dec;9(12):963-8.

Toxoplasma gondii microneme proteins: gene cloning and possible function. Author(s): Fourmaux, M N; Garcia-Reguet, N; Mercereau-Puijalon, O; Dubremetz, J F Language: eng Source Curr Top Microbiol Immunol 1996;219:55-8.

Transformation and cloning systems in non-Saccharomyces yeasts. Author(s): Klein, R D; Zaworski, P G Language: eng Source Bioprocess Technol 1990;8:245-310.

Trends in molecular genetics / edited by Umakant Sinha and Walter Klingmüller. Published/Created: Patna: Spectrum Pub. House, c1985. Related Names: Sinha, Umakant. Klingmüller, W., 1929- UGC Centre of Special Assistance. Patna University. Dept. of Botany. International Symposium on Gene Cloning and Recombinant DNA Technology (1983: Patna, India) Description: 158 p.: ill.; 25 cm. Summary: Selected papers of the International Symposium on Gene Cloning and Recombinant DNA Technology, 1983, Patna, India, organized by the UGC Centre of Special Assistance, and the Dept. of Botany, Patna University. Notes: Includes bibliographies. Subjects: Recombinant DNA--Congresses. Genetic engineering--Congresses. Molecular cloning--Congresses. LC Classification: QH442 .T74 1985 Dewey Class No.: 574.87/328 19

Triglyceride synthesis: insights from the cloning of diacylglycerol acyltransferase. Abstract: Although the biochemistry of triglyceride synthesis has been studied for decades, an understanding of the molecular processes involved has been lacking. The recent cloning of a gene encoding acyl coenzyme A : diacylglycerol acyltransferase, an enzyme that catalyses the final step in triglyceride synthesis, has opened this area to molecular investigation and has begun to provide new insights into triglyceride metabolism. Author(s): Farese, R V Jr; Cases, S; Smith, S J Language: eng Subject(s): Source Curr Opin Lipidol 2000 Jun;11(3):229-34.

Type C retroviruses as vectors for cloning cellular genes with probable transforming function. Author(s): Stephenson, J R; Khan, A S; van de Ven, W J; Reynolds, F Source J Natl Cancer Inst 1979 Nov;63(5):1111-9.

Understanding cloning / from the editors of Scientific American; compiled and with introductions by Sandy Fritz; foreword by William A. Haseltine. Published/Created: New York, NY: Warner Books, c2002. Related Names: Fritz, Sandy. Description: viii, 136 p.: ill.; 21 cm. ISBN: 0446678740 Subjects: Cloning--Popular works. Series: Science made accessible LC Classification: QH442.2 .U475 2002 Dewey Class No.: 660.6/5 21

Universal TA cloning. Abstract: TA cloning is one of the simplest and most efficient methods for the cloning of PCR products. The procedure exploits the terminal transferase activity of certain thermophilic DNA polymerases, including Thermus aquaticus (Taq) polymerase. Taq polymerase has non-template dependent activity which preferentially adds a single adenosine to the 3'-ends of a double stranded DNA molecule, and thus most of the molecules PCR amplified by Taq polymerase possess single 3'-A overhangs. The use of a linearized "T-vector" which has single 3'-T overhangs on both ends allows direct, high-efficiency cloning of PCR products, facilitated by complementarity between the PCR product 3'-A overhangs and vector 3'-T overhangs. The TA cloning method can be easily modified so that the same T-vector can be used to clone any double-stranded DNA fragment, including PCR products amplified by any DNA polymerase, as well as all blunt- and sticky-ended DNA species. This technique is especially useful when compatible restriction sites are not available for the subcloning of DNA fragments from one vector to another. Directional cloning is made possible by appropriate hemi-phosphorylation of both the T-vectors and the inserts. With a single T-vector at hand, any DNA fragment can be cloned without compromising the cloning efficiency. The universal TA cloning method is thus both convenient and labor-saving. Author(s): Zhou, M Y; Gomez-Sanchez, C E Language: eng Source Curr Issues Mol Biol 2000 Jan;2(1):1-7.

Use of a cDNA expression-cloning vector and a secretion vector for mammalian gene expression in Saccharomyces cerevisiae. Author(s): Miyajima, A; Arai, K Language: eng : Amino Acid Source Biotechnology 1989;13:281-304.

Use of cDNA cloning to study the mechanism of action of glucocorticoid hormones at the molecular level. Abstract: We have previously described a dramatic phenomenon of phenotypic reversion (tumor to normal) caused by glucocorticoid hormones in C6/ST1 rat glioma cells, but not in their hormone-resistant C6/P7 counterpart. Blind cDNA cloning was adopted to search for glucocorticoid-regulated gene sequences responsible for this phenotype reversion. Differential hybridization and differential display of RNA were used in parallel to isolate a number of cDNA clones that were characterized by DNA sequencing and Northern blot analysis. This approach was coupled to the analysis of known growth control genes (oncogenes, tumor suppressor genes, cyclins, cyclin-dependent kinases, other kinases). Glucocorticoid target genes isolated from this cell system are likely to be good anti-tumor candidate molecules which can be used in tumor therapy and anti-tumor drug design. Author(s): Armelin, M C; Sasahara, R M; Flatschart, R; Vedoy, C Language: eng Source Braz J Med Biol Res 1996 Dec;29(12):1751-7.

Use of cDNA library expression cloning to identify components of heterodimeric receptor complexes. Author(s):Chang, C; Lanier, L L Language: eng Subject(s): Animal; Cloning, Molecular; DNA, Complementary; Dimerization; Gene Expression; Human; Killer Cells, Natural/immunology; Receptors, Immunologic/genetics Source Methods Mol Biol 2000;121:273-81.

Use of mammalian cell expression cloning systems to identify genes for cytokines, receptors, and regulatory proteins. Abstract: Mammalian cell expression cloning has become a standard technique for the isolation of mammalian genes or cDNAs. Its advantage is that the biological functions of the gene of interest are used for cloning. Therefore, the identified cDNAs or genes should be functional in vivo, and there is no need for physical or chemical information about the gene products, so that protein purification in sufficient quantity to raise antibodies or to obtain amino acid sequences is not necessary. Here, we summarize recent progress in mammalian cell cloning systems, and discuss the possible directions in which this technique will lead. Author(s): Nakayama, N; Yokota, T; Arai, K Language: eng Subject(s): Animal; Biotechnology; Cell Line; Cloning, Molecular/methods; Cytokines; DNA/genetics; Gene Expression; Genetic Engineering; Genetic Vectors; Human; Plasmids; Receptors, Cell Surface/genetics; Recombinant Proteins/genetics Source Curr Opin Biotechnol 1992 Oct;3(5):497-505.

Use of plasmids from Staphylococcus aureus for cloning of DNA in Bacillus subtilis. Author(s): Ehrlich, S D; Niaudet, B; Michel, B Language: eng Subject(s): Bacillus subtilis/genetics; Cloning, Molecular; DNA, Recombinant; Genetic Vectors; Plasmids; Staphylococcus aureus/genetics; Transformation, Genetic Source Curr Top Microbiol Immunol 1982;96:19-29.

Use of recombinant DNA techniques in cloning DNA repair genes and in the study of mutagenesis in mammalian cells. Author(s): Lehmann, A R Language: eng Subject(s): Animal; Ataxia Telangiectasia/genetics; Cells, Cultured; Cloning, Molecular; Cricetulus; DNA Repair; DNA, Recombinant; Female; Genes, Bacterial; Hamsters; Human; Mutation; Ovary; Source Mutat Res 1985 Jun-Jul;150(1-2):61-7.

Use of vaccinia virus as an infectious molecular cloning and expression vector. Author(s): Moss, B; Smith, G L; Mackett, M Language: eng Subject(s): Chimera; Cloning, Molecular/methods; Genes, Viral; Genetic Vectors; Plasmids; Transfection; Vaccinia virus/genetics/isolation & purification; Viral Vaccines Source Gene Amplif Anal 1983;3:201-13.

Vaccine development for P. aeruginosa: based on antigen cloning of clinical vaccines][. Abstract: For the development of a clinical vaccine against P. aeruginosa infections we cloned the genes for the main outer membrane proteins F and I of P. aeruginosa and characterized protective epitopes by monoclonal antibodies. A recombinant hybrid protein (OprFaa190-342-OprIaa21-83) was expressed in E. coli which represents the protective epitopes. The vaccine showed to be highly protective in mice against experimental P. aeruginosa infections. A phase I trial in human volunteers showed that the vaccine is well tolerated and that high antibody titers against P. aeruginosa were induced. Author(s): von Specht, B U Language: ger; Antigens, Bacterial/immunology; Bacterial Vaccines/immunology; Cloning, Molecular; English Abstract; Human; Pseudomonas aeruginosa/immunology Source Zentralbl Chir 1999;124(4):298-302.

Vector systems for the expression, analysis and cloning of DNA sequences in S. cerevisiae. Author(s): Parent, S A; Fenimore, C M; Bostian, K A Language: eng Subject(s): Cloning, Molecular; DNA, Fungal; Gene Expression Regulation; Genetic Vectors;

Saccharomyces cerevisiae/genetics Source Yeast 1985 Dec;1(2):83-138.

Vectors for cloning in yeast. Author(s): Hinnen, A; Meyhack, B Language: eng Subject(s): Acid Phosphatase/genetics; Cloning, Molecular; DNA, Fungal/genetics; Genetic Complementation Test; Genetic Vectors; Plasmids; Saccharomyces cerevisiae/genetics; Transformation, Genetic Source Curr Top Microbiol Immunol 1982;96:101-17.

Vectors for cloning promoters and terminators. Author(s): Nierman, W C Language: eng Subject(s): Drug Resistance, Microbial; Escherichia coli; Galactokinase; Bacterial; Genes, Regulator; Genes, Structural; Genetic Vectors; Gram-Negative Bacteria/genetics; Plasmids; Promoter Regions; Terminator Regions; beta-Galactosidase/genetics Source Biotechnology 1988;10:153-77.

Vectors for gene cloning in Pseudomonas and their applications. Author(s): Sakaguchi, K Language: eng Subject(s): Cloning, Molecular; Escherichia coli/genetics; Genes, Bacterial; Genetic Engineering; Genetic Vectors; Leucine/genetics; Plasmids; Pseudomonas/genetics; Tryptophan/genetics Source Curr Top Microbiol Immunol 1982;96:31-45.

Vectors: a survey of molecular cloning vectors and their uses / edited by Raymond L. Rodriguez, David T. Denhardt. Published/Created: Boston: Butterworths, c1988. Related Names: Rodriguez, Raymond L. Denhardt, David T. Description: xiv, 578 p.: ill.; 25 cm. ISBN: 0409900427 Notes: Includes bibliographies and index. Subjects: Genetic vectors. Molecular cloning. Series: Biotechnology (Reading, Mass.); 10. LC Classification: QH442.2 .V43 1988 Dewey Class No.: 574.87/322 19

Vectors: cloning applications / edited by P. Jones. Published/Created: Chichester; New York: Wiley, c1998. Related Names: Jones, P. (Paul) Description: xiii, 157 p.: ill.; 22 cm. ISBN: 047196266X (alk. paper) Notes: "Published in association with BIOS Scientific Publishers Limited." Includes bibliographical references and index. Subjects: Molecular cloning--Laboratory manuals. Genetic vectors--Laboratory manuals. Series: Essential techniques series LC Classification: QH442.2 .V433 1998 Dewey Class No.: 572.8/6 21

Vectors: expression systems / edited by P. Jones. Published/Created: Chichester; New York: John Wiley & Sons, 1998. Related Names: Jones, P. (Paul) Description: xiii, 153 p.; 22 cm. ISBN: 0471962678 (alk. paper) Notes: "Published in association with BIOS Scientific Publishers Limited." Includes index. Subjects: Genetic vectors. Gene expression. Molecular cloning. Series: Essential techniques series LC Classification: QH442.2 .V446 1998 Dewey Class No.: 572.8/65 21

Vorilhon, Claude. Yes to human cloning: immortality thanks to science / Raël. Published/Created: London; Los Angeles: Tagman, 2001 Description: 172 p.; 24 cm. ISBN: 1903571049 (hardcover) 1903571057 (pbk.) Notes: Translated from the French. Subjects: Human cloning. LC Classification: QH442.2 .V675 2001

Wahrman, Miryam Z. Brave new Judaism: when science and scripture collide / Miryam Z. Wahrman. Published/Created: Hanover: University Press of New England [for] Brandeis University Press, c2002. Description: xix, 287 p.; 23 cm. ISBN: 1584650311 Notes: Includes bibliographical references (p. [239]-275) and index. Subjects: Judaism and science. Bioethics--Cloning--Medicine--Religious aspects--Judaism. Technology and Jewish law. LC Classification: BM538.S3 W34 2002 Dewey Class No.: 296.3/4957 21

Watson, James D., 1928- The DNA story: a documentary history of gene cloning / James D. Watson, John Tooze. Published/Created: San Francisco: W.H. Freeman and Co., c1981. Related Names: Tooze, John. Description: xii, 605 p.: ill.; 29 cm. ISBN: 071671292X Notes:

Includes index. Bibliography: p. 589-594. Subjects: Recombinant DNA--Cloning--Research--History--Addresses, essays, lectures. LC Classification: QH442 .W37 Dewey Class No.: 574.87/3282 19

Weldon, Fay. The cloning of Joanna May / Fay Weldon. Edition Information: 1st American ed. Published/Created: New York, N.Y., U.S.A.: Viking, 1990. Description: 265 p.: ill.; 24 cm. ISBN: 0670830909 Notes: "First published in Great Britain by W. Collins in 1989"--T.p. verso. Subjects: Widows--Fiction. Cloning--Fiction. LC Classification: PR6073.E374 C57 1990 Dewey Class No.: 823/.914 20

Weldon, Fay. The cloning of Joanna May / Fay Weldon. Published/Created: New York, N.Y., U.S.A.: Penguin Books, 1991. Description: 265 p.; 20 cm. ISBN: 0140128514 : Notes: "First published in Great Britain by William Collins 1989"--T.p verso. Subjects: Cloning--Fiction. Widows--Fiction. LC Classification: PR6073.E374 C57 1991 Dewey Class No.: 823/.914 20

Weldon, Fay. The cloning of Joanna May / Fay Weldon. Published/Created: London: Collins, 1989. Description: 265 p.; 24 cm. ISBN: 0002233495 : Subjects: Cloning--Fiction. Widows--Fiction. LC Classification: PR6073.E374 C57 1989b Dewey Class No.: 823/.914 20

West, Michael (Michael D.) The Immortal cell: one scientist's daring search for the keys to extending life / Michael West. Edition Information: 1st ed. Published/Created: New York: Doubleday, 2003. Description: p. cm. ISBN: 0385509286 Notes: Includes index. Subjects: West, Michael (Michael D.) Human cloning--Popular works. Cells--Aging--Popular works. Gerontologists--United States--Biography. LC Classification: QH442.2 .W47 2003 Dewey Class No.: 612.6 22

What has the cloning of the ATM gene told us about ataxia telangiectasia? Abstract: PURPOSE: To review what the cloning of the ATM gene has told us about ataxia telangiectasia (A-T). Conclusions: Perhaps the most interesting and potentially useful finding following the cloning of the gene has been the discovery of homologies with various previously known PIK related kinases. There is also a very wide range of mutations associated with A-T and it is clear that some ATM mutations can lessen the severity of both the clinical and cellular features of the disorder. The function(s) of ATM remain an enigma. There are several indications of its involvement in protecting the cell from the effects of ionizing radiation and radiomimetic agents, although the pathways that involve p53 and c-Abl are far from clear. Expression of AIM in tissue culture cells can reverse the cellular features of A-T, particularly the increased level of radiosensitivity. The localization of ATM on meiotic chromosomes and its speculated role in meiotic recombination is an important finding. Finally, cloning the ATM gene has allowed the development of mouse models, which are providing information about A-T and will be crucial for testing future treatments for the disorder. Author(s): Taylor, A M Language: eng Source Int J Radiat Biol 1998 Apr;73(4):365-71.

Williams, J. G. (Jeff G.) Genetic engineering / J. Williams, A. Ceccarelli, and A. Wallace. Edition Information: 2nd ed. Published: Oxford: Bios; New York: Springer, 2001. Related Names: Ceccarelli, A. Wallace, A. (Andrew), 1963- Description: ix, 149 p.: ill. (some col.); 24 cm. ISBN: 0387915753 Notes: Includes bibliographical references and index. Subjects: Genetic Engineering--Cloning, Organism--methods. Escherichia coli--genetics. Gene Expression--physiology. Series: The medical perspectives series LC Classification: QH442 .W55 2001 Dewey Class No.: 660.65 21

Williams, J. G. (Jeff G.) Genetic engineering / J.G. Williams, R.K. Patient. Published/Created: Oxford; Washington, DC: IRL Press, 1988. Related Names: Patient, R. K. (Roger K.) Description: x, 72 p.: ill.; 22 cm. ISBN: 1852210710 : Notes:

Includes bibliographies and index. Subjects: Cloning, Molecular. DNA, Recombinant. Series: In focus (Oxford, England) LC Classification: QH442 .W55 1988 Dewey Class No.: 574.87/3282 19

Wilmut, Ian. The second creation: Dolly and the age of biological control / Ian Wilmut, Keith Campbell, Colin Tudge. Edition Information: 1st ed. Published: New York: Farrar, Straus and Giroux, 2000. Related Names: Campbell, Keith, 1954- Tudge, Colin. Description: xvii, 333 p.: ill.; 23 cm. ISBN: 0374141231 (alk. paper) Notes: Includes index. Subjects: Cloning--Moral and ethical aspects. Dolly (Sheep) LC Classification: QH442.2 .W545 2000 Dewey Class No.: 174/.957 21

Wilson, F. Paul (Francis Paul) Masque / F. Paul Wilson and Matthew J. Costello. Published/Created: New York: Warner Books, c1998. Related Names: Costello, Matthew J. (Matthew John), 1948- Description: 342 p.; 24 cm. ISBN: 0446519774 Subjects: Genetic engineering--Fiction. Cloning--Fiction. LC Classification: PS3573.I45695 M37 1998 Dewey Class No.: 813/.54 21

Winnacker, Ernst L. From genes to clones: introduction to gene technology / Ernst-L. Winnacker; translated by Horst Ibelgaufts. Published/Created: Weinheim, Federal Republic of Germany; New York, NY, USA: VCH, c1987. Description: xiii, 634 p.: ill.; 24 cm. ISBN: 0895736144 (U.S.: pbk.) : Notes: Translation of: Gene und Klone. Includes bibliographies and index. Subjects: Molecular cloning. Recombinant DNA. LC Classification: QH442.2 .W5613 1987 Dewey Class No.: 574.87/322 19

Wong, Dominic W. S. The ABCs of gene cloning / Dominic W.S. Wong. Published/Created: New York: Chapman & Hall, c1997. Description: xiv, 213 p.: ill.; 23 cm. ISBN: 0412077914 (alk. paper) Notes: Includes bibliographical references (p. 193-203) and index. Subjects: Molecular cloning. LC Classification: QH442.2 .W66 1997 Dewey Class No.: 574.87/328 20

Worldviews and philosophical basis of human cloning Abstract: The article presents three standpoints on the question of moral permissibility of human cloning and shows the philosophical principles of it. 1. The moral consent to human cloning with the purposes of reproduction and therapy. The followers of human cloning refer to materialistic anthropology also to subjectivistic, relativistic and utilitarian ethics. 2. Those, who are adverse to human cloning with the purpose of reproduction, but they acquiesce to the so-called therapeutic cloning. They reject that human embryos and foetuses are individuals who come under protection of law. 3. Those, who reject human cloning for the purposes of reproduction and therapy alike. They assent to a personalistic anthropology and Christian ethics. A human being was created by God and human life begins at the moment of insemination. All three groups use various argumentation. The arguments for and against cloning are extracted from biology as well as psychology, philosophy, law and religion. The author of the article takes the last standpoint, but she does not see such arguments, that might convince the opposite parties to a suit. Author(s): Lukowska, A T Language: pol Subject(s): Bioethics; Cloning, Organism; English Abstract; Ethical Theory; Human; Moral Obligations; Personal Autonomy; Personhood; Philosophy, Medical; Religion and Science; World Health Source Med Wieku Rozwoj 2001;5(1 Suppl 1):27-37.

Xeroderma pigmentosum and molecular cloning of DNA repair genes. Abstract: Human cells from patients suffering with xeroderma pigmentosum (XP) characterized by extreme sensitivity to UV light and a high incidence of skin tumors fall into seven complementation groups, XPA to XPG, and are lacking a functional helicase, endonuclease, or lesion-recognizing protein involved in the initial steps during nucleotide excision repair (NER); a number of proteins involved in DNA repair are termed XPA to XPG depending on which one is defective in a particular complementation group of XP and include: (i) proteins involved in the recognition of (6-4) photoproducts (XPE) and of a broad range of lesions such as pyrimidine dimers (XPA); (ii) proteins that are DNA helicases and integral parts of the general transcription factor TFIIH functioning in both transcription and repair (XPB, XPD); (iii) endonucleases that perform the two incisions, the XPG incising six nucleotides (nt) to the 3' side from a photodimer and the ERCC1-XPF protein complex incising 22 nt to the 5' side of the lesion; and (iv) single-strand DNA-binding proteins (XPC). The ERCC6 helicase is largely responsible for coupling transcription to repair whereas XPC seems to be responsible for the repair of the inactive parts of the genome as well as for the repair of the nontranscribed strand in active genes. p53 recognizes insertion/ deletion mismatches as well as free ends of DNA produced by ionizing radiation to arrest the cell cycle. Most of the human DNA repair proteins have their counterparts in both budding and fission yeasts and some of them also in E. coli evoking an evolutionary conservation of DNA repair pathways. Accumulation of mutations within repair genes in single cells followed by their escape from immune surveillance & in clonal expansion may greatly contribute to the appearance & development of human cancers. Author(s): Boulikas, T Language: eng Source Anticancer Res 1996 Mar-Apr;16(2):693-708.

X-linked agammaglobulinemia--gene cloning and future prospects. Abstract: The btk gene has recently been identified as the causative gene in X-linked agammaglobulinemia (XLA). This has opened up many new possibilities for the treatment of this B-cell immunodeficiency. Christine Kinnon and colleagues review the high degree of sequence of homology of btk to the non-receptor tyrosine kinases and speculate on putative roles for this gene in B-cell development. Author(s): Kinnon, C; Hinshelwood, S; Levinsky, R J; Lovering, R C Language: eng Source Immunol Today 1993 Nov;14(11):554-8.

YAC cloning of DNA embedded in an agarose matrix. Abstract: Yeast artificial chromosome (YAC) cloning of DNA in agarose is an alternative method to cloning from aqueous solutions. It minimizes any shearing that may result from handling of high molecular weight DNA and can be done with nanogram to microgram amounts of material, which facilitates construction of YACs from sources of DNA other than genomic DNA isolated from cells. The average size of the YACs recovered (200-1000 kb) and efficiency of transformation of ligation products (200-1000 cfu/micrograms) are similar to those reported using aqueous protocols. This method has been used to construct chromosome specific YACs, and it should be possible to apply the technique to the construction of chromosome specific libraries using flow sorted chromosomes as source material, and the cloning of restriction fragments isolated by preparative pulsed field gel electrophoresis. Author(s): McCormick, M K; Antonarakis, S E; Hieter, P Language: eng Source Genet Anal Tech Appl 1990 Sep;7(5):114-8.

YAC cloning: options and problems. Author(s): Burke, D T Language: eng Subject(s): Chromosomes, Fungal; Cloning, Molecular; Gene Library; Plasmids; Saccharomyces cerevisiae/genetics; Transformation, Genetic Source Genet Anal Tech Appl 1990 Sep;7(5):94-9.

YAC protocols / edited by David Markie. Published/Created: Totowa, N.J.: Humana Press, c1996. Related Names: Markie, David. Description: xii, 378 p.: ill.; 23 cm. ISBN: 0896033139 (alk. paper) Notes: Includes bibliographical references and index. Subjects: Molecular cloning--Methodology. Genetic vectors. Yeast fungi--Genetic engineering. Saccharomyces cerevisiae. Series: Methods in molecular biology (Clifton, N.J.); v. 54. LC Classification: QH442.2 .Y33 1996 Dewey Class No.: 575.1/0724 20

Yeast artificial chromosome cloning. Abstract: Yeast artificial chromosome (YAC) cloning systems enable the cloning of DNA stretches of 50 to well over 2000 kb. This makes it possible to study large intact regions of DNA in detail, by restriction mapping the YAC to produce a physical map and by examining the YAC for coding sequences or genes. YACs are important for their ability to clone the complete sequences of large genes or gene complexes that exceed the size limit for cloning in conventional bacterial cloning vectors like plasmids (up to 10 kb), bacteriophage (15 kb), and cosmids (50 kb). A major advantage of cloning in yeast, a eukaryote, is that many sequences that are unstable, underrepresented, or absent when cloned into prokaryotic systems, remain stable and intact in YAC clones. It is possible to reintroduce YACs intact into mammalian cells where the introduced mammalian genes are expressed and used to study the functions of genes in the context of flanking sequences. The correct protein processing mechanisms are present in the mammalian cells to ensure that a viable protein product is produced. Author(s): Ramsay, M Language: eng Subject(s): Animal; Biotechnology; Cell Line, Transformed; Chromosome Mapping; Chromosomes, Artificial, Yeast/genetics; Cloning, Molecular/methods; DNA/genetics; Gene Library; Genetic Markers; Genetic Techniques; Genetic Vectors; Sequence Tagged Sites; Transformation, Genetic Source Mol Biotechnol 1994 Apr;1(2):181-201.

Yeast systems for expression cloning and analysis of genes of higher eukaryotes; Okazaki, K Language: jpn Subject(s): Eukaryotic Cells; Gene Expression ; Human; Saccharomyces cerevisiae ; Signal Transduction. Source Tanpakushitsu Kakusan Koso 1994 Mar;39(4):439-46.

AUTHOR INDEX

C

D

E

F

G

H

L

M

N

O

P

R

S

T

U

V

W

X

Y

Z

TITLE INDEX

A

SUBJECT INDEX

A

B

C

D

E

H

N

S

T

U

V

W

X

Y

Z